Biofiller-Reinforced Biodegradable Polymer Composites

Biofiller-Reinforced Biodegradable Polymer Composites

Edited by
R. Jumaidin, S.M. Sapuan, and H. Ismail

CRC Press
Taylor & Francis Group
Boca Raton London New York

CRC Press is an imprint of the
Taylor & Francis Group, an **informa** business

First edition published 2021
by CRC Press
6000 Broken Sound Parkway NW, Suite 300, Boca Raton, FL 33487-2742

and by CRC Press
2 Park Square, Milton Park, Abingdon, Oxon, OX14 4RN

ISBN: 978-0-367-27264-7 (hbk)
ISBN: 978-0-367-55569-6 (pbk)
ISBN: 978-0-429-32211-2 (ebk)

Typeset in Times
by codeMantra

Contents

Preface

Biodegradable polymer can be referred to any polymer that can be naturally decomposed by the action of living organisms in soil or water. In recent times, the field of biodegradable material has attracted the attention of global researchers due to the needs to encounter the environmental problem arising from nonbiodegradable material. In general, biodegradable polymer is produced from natural origins such as plants, animals, or microorganisms. Nevertheless, the neat biodegradable polymer has several limitations, which limit the potential application in the real industry. Hence, modification of biodegradable polymer with biofiller has been carried out by researchers in order to reinforce the material or to enhance the desirable properties of the material. Biofiller can be referred to any types of material which is derived from natural resources and has great potential to improve the desirable properties of the biodegradable polymer. The approach of using biofiller to improve the properties of biodegradable polymer is a totally green solution which is capable of enhancing the properties of material while preserving the environmentally friendly characteristics of the material. This phenomenon has created new opportunities for the development of new materials for food packaging, miscellaneous packaging, pharmaceutical, and other potential applications which are commonly associated with single-use application. Biofiller-reinforced biodegradable polymer composites present a comprehensive review and findings on recent development in bio-based composites and their use in various applications. The information contained in this book is vital for the adaptation of new technology related to biodegradable materials since detailed and up-to-date information is a highly important requirement. This book might attract the attention of global researchers to advance their investigation and innovation in the field of biodegradable material. This book covers a broad range of different research fields, including the introduction of fully bio-based composites (biocomposites) in general perspective, the processing and characteristics of a new composite material, and the introduction of new combination of biocomposites such as soy protein, nanocellulose, and many other types of bio-based material. New modification on the most popular biopolymer, namely, polylactic acid (PLA), was also covered in this book. Recent findings on a new biodegradable polymer, namely, thermoplastic starch (TPS), were comprehensively discussed. The introduction of new materials to develop biodegradable polymer like Ubi gadong (*Dioscorea daemona*) was also discussed in one of the chapters, and the modification of natural fiber for further enhancement of composites and the compatibility of the natural filler like protein into biocomposites were elaborated as well. Various kinds of new approaches and advanced techniques in the development and modification of biocomposites were discussed in this book. As such, it offers a valuable reference guide for scholars interested in the bio-based polymer and biofiller as well as the end application as the biocomposite materials.

This book also provides interesting vision for the graduate students and scientists pursuing research in the fields of environmentally friendly material, biodegradable polymer, biofiller, or the combination of both, namely, the biocomposites. It is hoped that this book will give a positive insight to the reader and will nurture more new research and innovation in the field of biodegradable materials.

Editors

R. Jumaidin is the Head of Department of Manufacturing at Faculty of Mechanical and Manufacturing Engineering Technology, Universiti Teknikal Malaysia Melaka. He earned his BEng degree in Mechanical Engineering (Major in Structure and Materials) from Universiti Teknikal Malaysia Melaka in 2009 and MEng in Mechanical Materials from Universiti Teknologi Malaysia in 2012. He obtained his PhD in Material Engineering from Universiti Putra Malaysia in 2017 and received Graduate on Time Award upon graduation. His research interests include biopolymer, cellulose materials, biocomposites, and thermoplastic starch (TPS). New modification of TPS using cogon grass and seaweed waste is one of the interesting pieces of research that has been published in *International Journal of Biological Macromolecules*. To date, he has published more than 35 scientific articles, 2 chapters in book, and 5 proceedings. His citation Index (H-Index) is 11 with a citation number of 458 and counting. He has experience in reviewing numerous scientific papers from various international journals and conference. He has won several awards for the invention and innovation of bio-based composites, such as Gold Medal Award at UTeMEX Innovation 2020 and Silver Award at Malaysia Technology Expo 2020. He has also received International Merit Award from Republic of Croatia for his innovation during Malaysia Technology Expo. He has also received Journal Publication Incentive Award on 2018 and 2019 for the excellent publication achievement in his university. He has received Best Paper Award during Engineering Technology International Conference 2018 and Best Technical Paper during Sugar Palm National Seminar 2019. To date, he is an Honorary Member of Society of Sugar Palm Development.

S.M. Sapuan is a professor of composite materials at Universiti Putra Malaysia. He earned his BEng degree in Mechanical Engineering from University of Newcastle, Australia, in 1990; MSc from Loughborough University, the United Kingdom, in 1994; and PhD from De Montfort University, the United Kingdom, in 1998. His research interests include natural fiber composites, materials selection, and concurrent engineering. To date, he has authored or co-authored more than 1,521 publications (730 papers published/accepted in national and international journals, 16 authored books, 25 edited books, 153 chapters in books, and 597 conference proceedings/seminar papers/presentation (26 of which are plenary and keynote lectures, and 66 of which are invited lectures)). S.M. Sapuan was the recipient of Rotary Research Gold Medal Award 2012; The Alumni Medal for Professional Excellence Finalist; 2012 Alumni Awards, University of Newcastle, NSW, Australia; and Khwarizmi International Award (KIA). In 2013, he was awarded with 5 Star Role Model Supervisor award by UPM. He has been awarded "Outstanding Reviewer" by Elsevier for his contribution in reviewing journal papers. He received Best Technical Paper Award in UMIMAS STEM International Engineering Conference in Kuching, Sarawak, Malaysia. S.M. Sapuan was recognized as the first Malaysian to be conferred Fellowship by the US-based Society of Automotive Engineers International (FSAE) in 2015. He was the 2015/2016 recipient of SEARCA Regional Professorial Chair. In 2016 ranking of UPM researchers based on the number of citations and H-index by SCOPUS, he is ranked the sixth from 100 researchers. In 2017, he was awarded with IOP Outstanding Reviewer Award by Institute of Physics, the United Kingdom; National Book Award; The Best Journal Paper Award, UPM; Outstanding Technical Paper Award; Society of Automotive Engineers International, Malaysia; and Outstanding Researcher Award, UPM. He also received in 2017 Citation of Excellence Award from Emerald, the United Kingdom, SAE Malaysia; the Best Journal Paper Award; IEEE/TMU Endeavour Research Promotion Award; Best Paper Award by Chinese Defence Ordnance; and Malaysia's Research Star Award (MRSA), from Elsevier.

Prof. Ts. Dr. H. Ismail is an educator, researcher, and distinguished scientist. He received his doctorate in Polymer Technology from Loughborough University of Technology, the United Kingdom, in 1995. He is a Fellow of the Academy of Science Malaysia (ASM) and Professional Technologist of Malaysian Board of Technologies (MBOT). Professor Hanafi is a Chief Editor for Progress in Rubber, Plastics, and Recycling Technology (Sage) and Editorial Boards for Polymer Plastic Technology and Materials (Taylor & Francis), *Iranian Polymer Journal (Springer), Journal of Vinyl and Additive Technology* (Wiley), Polymer Testing (Elsevier), and *Journal of Rubber Research* (Springer). In 2000, Professor Hanafi received Khwarizmi International Award (KIA) from Iranian Research Organization for Science and Technology (IROST) and Asian and Pacific Centre for Transfer of Technology (APCTT) International Award from the United Nations Economic and Social Commission for Asia and the Pacific (ESCAP). In 2001, he was conferred with International Award from Islamic Educational, Scientific and Cultural Organization (ISESCO). In 2004, Professor Hanafi received Malaysian Excellent Scientist Award by Malaysian Ministry of Higher Education. In 2012, ASM has conferred Professor Hanafi Fellow of Academy Science (FASc) and the 2012 Top Research Scientists Malaysia (TRSM) Award. In 2016, he was listed as "The Most Cited Researchers 2016" by Shanghai Academic Ranking of World Universities (ARWU). In addition to his personal achievements, his research products have unfailingly won over 60 invention awards, including gold medals and special awards in many exhibitions nationally and internationally such as exhibitions in Kuwait, Switzerland, Germany, Belgium, Poland, Korea, Romania, Canada, Croatia, and the United Kingdom. Professor Ts. Dr. Hanafi's years of experience has made him a well-known figure in Polymer Science and Technology. To date, Professor Hanafi has successfully acquired 722 international journal publications with H-Index (SCOPUS): 52 and Total Citation Index: 12,212 and H-Index (GOOGLE SCHOLAR): 62 and Total Citation Index: 17,251. For the past 3 years (2017–2019), Professor Hanafi was awarded many international awards such as World Golden Scientist Award and Outstanding Leaders of Global Inventors by Korea Inventor Academy (2016), Scientist Recognition Award by Romanian Inventors Society (2018), Innovation in Polymer Science and Technology Award by Indonesia Polymer Society (2019), and Excellent Research Award by Croatian Inventors Network (2019).

Contributors

N.W. Adam
Faculty of Mechanical and
  Manufacturing Engineering
  Technology
Universiti Teknikal Malaysia Melaka
Melaka, Malaysia

M.R.M. Asyraf
Department of Aerospace Engineering
Universiti Putra Malaysia
Seri Kembangan, Malaysia

M.S.N. Atikah
Department of Chemical and
  Environmental Engineering
Universiti Putra Malaysia
Seri Kembangan, Malaysia

A.B. Azhar
School of Materials and Mineral
  Resources Engineering
Universiti Sains Malaysia
George Town, Malaysia

W.S. Chow
School of Materials and Mineral
  Resources Engineering
Universiti Sains Malaysia
George Town, Malaysia

T.T. Dele-Afolabi
Advanced Engineering Materials and
  Composites (AEMC)
Universiti Putra Malaysia
Seri Kembangan, Malaysia

A. Edhirej
Department of Mechanical and
  Manufacturing Engineering
Universiti Putra Malaysia
and
Department of Mechanical Engineering
University of Sabha, Libya

Z.A.A. Hamid
School of Materials and Mineral
  Resources Engineering
Universiti Sains Malaysia
George Town, Malaysia

K.Z. Hazrati
Department of Mechanical and
  Manufacturing Engineering
Universiti Putra Malaysia
Seri Kembangan, Malaysia

M.D. Hazrol
Department of Mechanical and
  Manufacturing Engineering
Universiti Putra Malaysia
Seri Kembangan, Malaysia

M.I.J. Ibrahim
Department of Mechanical and
  Manufacturing Engineering
Universiti Putra Malaysia
Seri Kembangan, Malaysia
and
Department of Mechanical Engineering
University of Sabha, Libya
Sabha, Libya

R. Ibrahim
Innovation & Commercialization
    Division
Forest Research Institute Malaysia
Kuala Lumpur, Malaysia

R.A. Ilyas
Advanced Engineering Materials and
    Composites (AEMC)
Universiti Putra Malaysia
Seri Kembangan, Malaysia

H. Ismail
School of Materials and Mineral
    Resources Engineering
Universiti Sains Malaysia
George Town, Malaysia

N.Z. Ismarrubie
Department of Mechanical and
    Manufacturing Engineering
Universiti Putra Malaysia
Seri Kembangan, Malaysia

M. Jawaid
Institute of Tropical Forestry and Forest
    Products
Universiti Putra Malaysia
Seri Kembangan, Malaysia

R. Jumaidin
Faculty of Mechanical and
    Manufacturing Engineering
    Technology
Universiti Teknikal Malaysia Melaka
Melaka, Malaysia

A.W.M. Kahar
School of Materials Engineering
Universiti Malaysia Perlis
Arau, Malaysia

R.Z. Khoo
Universiti Sains Malaysia
George Town, Malaysia

T.P. Leng
Universiti Malaysia Perlis
Arau, Malaysia

K.Y. Low
Universiti Malaysia Perlis
Arau, Malaysia

K.I. Ku Marsilla
Universiti Sains Malaysia
George Town, Malaysia

M.T. Mastura
Universiti Teknikal Malaysia Melaka
Melaka, Malaysia

A.L. Pang
Universiti Sains Malaysia
George Town, Malaysia

R.K.C. Pani Sellivam
Universiti Malaysia Perlis
Arau, Malaysia

M.S. Rasidi
Universiti Malaysia Perlis
Arau, Malaysia

A. Rusli
Universiti Sains Malaysia
George Town, Malaysia

Z.A.S. Saidi
Universiti Teknikal Malaysia Melaka
Melaka, Malaysia

H.N. Salwa
Universiti Putra Malaysia
Seri Kembangan, Malaysia

S.T. Sam
Universiti Malaysia Perlis
Arau, Malaysia

S.M. Sapuan
Universiti Putra Malaysia
Seri Kembangan, Malaysia

R. Syafiq
Universiti Putra Malaysia
Seri Kembangan, Malaysia

P.L. Teh
Universiti Malaysia Perlis
Arau, Malaysia

N.D. Yaacob
Universiti Malaysia Perlis
Arau, Malaysia

N.F. Zaaba
Universiti Sains Malaysia
George Town, Malaysia

E.S. Zainudin
Universiti Putra Malaysia
Seri Kembangan, Malaysia

M.Y.M. Zuhri
Universiti Putra Malaysia
Seri Kembangan, Malaysia

1 Introduction to Biofiller-Reinforced Degradable Polymer Composites

R.A. Ilyas, S.M. Sapuan, M.R.M. Asyraf, and M.S.N. Atikah
Universiti Putra Malaysia

R. Ibrahim
Forest Research Institute Malaysia

T.T. Dele-Afolabi and M.D. Hazrol
Universiti Putra Malaysia

CONTENTS

1.1 INTRODUCTION

In recent, high performance product derived from bio-based materials has become growing interest for the researchers all around the world [1]. For the past several decades, the conventional-based product was shown an increasing trend due to the significant growth of the global demand towards the material [2–5], which was attributable to its diverse and universal properties to be used in the development of science and technology [6–8]. The materials are widely used, especially, for structural purposes in the automotive, marine, aerospace, food packaging, and construction sectors due to their lightweight properties such as higher strength and durability [9–16]. Around 140 million tons of petroleum-based polymers is manufactured globally [17]. However, petroleum-based polymers show negative effects on the

1

environment, which leads to several issues, including health problems and pollution. Moreover, the majority of conventional plastics that are used end up in landfill areas. These observations showed that the synthetic plastics contribute to the major solid waste environmental pollutants [18]. Subsequently, the increased costs and the negative effect on environment make the engineers and researchers show high interest in polymer biodegradation [19].

Generally, most of the synthetic plastics are nonbiodegradable, which requires thousands of years to be disposed. This situation is undesirable for the environment, and the subsequent plan should be made to find alternative materials with biodegradable properties [20]. Based on the previous studies, the awareness of global communities is currently being centered on eco-friendly composites (eco-composites), which are made up of natural-based fiber and biopolymer materials [21]. The materials are produced from plants and animal wastes, which are biodegradable and nontoxic. Moreover, they are attractive because they are safe to be processed, and produced neutral and nonpolluted by-products [22]. Currently, it is shown an increasing demand on the eco-composites, which tends to have higher production than before. Several material engineering studies enticed many benefits and advantageous of the bio-based plastics compared to those of man-made plastics [23].

This chapter explains and elaborates the developments of science and technology related to biocomposites from the current research and industrial perspectives. The aim is to further discuss the biofillers incorporated with various types of polymeric matrices such as elastomers, thermoplastic, and thermoset. Moreover, this chapter also explains the mechanical and physical properties of the biocomposites, and the recent developments and applications in this field. Last, this chapter documents the challenges and limitations of the incorporation of biofillers in biocomposites.

1.2 CLASSIFICATION OF FIBERS IN COMPOSITE MATERIALS

Basically, a composite material is made up of two or more individual constituents, which are combined together to form a new developed material with improved physical, thermal, chemical, and mechanical properties rather than those individual components acting alone. The reinforcement component in the composite materials usually provides the structure with stiffer and stronger behavior when embedded in the matrix component [24]. The benefits of such composites are that they have higher strength and stiffness [25]. The reinforcement fiber can be obtained from the chemical extraction of rare earth materials, and plant and animal wastes. For the plant-based fibers, the materials are obtained from various parts of the plant, such as bast, seed, fruit, stalk, leaf, and grass. Figure 1.1 shows the classification of resources of plant-based fibers [26]. The plant-based fibers were previously categorized as nonwood fibers. In the recent years, researchers and engineers are focusing on these nonwood fibers since the materials can reduce the dependency on the hard wood fibers, which leads to deforestation. Later, the deforestation will cause the loss of biodiversity [27].

Recently, the increasing use of plant-based fibers can be attributed to their affordability, availability, biodegradability, renewability, recyclability, and

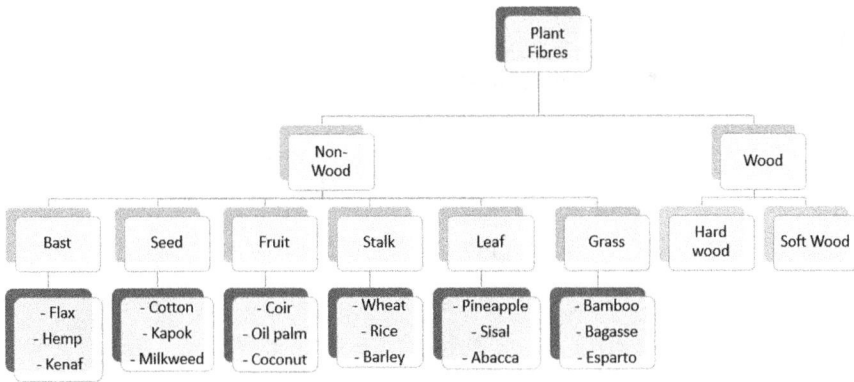

FIGURE 1.1 Classification of plant fibers from the distinct parts of plant [26].

processability [28,29]. In addition, the plant-based fibers present many benefits, such as good tensile strength, good insulation properties, lean in term of health risk, low density, and less energy consumption during manufacturing process compared to the conventional petroleum-based fibers [30]. Most of these natural fibers are suitable and have potential to be fabricated into the composite as well as other value-added products. Table 1.1 shows the comparison between plant-based fibers and conventional fibers.

In terms of mechanical properties, the fibers vary with each other based on the types and the resources that they are obtained. Table 1.2 depicts the mechanical behaviors in tensile mode of synthetic and natural fibers. From Table 1.2, the conventional fibers (E-glass and S-glass) show greater values of tensile strength compared to the majority of plant-based fibers. In terms of tensile modulus, three plant-based fibers are compared to the conventional fibers.

TABLE 1.1
Benefits of Plant-Based Fibers Compared to Conventional Fibers [31,32]

Attribute	Plant-Based Fiber	Conventional Fiber
Cost	Low	High
Carbon emission effect	Less	More
Density	Light	Heavy than bio-based fiber
Distribution	Wide-ranging	High
Disposal	Highly degradable	Consume thousands of years to degrade
Recyclability	Yes	No
Renewability	Yes	No
Energy consumption	Low	High
Health risk when inhaled	No	Yes

TABLE 1.2
Mechanical Performance of Various Types of Fibers [28,33–41]

Fiber	Tensile Modulus (GPa)	Tensile Strength (MPa)	Elongation at Break
E-glass	86	4,570	2.8
S-glass	86	4,570	2.8
Bagasse	19.7–27.1	20–290	1.1
Coir	6	220	15–25
Abaca	–	980	–
Cotton	12	400	3–10
Banana	33.8	355	5.3
Jute	10–30	400–800	1.8
Flax	60–80	800–1,500	1.2–1.6
Kenaf (bast)	–	295	2.7–6.9
Pineapple	82	170–1,627	1–3
Sisal	38	6–700	2–3
Ramie	44	500	2
Oil palm (empty fruit bunch)	3.2	248	2.5
Sugar palm(frond)	10.4	42.14	9.8
Hemp	70	550–900	1.6
Henequen	–	3–4.7	–

1.3 POLYMER MATRICES AS THE MAIN CONSTITUENTS IN COMPOSITE MATERIALS

Polymer-based materials can be classified into several varieties, such as thermoplastics, thermosetting, and elastomers. Thermoplastic materials are currently used as matrices for biofillers; the most commonly used thermoplastics for this purpose are polyethylene, polyvinyl chloride (PVC), and polypropylene (PP). Meanwhile, thermosetting polymers that are usually used in plastic industries are phenolics, epoxy, and polyester. At the same time, natural rubber is an elastomer that is included in the polymer classification [42].

1.4 BIOFILLERS IN POLYMER MATRICES

Polymers play a dominant role in manufacturing sectors because of their extraordinary flexibility along with their ultimate properties [43]. Their properties are high durability, mechanical and thermal performance, and chemical resistance [44]. This section further highlights the latest development in thermoset matrices reinforced with bio-based fibers. Table 1.3 gives the combination of renewable fibers, which are extracted from plants, with the thermosetting materials such as polyester, polyurethane, phenolic, and epoxy polymers.

As shown in Table 1.3, the combination of eggshell biofiller with polyester to develop a composite material was made by Toro et al. [47]. The research found that

TABLE 1.3

Biofillers/Fibers Incorporated in Polymer Matrices

Biofiller/Fiber	Polymeric Matrix	Experimental Condition	Results	References
Banana/sisal	Polyester matrix	Size of the particle was 30 mm. Sisal and banana loading of 40% vol. was utilized	Reduced the effectiveness of thermal conductivity because of the incorporation of banana and sisal	[45]
PALF		Size of the particle was 30 mm. Hybrid glass fiber and pineapple leaf loading of 0.40% vol. was implemented	Effective in terms of thermal conductivity	[45]
Spent coffee ground	Phenol–formaldehyde matrix	Size of the particle was 149 mm. 100 parts of the spent coffee fillers was combined with 100 parts of a phenol–formaldehyde resin	Good surface finish. The electrical insulation of the material was as good as the wood-filled plastic	[46]
Eggshells	Polyester matrix	0%–50% of the w/w eggshell filler. 2% of catalyst and accelerator was applied, respectively	At 40 wt% of eggshell fillers, the tensile and flexural strength increased. At 50 wt% of eggshell fillers, the compressive strength and hardness increased	[47]
Coconut shell	Epoxy matrix	Size of the particle was less than 100 μm. 30% of the w/w filler was implemented	The combination of epoxy resin and coconut shell fillers is effective in terms of the interfacial adhesion between them. Significantly improved with respect to hardness, tensile strength, and modulus. However, the material showed a decreasing trend in terms of impact strength, as compared to pure epoxy resin	[21]

(Continued)

TABLE 1.3 (*Continued*)
Biofillers/Fibers Incorporated in Polymer Matrices

Biofiller/Fiber	Polymeric Matrix	Experimental Condition	Results	References
Coconut shell	Polyester matrix	Size of the particle was 304 μm Filler contents were selected at 0, 15, 30, 45, and 60 php. The catalyst used was Butanox M-60	The analysis showed an increasing pattern in coconut shell content, causing the increase in Young's modulus and tensile strength. However, the water absorption decreased along with the elongation at break. The tendency of filler increased due the increase in the filler content	[48]
CSA	Epoxy matrix	Size of the particle was 53 μm. Resin-to-hardener weight ratio was 100:50	The increment in CSA fillers leads to an increase in elastic modulus, tensile strength, and microhardness of the composite laminate. Yet an increase in CSA contents tends to reduce the elongation and load values at break	[49]

the performance of eggshell/polyester composites has significantly been improved in terms of mechanical properties and density with the addition of eggshell fillers. In this case, the tensile and compressive strengths of the composite are significantly increased with the addition of uncarbonized fillers up to 40 wt%. Moreover, the maximum value of the mechanical performance is 28.378 MPa at 20 wt% of carbonized eggshell filler. This indicates that the higher the carbonized eggshell filler, the better the stress transfer between the matrices, thus leading to the increase in mechanical performance.

Idicula et al. [45] carried out a research on the incorporation of hybrid banana/sisal and pineapple leaf fibers (PALF) in polyester polymer matrix. They found out that the addition of the natural fibers effectively reduced the thermal conductivity of the composites. The addition of the biofillers reduced the thermal conductivity as low as 50% compared to the glass/polyester composites.

Eugenie [46] established that spent coffee ground was one of the versatile fillers used in various types of polymer matrices. The filler can be successfully applied for different purposes, where in most cases the function of colorization of dark color can be included in. This study focused on the addition of spent coffee ground fillers in phenol–formaldehyde resin with the major advantage of having significantly good surface finish. The research further revealed that the electrical insulation of the filler in a composite is comparable to the wood-filled plastic composites. It is also noticed that there was no need to include additional lubricant like stearic acid since the filler powder itself contains lubricant during the fabrication of composites.

In another study, Husseinsyah and Mostapha [48] performed an experimental work on the performance of coconut filler-reinforced polyester composites. The incorporation of coconut shells at different concentrations (0, 15, 30, 45, and 60 php (per hundred parts)) in polyester resin was assisted by Butanox M-60 added as a catalyst to induce polymerization reaction. The results showed that the higher the coconut shell content, the higher the tensile strength, Young's modulus, and the water absorption capacity. This was due to the fact that coconut shell was hydrophilic in nature, which caused the water uptake by the lignocellulosic components, and subsequently formed hydrogen bond between fillers and water.

A study by Sarki et al. [21] was conducted with agricultural wastes, i.e., coconut shells as alternative fillers to reinforce epoxy composites. The performance of 30 wt% of coconut shell-reinforced epoxy composites was investigated. SEM (scanning electron microscopy) micrographs showed that the composite experienced better interfacial adhesion with the inclusion of fillers in epoxy composites. Tensile, hardness, and impact performance of the composites were found to be significantly increased compared to the performance of virgin epoxy due to the increased filler–matrix interaction. The addition of coconut shell fillers into epoxy composites displayed the potential of the composites to be used in eco-buildings.

Imoisili et al. [49] performed an experimental work on the mechanical properties of coconut shell ash (CSA)-reinforced polyester composites. In their study, five filler concentrations were used, i.e., from 5 to 25 wt%. The test results showed that higher filler concentrations resulted in the increase in mechanical performance of the composites (such as elastic modulus, tensile strength, and microhardness). However, the elongation at break decreased with the increase in filler concentration.

1.4.1 BIOFILLERS AS FILLERS/FIBERS IN BIOPOLYMER COMPOSITES

The use of nonbiodegradable plastics originally derived from petroleum raised several concerns involving environmental problems [50–55]. Therefore, the awareness towards environmental sustainability and "green" movement is currently guiding the development of innovative eco-composite materials comprising bio-based filler materials and biodegradable polymers [56–58]. This section highlights the latest development in biopolymer matrices reinforced or filled with biofillers. Table 1.4 summarizes some studies on biofillers/fibers incorporated into biopolymer matrices.

Modi et al. [58] conducted a research on mechanical and rheological properties of a bacterial polyester-based poly-(3-hydroxybutyrate-*co*-3-hydroxyvalerate) (PHBV) biopolymer composites reinforced with selected plant fibers. The composites used different biofibers, including common reed, reed canary grass, and water celery with

TABLE 1.4

Summary of Studies on Biofillers/Fibers Incorporated into Biopolymer Matrices

Biofiller/Fiber	Biopolymer Matrix	Experimental Condition	Results	References
Common reed/ reed canary grass/water celery	PHBV. Wheat gluten-bio-based plasticizers (glycerol, octanoic acid, and 1,4-butanediol)	Fiber size of less than 1 mm was used, while 2%, 5%, and 10% fiber loading was implemented	In all cases, a significant improvement was observed in the Young's modulus of the composites relative to the pure PHBV. At 10% biofiber addition, viscosity at 170°C significantly improved	[45]
Microalgal biomass		Microalgal biomass in amounts of 10, 20, and 30 php relative to the total amount of gluten and plasticizer was employed	Plasticized samples with 1,4-butanediol (30 php) increased the tensile modulus and tensile strength from 36.5 to 273.1 and 3.3 to 4.9 MPa, respectively	[59]
Clamshell powder	PHB	Ratio of PHB to $CaCO_3$ was fixed at 70:30 while the binder contents varied between 3 and 9 wt%	Tensile strength and tensile modulus of approximately 7.32 and 208.8 MPa, respectively, were measured for the clamshell-reinforced PHB biocomposite	[46]
Cellulose nanofibrils/ nanoclays	PHB, PLA	Four grades of the biocomposites, namely, PHB/NC, PHB/CNF, PLA/NC, and PLA/CN, were prepared using 0.5%, 2%, and 4% of the biofillers	Generally, the mechanical properties increased with increasing biofiller content in both the PHB grade and the PLA grade	[47]
BF/WF	PLA and PBS	PBS/PLA: natural flour (BF, WF) = 70:30. Average particle sizes of BF and WF were 580 and 160 μm, respectively	Tensile strength and heat deflection temperature of the biopolymer-based biocomposites were superior to those of the polymer-based composites	[21]

(Continued)

TABLE 1.4 (*Continued*)
Summary of Studies on Biofillers/Fibers Incorporated into Biopolymer Matrices

Biofiller/Fiber	Biopolymer Matrix	Experimental Condition	Results	References
Canola meal, corn gluten meal, soy meal, switch grass	PBS	Polymer and biofiller weight ratios were 75 and 25 wt%	Biodegradation rate significantly increased in the meal-based composites as compared to both the pure PBS and the biocomposite prepared with switch grass	[48]
Treated eggshell	Bio-based polyethylene from sugarcane	Six different concentrations were used between the range of 0%–40%	Improved stiffness, hardness, flexural modulus, and tensile modulus were exhibited by the developed biocomposites	[49]
Sugar palm nanocrystalline celluloses (SPNCCs)	Sugar palm starch	Seven different concentrations were used between the range of 0–1.0 wt%	Improved mechanical and water barrier properties of bionanocomposites	[60,61]
Sugar palm nanofibrillated celluloses (SPNFCs)	Sugar palm starch	Seven different concentrations were used between the range of 0–1.0 wt%	Improved mechanical and water barrier properties of bionanocomposites	[9,62–65]
Corn straw, soy stalk, wheat straw	PHBV	PHBV/agro-residue filler ratios were fixed at 70:30 and 60:40	The tensile modulus and storage modulus of PHBV were improved by the maximum of 256% and 308%, respectively, with the reinforcement of 30 wt% agro-residue	[46]

2%, 5%, and 10% of fiber loadings, and the research exhibited an improvement in Young's modulus, while slight changes in the tensile strength and elongation at break were evident compared to the neat PHBV. In addition, an improvement in viscosity at 170°C was recorded with the addition of 10% biofibers due to fiber–fiber and fiber–matrix interactions.

Ciapponi et al. [59] proposed the microalgal biomass as a suitable filler for developing new bioplastic compounds with high thermal stability from wheat gluten, and bio-based plasticizers (glycerol, octanoic acid, and 1,4-butanediol). This was empirically confirmed in samples plasticized with 1,4-butanediol, where the 30 php increased the tensile modulus and tensile strength from 36.5 to 273.1MPa and 3.3 to 4.9 MPa, respectively. However, the addition of the microalgal biomass slightly raised the surface sensitivity against water, with 30 php of biomass reducing the water contact angle from 41°C to 22°C.

A successful development of a totally "green" biocomposite of polyhydroxybutyrate (PHB) filled with clamshell biofiller was reported by Othman et al. [55]. Their findings showed that tensile strength and tensile modulus of approximately 7.32 and 208.8 MPa, respectively, were measured for the clamshell fiber-reinforced PHB biocomposites.

Al et al. [66] investigated the effects of cellulose nanofibrils and nanoclays on the mechanical and thermal properties of PHB and polylactic acid (PLA) biopolymer composites and showed that there was a great potential of biofillers in the development of highly reliable biocomposites. Relative to the plain biopolymer nanocomposites, the mechanical properties, including tensile strength, tensile modulus, flexural strength, and flexural modulus, increased with the increase in the contents of the biofillers for both the PHB and PLA biopolymers. The addition of the biofillers in both the PHB and PLA biopolymers improved the thermal stability.

Kim et al. [67] reported an increase in the interfacial adhesion, mechanical and thermal properties of biocomposites prepared from biodegradable polymers, including PLA and poly(butylene succinate) (PBS), filled with natural flour fillers (bamboo flour (BF) and wood flour (WF)). Generally, tensile strength and heat deflection temperature of both the PLA- and PBS-based biocomposites were higher than those of poly(styrene-b-ethylene-co-butylene-b-styrene) triblock copolymer and PP-based biocomposites.

Anstey et al. [68] developed biocomposites from (PBS biopolymer and bio-based fillers derived from coproducts of biofuel production. From the analysis, it was observed that the addition of meal-based fillers (canola meal, soy meal, and corn gluten meal) improved the biodegradability of the biopolymer matrix and the meal-based fillers degraded the biopolymer matrix faster than both PBS polymer and the biocomposites prepared using switch grass biofillers. The high concentration of proteins in the meal-based composites, which facilitated microorganism growth, was attributed to the improved biodegradability exhibited by the developed biocomposites.

A biocomposite based on "green" polyethylene biopolymer and treated eggshell biofillers was developed by Boronat et al. [69]. The results obtained showed that $CaCO_3$ from the treated eggshell effectively enhanced the mechanical properties (stiffness, hardness, and tensile strength) of the developed biocomposites. Ahankari et al. [70] successfully prepared new "green" composites by adding fillers from agro-residues (corn straw, soy stalk, and wheat straw) into the bacterial polyester PHBV. From a comparative analysis, the tensile strength and the storage modulus of PHBV were improved by the maximum of 256% and 308%, respectively, with the reinforcement of 30 wt% agro-residues.

1.5 POTENTIAL APPLICATIONS OF BIOFILLER-REINFORCED DEGRADABLE POLYMER COMPOSITES

Biofiller-reinforced degradable polymer composites demonstrate huge prospects among other categories of particulate filler-reinforced composites based on the numerous findings emphasized in this chapter. Among the interesting innovations of the biofiller-reinforced degradable polymer composites are the ease of processing with different types of polymers and biopolymer matrices, and the applicability of several manufacturing procedures that have been used for developing other particulate filler-reinforced composites. Therefore, according to the findings highlighted in this chapter, and due to their lightweight and low cost, biofiller-reinforced degradable polymer composites have great potential as alternative composite materials, especially in industrial applications such as food packaging [71–73] (containers) (Figure 1.2), automotive (Figures 1.3 and 1.4) and aerospace components (interior–exterior panels, high-performance components, and gas tanks), construction (flooring, structural panels, and building sections) (Figure 1.5), composite frame (Figure 1.6), table and chair composites (Figures 1.7–1.9), solar module (Figure 1.10), safety helmet (Figure 1.11), profiling and roof (Figure 1.12), and electrical and electronics (printed circuit boards and other integrated circuit components) [74,75]. Besides, bamboo strip/epoxy composites were found to be remarkable materials to be used in marine industry. Siti et al. [76] have manufactured bamboo boat hull using vacuum bagging and compression molding technique, as shown in Figure 1.13. Moreover, other potential applications of biocomposites included hoses, V belts, gloves, tires, and complex-shaped mechanical goods. For manufacturing of tires, carbon black has been extensively used for obtaining improved initial modulus and durability [76]. As scientists and engineers pay more and more attention to the environmental protection and sustainability, the utilization of various natural fibers, especially bamboo, as

FRPC MADE FROM
KENAF FIBRE + HDPE
[40% : 60%]

INSTITUT PERHUTANAN TROPIKA & PRODUK HUTAN
(INTROP) UPM

PRODUK KENAF

FIGURE 1.2 Kenaf composite plate and food container.

FIGURE 1.3 Kenaf composite automotive package tray.

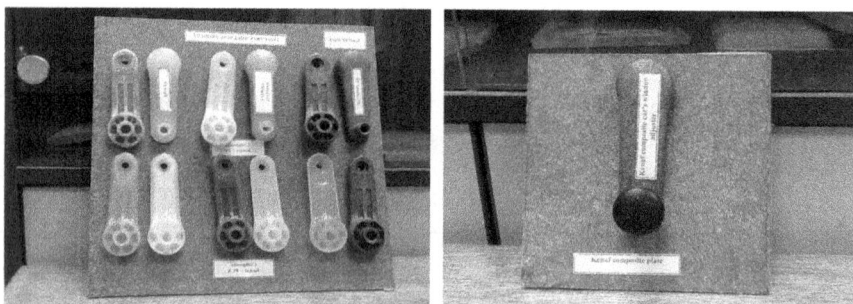

FIGURE 1.4 Kenaf composite car window regulators.

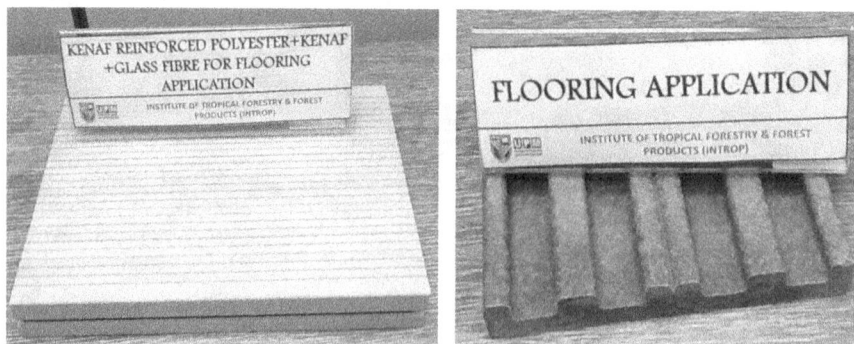

FIGURE 1.5 Kenaf composite flooring.

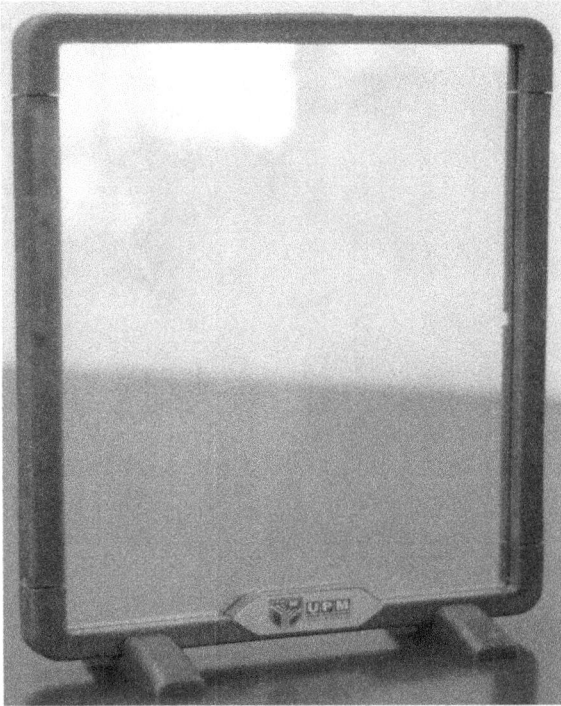

FIGURE 1.6   Frame from kenaf fiber composites.

FIGURE 1.7   Kenaf composite table and kenaf composite chair.

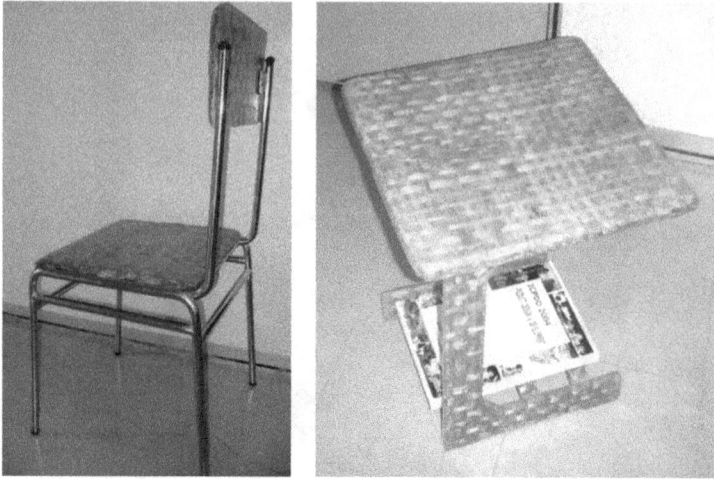

FIGURE 1.8 Banana stem composite chair and table.

FIGURE 1.9 Sugar palm fiber composite table.

biofiller, which is an alternative for petroleum fossil fuels, in natural rubber polymer matrix creates greener tires, as displayed in Figure 1.14. Biocomposites are also utilized in medical industry. The cockle (*Anadara granosa*) is by far the most ample species cultured in Malaysia. A possible advantage of using cockle shell as a biomineral is that they may act as analogs of calcium carbonate present in vivo. Cockle shell-based biocomposites have high potential to be used as a scaffold for bone tissue engineering [77] (see Figures 1.15 and 1.16).

Biocomposites in plastic packaging bring the benefit of using less fossil-based materials in current packaging and reducing the carbon footprint. In the best case, packaging can be 100% and made of renewable material. Research that had been

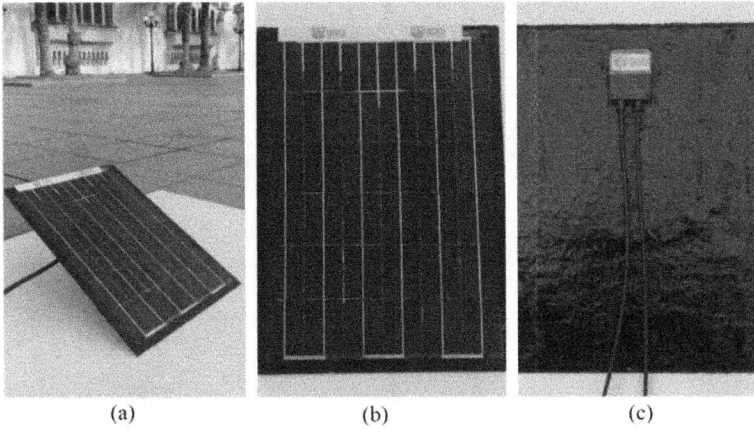

FIGURE 1.10 Fabricated solar module: (a) module while exposed to sun radiation; (b) front view of the module; (c) back view of the module showing the sealed backsheets.

FIGURE 1.11 Safety helmet from kenaf composites.

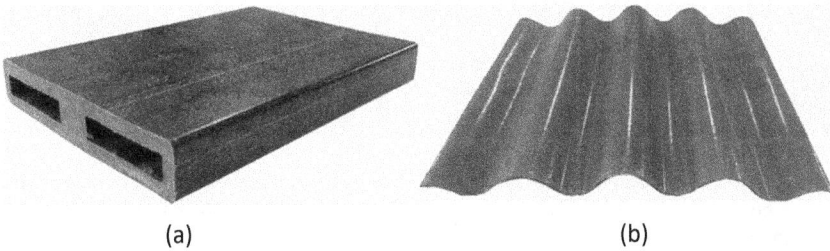

FIGURE 1.12 Profiling (a) and roof (b) made from bamboo composites reinforcement thermoplastic. (Reproduced with permission from [76], Creative Commons Attribution 3.0 License 2012, IntechOpen.)

FIGURE 1.13 Manufacturing process of bamboo boat hull for water sports activities. (Reproduced with permission from [76], Creative Commons Attribution 3.0 License 2012, IntechOpen.)

Bamboo carbon black   Green tire

FIGURE 1.14 Green tire made from bamboo carbon black. (Reproduced with permission from [76], Creative Commons Attribution 3.0 License 2012, IntechOpen.)

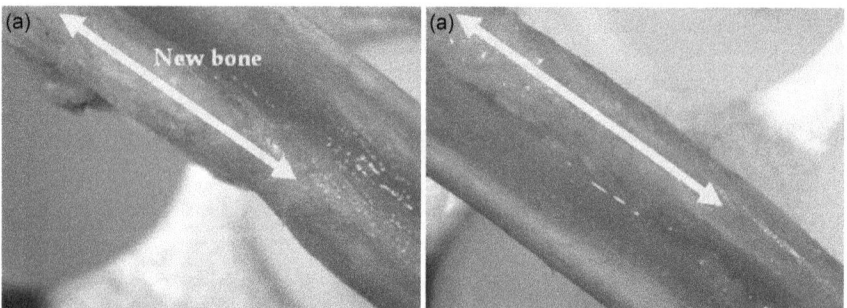

FIGURE 1.15 Photographs of the radial bone implanted with seeded scaffold in group A at 8 weeks postimplantation showing the new bone formation that completely filled the defect area and restored the continuity of the radial bone (arrows). (a) Lateral view and (b) medial view cockle shell-based biocomposite scaffold for bone tissue engineering. (Reproduced with permission from [77], Creative Commons Attribution 3.0 License 2012, IntechOpen.)

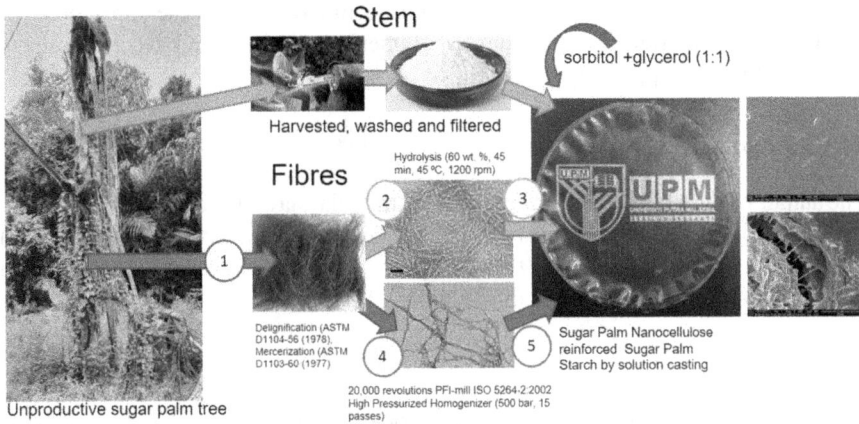

FIGURE 1.16 Sugar palm nanocellulose-reinforced sugar palm starch biopolymer composite.

conducted by Ilyas et al. [60] proved that plastic packaging can be made using 100% renewable materials such as sugar palm starch and sugar palm fiber. In their work, sugar palm fiber was subjected to mechanical (high-pressurized homogenizer) [78] and chemical treatments (mercerization, delignification, and bleaching) [79,80] to obtain sugar palm nanocellulose. The nanocellulose thus obtained was then reinforced with sugar palm starch to produce strong biofilm for packaging applications. Exhaustive discussion of the processing, characterization, and potential applications of these bio-based materials can be found in Ilyas's work [9,19,57,60–65,78,81–87].

1.6 CONCLUSIONS

As presented in this chapter, investigations on the biofiller-reinforced degradable polymer composites have gained attention due to their excellent properties and positive environmental impact. In most of the studies reviewed, the addition of biofillers significantly improved the interfacial adhesion and thermal and mechanical properties. In particular, mechanical properties such as tensile strength, tensile modulus, flexural strength, and flexural modulus increased with the increase in filler loading until the optimal loading was reached. Meanwhile, the treatment of the biofillers, especially the agro-waste biofillers, is useful in enhancing their intrinsic properties and in turn significantly improving the biocomposites developed using the treated biofillers. However, the biofiller-reinforced degradable polymer composites were marred by some limitations such as high sensitivity to moisture and temperature, and the difficulty in controlling the biodegradability rate. Therefore, further exploration on more advanced processing techniques for the biofillers and their composites is required to overcome these limitations so that the biofiller-reinforced degradable polymer composites can have a promising future in different applications.

ACKNOWLEDGMENTS

The authors would like to thank Universiti Putra Malaysia and Ministry of Education, Malaysia, for the financial support through the Graduate Research Fellowship (GRF) scholarship, HICoE Grant (6369107), and Fundamental Research Grant Scheme (FRGS/1/2017/TK05/UPM/01/1) (5540048).

REFERENCES

1. C. N. A. Jaafar, M. A. M. Rizal, and I. Zainol, "Effect of kenaf alkalization treatment on morphological and mechanical properties of epoxy/silica/kenaf composite," *International Journal of Engineering and Technology*, vol. 7, pp. 258–263, 2018.
2. N. H. Hamid, N. Hassan, P. M. Tahir, and S. Ujang, "Effectiveness of acetic, propionic and butyric anhydrides to protect rubberwood from decayed by white rot Trametes versicolor," *IOP Conference Series: Materials Science and Engineering*, vol. 368, no. 012037, pp. 1–13, 2018.
3. N. H. Hamid, W. B. Hisan, W. S. Izzati. U. H. Abdullah, A. A. A. Azim, and P. M. Tahir, "Mechanical properties and moisture absorption of epoxy composites mixed with amorphous and crystalline silica from rice husk," *BioResources*, vol. 14, no. 3, pp. 7363–7374, 2019.
4. H. N. Hisham, M. Hale, and A. L. Norasikin, "Equilibrium moisture content and moisture exclusion efficiency of acetylated rattan (*Calamus manan*)," *Journal of Tropical Forest Science*, vol. 26, no. 1, pp. 32–40, 2014.
5. H. Hamid, M. F. Yazid, N. Hassan, and M. D. Hale, "The selected properties of hybrid particleboard made from oil palm empty fruit bunch (EFB) and merpauh (Irvingla Malayanan Olive) sawdust wastes," *Asian Journal of Applied Sciences*, vol. 5, no. 04, pp. 615–623, 2015.
6. W. Ashraf, M. R. Ishak, M. Y. M. Zuhri, N. Yidris, A. M. B. Yaacob, and M. R. M. Asyraf, "Investigation of different facesheet materials on compression properties of honeycomb sandwich composite," in Seminar Enau Kebangsaan, pp. 129–132, 2019.
7. C. N. A. Jaafar, I. Zainol, and M. A. M. Rizal, "Preparation and characterisation of epoxy/silica/kenaf composite using hand lay-up method," in *27th Scientific Conference of the Microscopy Society Malaysia (27th SCMSM 2018)*, pp. 2–6, 2018.
8. H. A. Aisyah *et al.*, "Thermal properties of woven kenaf/carbon fibre-reinforced epoxy hybrid composite panels," *International Journal of Polymer Science*, vol. 2019, no. December, pp. 1–8, Dec. 2019.
9. R. A. Ilyas *et al.*, "Sugar palm (Arenga pinnata [Wurmb.] Merr) starch films containing sugar palm nanofibrillated cellulose as reinforcement: water barrier properties," *Polymer Composites*, vol. 41, no. 2, pp. 459–467, Feb. 2020.
10. M. L. Sanyang, S. M. Sapuan, M. Jawaid, M. R. Ishak, and J. Sahari, "Development and characterization of sugar palm starch and poly(lactic acid) bilayer films," *Carbohydrate Polymers*, vol. 146, pp. 36–45, 2016.
11. A. N. Johari *et al.*, "Fabrication and cut-in speed enhancement of savonius vertical axis wind turbine (SVAWT) with hinged blade using fiberglass composites," in *Seminar Enau Kebangsaan*, no. 1997, pp. 978–983, 2019.

12. M. R. M. Asyraf, M. R. Ishak, M. R. Razman, and M. Chandrasekar, "Fundamentals of creep, testing methods and development of test rig for the full-scale crossarm: a review," *Jurnal Teknologi*, vol. 81, no. 4, pp. 155–164, 2019.

13. N. M. Nurazzi, A. Khalina, S. M. Sapuan, R. A. Ilyas, S. A. Rafiqah, and Z. M. Hanafee, "Thermal properties of treated sugar palm yarn/glass fiber reinforced unsaturated polyester hybrid composites," *Journal of Materials Research and Technology*, vol. 9, no. 4, pp. 1606–1618, 2020.

14. M. N. Norizan, K. Abdan, R. A. Ilyas, and S. P. Biofibers, "Effect of fiber orientation and fiber loading on the mechanical and thermal properties of sugar palm yarn fiber reinforced unsaturated polyester resin composites," *Polimery*, vol. 65, no. 2, pp. 34–43, 2020.

15. N. M. Nurazzi, A. Khalina, S. M. Sapuan, and R. A. Ilyas, "Mechanical properties of sugar palm yarn/woven glass fiber reinforced unsaturated polyester composites : effect of fiber loadings and alkaline treatment," *Polimery*, vol. 64, no. 10, pp. 12–22, 2019.

16. M. R. M. Asyraf, M. R. Ishak, S. M. Sapuan, N. Yidris, and R. A. Ilyas, "Woods and composites cantilever beam: a comprehensive review of experimental and numerical creep methodologies," *Journal of Materials Research and Technology*, vol. 9, no. 3, pp. 6759–6776, 2020.

17. A. S. M. Bashir and Y. Manusamy, "Recent developments in biocomposites reinforced with natural biofillers from food waste," *Polymer - Plastics Technology and Engineering*, vol. 54, no. 1, pp. 87–99, 2015.

18. M. Shimao, "Biodegradation of plastics," *Current Opinion in Biotechnology*, vol. 12, no. 3, pp. 242–247, 2001.

19. R. A. Ilyas *et al.*, "Sugar palm (Arenga pinnata (Wurmb.) Merr) cellulosic fibre hierarchy: a comprehensive approach from macro to nano scale," *Journal of Materials Research and Technology*, vol. 8, no. 3, pp. 2753–2766, May 2019.

20. L. Katarzyna and L. Grazyna, "Polymer biodegradation and biodegradable polymers—a review," *Polish Journal of Environmental Studies*, vol. 19, no. 2, pp. 255–266, 2010.

21. J. Sarki, S. B. Hassan, V. S. Aigbodion, and J. E. Oghenevweta, "Potential of using coconut shell particle fillers in eco-composite materials," *Journal of Alloys and Compounds*, vol. 509, no. 5, pp. 2381–2385, 2011.

22. K. tak Lau, P. yan Hung, M. H. Zhu, and D. Hui, "Properties of natural fibre composites for structural engineering applications," *Composites Part B: Engineering*, vol. 136, no. September 2017, pp. 222–233, 2018.

23. O. Faruk, A. K. Bledzki, H.-P. Fink, and M. Sain, "Biocomposites reinforced with natural fibers: 2000–2010," *Progress in Polymer Science*, vol. 37, no. 11, pp. 1552–1596, Nov. 2012.

24. R. M. Shahroze, M. Chandrasekar, K. Senthilkumar, T. Senthilmuthukumar, M. R. Ishak, and M. R. M. Asyraf, "A review on the various fibre treatment techniques used for the fibre surface modification of the sugar palm fibres," in *Seminar Enau Kebangsaan*, no. 1, pp. 48–52, 2019.

25. P. A. Fowler, J. M. Hughes, and R. M. Elias, "Biocomposites: technology, environmental credentials and market forces," *Journal of the Science of Food and Agriculture*, vol. 86, no. 12, pp. 1781–1789, Sep. 2006.

26. M. L. Sanyang, S. M. Sapuan, M. Jawaid, M. R. Ishak, and J. Sahari, "Recent developments in sugar palm (Arenga pinnata) based biocomposites and their potential industrial applications: a review," *Renewable and Sustainable Energy Reviews*, vol. 54, pp. 533–549, 2016.

27. T. Bou, "Malaysia has the world's highest deforestation rate, reveals Google forest map," *Aljazeera News*, 2019.

28. R. Burgueño, M. J. Quagliata, A. K. Mohanty, G. Mehta, L. T. Drzal, and M. Misra, "Hybrid biofiber-based composites for structural cellular plates," *Composites Part A: Applied Science and Manufacturing*, vol. 36, no. 5, pp. 581–593, May 2005.

29. S. K. Acharya and S. Dalbehera, "Study on mechanical properties of natural fiber rein-forced woven jute-glass hybrid epoxy composites," *Advances in Polymer Science and Technology*, vol. 4, no. 1, pp. 1–6, 2014.

30. M. Jawaid and H. P. S. Abdul Khalil, "Cellulosic/synthetic fiber reinforced polymer hybrid composites: a review." *Carbohydr Polymers*, vol. 86, pp. 1–18, 2011.

31. P. Wambua, J. Ivens, and I. Verpoest, "Natural fibres: can they replace glass in fibre rein-forced plastics?," *Composites Science and Technology*, vol. 63, no. 9, pp. 1259–1264, 2003.

32. P. A. Sreekumar, *Matrices for natural-fibre reinforced composites*. Brimingham: Woodhead Publishing Limited, 2008.

33. S. S. Bhagawan, D. K. Tripathy, and S. K. De, "Stress relaxation in short jute fiber-reinforced nitrile rubber composites," *Journal of Applied Polymer Science*, vol. 33, no. 5, pp. 1623–1639, 1987.

34. K. N. Law, W. R. W. Daud, and A. Ghazali, "Morphological and chemical nature of fiber strands of oil palm empty-fruit-bunch (OPEFB)," *BioResources*, vol. 2, no. 3, pp. 351–362, 2007.

35. N. Amir, K. A. Z. Abidin, and F. B. M. Shiri, "Effects of fibre configuration on mechanical properties of banana fibre/PP/MAPP natural fibre reinforced polymer composite," *Procedia Engineering*, vol. 184, pp. 573–580, 2017.

36. I. S. M. A. Tawakkal, R. A. Talib, K. Abdan, and C. N. Ling, "Mechanical and physical properties of kenaf-derived cellulose (KDC)-filled polylactic acid (PLA) composites," *BioResources*, vol. 7, no. 2, pp. 1643–1655, Feb. 2012.

37. D. Bachtiar, M. S. Salit, E. Zainudin, K. Abdan, and K. Z. H. M. Dahlan, "Effects of alkaline treatment and a compatibilizing agent on tensile properties of sugar palm fibre reinforced high impact polystyrene composites," *BioResources*, vol. 6, no. 4, pp. 4815–4823, 2011.

38. M. H. Mustafa and B. Dauda, "Unsaturated polyester resin reinforced with chemically modified natural fibre," *IOSR Journal of Polymer and Textile Engineering (IOSR-JPTE)*, vol. 1, no. 4, pp. 31–38, 2014.

39. M. Hosur, H. Maroju, and S. Jeelani, "Comparison of effects of alkali treatment on flax fibre reinforced polyester and polyester-biopolymer blend resins," *Polymers and Polymer Composites*, vol. 23, no. 4, pp. 229–242, 2015.

40. C. Y. Lai, S. M. Sapuan, M. Ahmad, N. Yahya, and K. Dahlan, "Mechanical and electrical properties of coconut coir fiber-reinforced polypropylene composites," *Polymer-Plastic Technology and Engineering*, vol. 44, no. 4, pp. 619–632, 2005.

41. M. S. Huda, L. T. Drzal, A. K. Mohanty, and M. Misra, "Effect of chemical modifica-tions of the pineapple leaf fiber surfaces on the interfacial and mechanical properties of laminated biocomposites," *Composite Interfaces*, vol. 15, no. 2–3, pp. 169–191, 2008.

42. R. Malkapuram, V. Kumar, and Y. Singh Negi, "Recent development in natural fiber reinforced polypropylene composites," *Journal of Reinforced Plastics and Composites*, vol. 28, no. 10, pp. 1169–1189, 2009.

43. J. M. Raquez, M. Deléglise, M. F. Lacrampe, and P. Krawczak, "Thermosetting (bio) materials derived from renewable resources: a critical review," *Progress in Polymer Science (Oxford)*, vol. 35, no. 4, pp. 487–509, 2010.

44. G. F. Fernando and B. Degamber, "Process monitoring of fibre reinforced composites using optical fibre sensors," *International Materials Reviews*, vol. 51, no. 2, pp. 65–106, 2006.

45. M. Idicula, A. Boudenne, L. Umadevi, L. Ibos, Y. Candau, and S. Thomas, "Thermophysical properties of natural fibre reinforced polyester composites," *Composites science and technology*, vol. 66, no. 15, pp. 2719–2725, 2006.

46. L. Eugenie, "Waste coffee grounds as a filler in thermosetting materials," US Patent 3499851, 1970.

47. P. Toro, R. Quijada, M. Yazdani-Pedram, and J. L. Arias, "Eggshell, a new bio-filler for polypropylene composites," *Materials Letters*, vol. 61, no. 22, pp. 4347–4350, 2007.

48. S. Husseinsyah and M. Mostapha, "The effect of filler content on properties of coconut shell filled polyester composites," *Malaysian polymer journal*, vol. 6, no. 1, pp. 87–97, 2011.

49. P. E. Imoisili, C. M. Ibegbulam, and T. I. Adejugbe, "Effect of concentration of coconut shell ash on the tensile properties of epoxy composites," *The Pacific Journal of Science and Technology*, vol. 13, no. 1, pp. 463–468, 2012.

50. E. Syafri *et al.*, "Effect of sonication time on the thermal stability, moisture absorption, and biodegradation of water hyacinth (Eichhornia crassipes) nanocellulose-filled bengkuang (Pachyrhizus erosus) starch biocomposites," *Journal of Materials Research and Technology*, vol. 8, no. 6, pp. 6223–6231, Nov. 2019.

51. H. Abral *et al.*, "Transparent and antimicrobial cellulose film from ginger nanofiber," *Food Hydrocolloids*, vol. 98, no. August 2019, p. 105266, 2020.

52. H. Abral *et al.*, "Highly transparent and antimicrobial PVA based bionanocomposites reinforced by ginger nanofiber," *Polymer Testing*, vol. 81, p. 106186, 2020.

53. H. Abral *et al.*, "Effect of ultrasonication duration of polyvinyl alcohol (PVA) gel on characterizations of PVA film," *Journal of Materials Research and Technology*, vol. 9, no.2, pp. 2477–2486, 2020.

54. H. Abral *et al.*, "A simple method for improving the properties of the sago starch films prepared by using ultrasonication treatment," *Food Hydrocolloids*, vol. 93, pp. 276–283, 2019.

55. M. Othman, M. Yusoff Hashim, I. N. Azowa, K. Khalid, and S. A. Mohamad, "Polyhydroxybutrate filled clamshell biofiller: effect of polyvinylpyrrolidone," *Advanced Materials Research*, vol. 1115, pp. 296–299, Jul. 2015.

56. M. J. Halimatul, S. M. Sapuan, M. Jawaid, M. R. Ishak, and R. A. Ilyas, "Water absorption and water solubility properties of sago starch biopolymer composite films filled with sugar palm particles," *Polimery*, vol. 64, no. 9, pp. 27–35, 2019.

57. M. J. Halimatul, S. M. Sapuan, M. Jawaid, M. R. Ishak, and R. A. Ilyas, "Effect of sago starch and plasticizer content on the properties of thermoplastic films: mechanical testing and cyclic soaking-drying," *Polimery*, vol. 64, no. 6, pp. 32–41, 2019.

58. S. J. Modi, K. Cornish, K. W. Koelling, and Y. Vodovotz, "Mechanical and rheological properties of PHBV bioplastic composites engineered with invasive plant fibers," *Transactions of the ASABE*, vol. 59, no. 6, pp. 1883–1891, Dec. 2016.

59. R. Ciapponi, S. Turri, and M. Levi, "Mechanical reinforcement by microalgal biofiller in novel thermoplastic biocompounds from plasticized gluten," *Materials*, vol. 12, no. 9, p. 1476, May 2019.

60. R. A. Ilyas, S. M. Sapuan, M. R. Ishak, and E. S. Zainudin, "Development and characterization of sugar palm nanocrystalline cellulose reinforced sugar palm starch bionanocomposites," *Carbohydrate Polymers*, vol. 202, pp. 186–202, Dec. 2018.

61. R. A. Ilyas, S. M. Sapuan, M. R. Ishak, and E. S. Zainudin, "Sugar palm nanocrystalline cellulose reinforced sugar palm starch composite: degradation and water-barrier properties," in *IOP Conference Series: Materials Science and Engineering*, vol. 368, no. 1, 2018.

62. M. S. N. Atikah *et al.*, "Degradation and physical properties of sugar palm starch/sugar palm nanofibrillated cellulose bionanocomposite," *Polimery*, vol. 64, no. 10, pp. 27–36, 2019.

63. R. A. Ilyas, S. M. Sapuan, M. R. Ishak, and E. S. Zainudin, "Water transport properties of bio-nanocomposites reinforced by sugar palm (Arenga Pinnata) nanofibrillated cellulose," *Journal of Advanced Research in Fluid Mechanics and Thermal Sciences Journal*, vol. 51, no. 2, pp. 234–246, 2018.

64. R. A. Ilyas *et al.*, "Thermal, biodegradability and water barrier properties of bio-nanocomposites based on plasticised sugar palm starch and nanofibrillated celluloses from sugar palm fibres," *Journal of Biobased Materials and Bioenergy*, vol. 14, pp. 1–13, 2020.

65. R. A. Ilyas *et al.*, "Effect of sugar palm nanofibrillated cellulose concentrations on morphological, mechanical and physical properties of biodegradable films based on agro-waste sugar palm (Arenga pinnata (Wurmb.) Merr) starch," *Journal of Materials Research and Technology*, vol. 8, no. 5, pp. 4819–4830, Sep. 2019.

66. G. Al, D. Aydemir, B. Kaygin, N. Ayrilmis, and G. Gunduz, "Preparation and characterization of biopolymer nanocomposites from cellulose nanofibrils and nanoclays," *Journal of Composite Materials*, vol. 52, no. 5, pp. 689–700, Mar. 2018.

67. H.-S. Kim, B.-H. Lee, S. Lee, H.-J. Kim, and J. R. Dorgan, "Enhanced interfacial adhesion, mechanical, and thermal properties of natural flour-filled biodegradable polymer bio-composites," *Journal of Thermal Analysis and Calorimetry*, vol. 104, no. 1, pp. 331–338, Apr. 2011.

68. A. Anstey, S. Muniyasamy, M. M. Reddy, M. Misra, and A. Mohanty, "Processability and biodegradability evaluation of composites from poly(butylene succinate) (PBS) bioplastic and biofuel co-products from Ontario," *Journal of Polymers and the Environment*, vol. 22, no. 2, pp. 209–218, Jun. 2014.

69. T. Boronat, V. Fombuena, D. Garcia-Sanoguera, L. Sanchez-Nacher, and R. Balart, "Development of a biocomposite based on green polyethylene biopolymer and eggshell," *Materials & Design*, vol. 68, pp. 177–185, Mar. 2015.

70. S. S. Ahankari, A. K. Mohanty, and M. Misra, "Mechanical behaviour of agro-residue reinforced poly(3-hydroxybutyrate-co-3-hydroxyvalerate), (PHBV) green composites: a comparison with traditional polypropylene composites," *Composites Science and Technology*, vol. 71, no. 5, pp. 653–657, Mar. 2011.

71. R. Jumaidin *et al.*, "Effect of cogon grass fibre on the thermal, mechanical and biodegradation properties of thermoplastic cassava starch biocomposite," *International Journal of Biological Macromolecules*, no. xxxx, 2019.

72. R. Jumaidin *et al.*, "Characteristics of cogon grass fibre reinforced thermoplastic cassava starch biocomposite: water absorption and physical properties," *Journal of Advanced Research in Fluid Mechanics and Thermal Sciences 62*, vol. 62, no. 1, pp. 43–52, 2019.

73. R. Jumaidin, R. A. Ilyas, M. Saiful, F. Hussin, and M. T. Mastura, "Water transport and physical properties of sugarcane bagasse fibre reinforced thermoplastic potato starch biocomposite," *Journal of Advanced Research in Fluid Mechanics and Thermal Sciences*, vol. 61, no. 2, pp. 273–281, 2019.

74. R. A. Ilyas and S. M. Sapuan, "The preparation methods and processing of natural fibre bio-polymer composites," *Current Organic Synthesis*, vol. 16, no. 8, pp. 1068–1070, Jan. 2020.

75. R. A. Ilyas and S. M. Sapuan, "Biopolymers and biocomposites : chemistry and technology," *Current Analytical Chemistry*, vol. 16, pp. 1–4, 2020.

76. S. Siti, H. P. S. Abdul, W. O. Wan, and M. Jawai, "Bamboo based biocomposites material, design and applications," in *Materials Science - Advanced Topics*. InTech, 2013.

77. Z. A. Bakar, B. F. Hussein, and N. M. Mustapha, "Cockle shell-based biocomposite scaffold for bone tissue engineering," in *Regenerative Medicine and Tissue Engineering - Cells and Biomaterials*. InTech, 2011.
78. R. A. Ilyas, S. M. Sapuan, M. R. Ishak, and E. S. Zainudin, "Sugar palm nanofibrillated cellulose (Arenga pinnata (Wurmb.) Merr): effect of cycles on their yield, physic-chemical, morphological and thermal behavior," *International Journal of Biological Macromolecules*, vol. 123, pp. 379–388, Feb. 2019.
79. R. A. Ilyas, S. M. Sapuan, M. R. Ishak, and E. S. Zainudin, "Effect of delignification on the physical, thermal, chemical, and structural properties of sugar palm fibre," *BioResources*, vol. 12, no. 4, pp. 8734–8754, 2017.
80. R. A. Ilyas, S. M. Sapuan, and M. R. Ishak, "Isolation and characterization of nanocrystalline cellulose from sugar palm fibres (Arenga Pinnata)," *Carbohydrate Polymers*, vol. 181, pp. 1038–1051, Feb. 2018.
81. M. D. Hazrol, S. M. Sapuan, R. A. Ilyas, M. L. Othman, and S. F. K. Sherwani, "Electrical properties of sugar palm nanocrystalline cellulose, reinforced sugar palm starch nanocomposites," *Polimery*, vol. 55, no. 5, pp. 33–40, 2020.
82. R. A. Ilyas *et al.*, "Production, processes and modification of nanocrystalline cellulose from agro-waste: a review," in *Nanocrystalline Materials*. London: IntechOpen, pp. 3–32, 2019.
83. A. M. N. Azammi *et al.*, "Characterization studies of biopolymeric matrix and cellulose fibres based composites related to functionalized fibre-matrix interface," in *Interfaces in Particle and Fibre Reinforced Composites*, 1st ed., no. November. London: Elsevier, pp. 29–93, 2020.
84. S. M. Sapuan *et al.*, "Development of sugar palm-based products: a community project," in *Sugar Palm Biofibers, Biopolymers, and Biocomposites*, 1st ed. Boca Raton, FL: CRC Press/Taylor & Francis Group, pp. 245–266, 2018.
85. R. A. Ilyas, S. M. Sapuan, M. L. Sanyang, M. R. Ishak, and E. S. Zainudin, "Nanocrystalline cellulose as reinforcement for polymeric matrix nanocomposites and its potential applications: a review," *Current Analytical Chemistry*, vol. 14, no. 3, pp. 203–225, May 2018.
86. R. A. Ilyas, S. M. Sapuan, M. R. Ishak, E. S. Zainudin, and M. S. N. Atikah, "Characterization of Sugar Palm Nanocellulose and Its Potential for Reinforcement with a Starch-Based Composite," in *Sugar Palm Biofibers, Biopolymers, and Biocomposites*, 1st ed. Boca Raton, FL : CRC Press/Taylor & Francis Group, pp. 189–220, 2018.
87. M. L. Sanyang, R. A. Ilyas, S. M. Sapuan, and R. Jumaidin, "Sugar palm starch-based composites for packaging applications," in *Bionanocomposites for Packaging Applications*. Cham: Springer International Publishing, pp. 125–147, 2018.

2 Processing and Properties of Poly(lactic acid)/ Nanocellulose Nanocomposites

R.Z. Khoo, W.S. Chow, and H. Ismail
Universiti Sains Malaysia

CONTENTS

2.1 INTRODUCTION: POLYMER NANOCOMPOSITES

Polymer nanocomposites have attracted much attention worldwide from both academia and industry. Polymer nanocomposites show many desirable properties, such as improved modulus and strength, increased heat distortion temperature, enhanced barrier characteristics, and reduced gas permeability [1]. Understanding the property changes as the filler dimensions decrease to the nanoscale level is essential to optimize the nanocomposites [2]. Polymer nanocomposites are a class of multi-phase materials containing polymer matrix resin and dispersion of an ultrafine phase that have at least one dimension in the range of 1–100 nm [3]. Often, the synergistic improvement in the properties of the polymer nanocomposites is attributed to the confinement of the polymer chain and specific polymer–nanofiller interaction [4]. Polymer nanocomposites have offered a great opportunity in automotive [5], electronics [6], optical and magnetic application [7], food packaging [8], organic solar cell [9], medicine [10], construction and building [11], and water purification [12].

Biopolymer nanocomposites open an opportunity for the use of new, high-performance, lightweight green nanocomposite materials to replace conventional nonbiodegradable petroleum-based plastic materials [13]. Poly(lactic acid) (PLA) is one of the most widely used biopolymers due to its renewability, biodegradability, and biocompatibility. PLA is produced either by direct polycondensation of lactic acid or by ring-opening polymerization of lactide [14]. Nanocellulose has been received significant research interest as potential nanofiller for the reinforcements in the polymer matrices due to its renewability, biocompatibility, and tunable surface properties. Nanocellulose-reinforced polymer nanocomposites have high potential in various applications, e.g., biomedical, packaging, electronics, environment, and water treatment [15]. PLA can be used in biomedical application, the short-term application (e.g., packaging), and the long-term application (e.g., automotive and electronics). Incorporation of nanofillers is an interesting way to extend and improve the properties of PLA [16]. The addition of cellulose nanocrystals (CNCs) can modify the mechanical, thermal, and antimicrobial properties of PLA [17,18].

2.2 POLY(LACTIC ACID) (PLA)

PLA has been one of the most anticipated materials in the category of biopolymers due to its attractive mechanical properties, biodegradability, and also renewability. Being a part of the aliphatic polyester family, PLA is a type of bio-based polymer, which is produced from the fermentation of agricultural-based materials such as corn and sugar beet [19]. Figure 2.1 shows the chemical structure of PLA.

The glass transition temperature and melting temperature of PLA are about 55°C and 155°C, respectively. As a result, the processing temperature of PLA is around 165°C–180°C [20]. The promising properties of PLA such as good transparency, high strength and stiffness, low toxicity, and biodegradability make it a suitable material for the production of thermoformed containers, stretch blown bottles, and biomedical devices [21]. PLA is produced at a large scale of over 140,000 tonnes per year, which is a large quantity compared to other available biopolymers [22]. However, there are still some limitations of PLA such as brittleness (i.e., low impact strength, low elongation at break, and low toughness) and slow crystallization, which limit its potential applications in automotive, electronics, and packaging

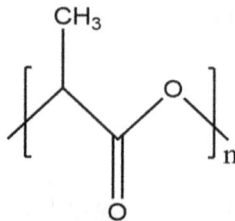

FIGURE 2.1   Chemical structure of PLA.

applications. Accordingly, adding filler or nanofiller into PLA is one of the good strategies in order to enhance its mechanical (especially in toughness and impact strength), thermal (especially crystallization rate, heat distortion temperature, and thermal stability), barrier, and flammability resistance properties [23].

2.3 NANOCELLULOSE

Cellulose is one of the most abundant biopolymers, occurring in wood, cotton, hemp, and other plant-based materials. It functions as the dominant reinforcing phase in plant structures and possesses good biocompatibility, remarkable mechanical properties, and low density. Plant fibers mainly consist of cellulose, hemicellulose, and lignin [24]. Native cellulose fibers is comprised of microfibrils. These microfibrils contain alternating amorphous and crystalline regions, which play an important role in producing CNC. CNC is a structure of densely organized crystalline particles engineered by nature in a way that makes it inherently strong [25]. In the open literature, CNC is also known as cellulose nanowhiskers (CNWs), crystalline nanocellulose (CNC), or nanocrystalline cellulose (NCC). CNC can be produced via chemical treatments of cellulose fiber. At nanometer sizes, nanocellulose can reinforce polymer composites even at low filler content. CNC can be extracted from naturally occurring polymers, mainly cellulose fibers. Cellulose is a natural linear biopolymer found in plant cells such as wood and cotton. It is the main building block in trees and plants. Figure 2.2 shows a repeating segment of glucose present in cellulose, also known as cellobiose.

Figure 2.3 shows the extraction method of CNC via acid hydrolysis from cellulose fibers. Generally, cellulose fibers first undergo alkaline treatment and bleaching to extract cellulose, and this process will also remove the soluble polysaccharides along with residual phenolic molecules (i.e., lignin). Then, cellulose is subjected to acid hydrolysis. The resulting suspension will be subsequently diluted with water and washed with successive centrifugation followed by ultrasonication in order to produce CNC. CNC obtained via acid hydrolysis is highly crystalline in nature and is of needle-like shape with 5–70 nm in diameter and 100–250 nm in length, as shown in Figure 2.4. The obtained particles have ~100% cellulose and high crystallinity (54%–88%). As mentioned earlier, cellulose consists of amorphous region and microcrystalline region. The amorphous region is easily hydrolyzed by strong acids such as sulfuric acid and hydrochloric acid [26], while leaving behind microcrystalline region with high aspect ratio (length-to-diameter ratio) in the form of an aqueous suspension.

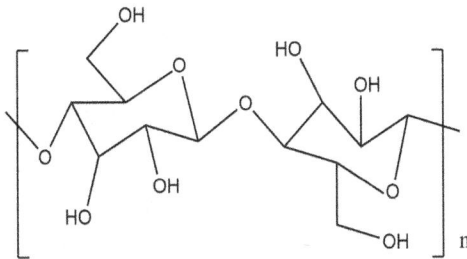

FIGURE 2.2 Chemical structure of cellobiose.

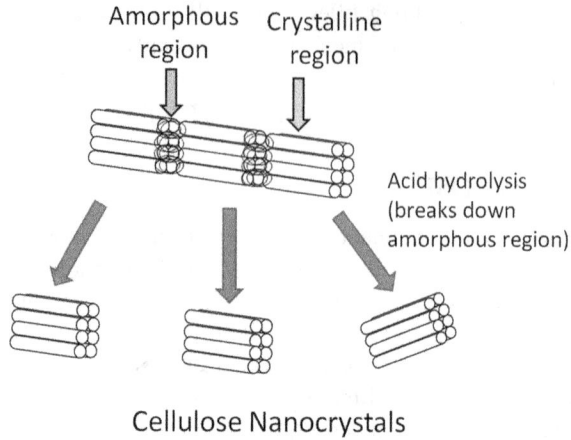

FIGURE 2.3   Synthesis of CNC via acid hydrolysis of cellulose.

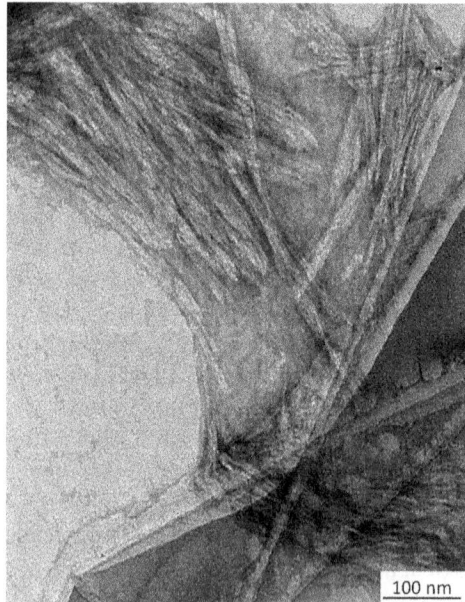

FIGURE 2.4   TEM micrograph of needle-like CNC extracted by acid hydrolysis.

It is important to note that different acids will have different effects on the CNC. For example, cellulose extracted using hydrochloric acid (HCl) presents a weak negatively charged surface, whereas cellulosed extracted using sulfuric acid (H_2SO_4) presents a more negatively charged surface functionality. Accordingly, H_2SO_4 has proven to be an effective hydrolyzing agent because of the esterification process that allows grafting of anionic sulfate ester groups [27]. As a result, a negative electrostatic layer will be formed on the outer layer of nanocrystals when these negatively charged groups are present, thus promoting better dispersion in water. Lu et al. [28] investigated the effects of temperature, sulfuric acid concentration, and mass of raw material on the yield and crystallinity of CNC extracted from filter paper, using ultrasonic wave and microwave-assisted technique simultaneously. In their study, they found that at the temperature of 70°C, 50% concentration of H_2SO_4 can convert 2 g of filter paper to CNC in 1.5 h. The yield of CNC is 78%, and the crystallinity of the CNC is approximately 80%.

2.4 PROCESSING OF PLA/CNC NANOCOMPOSITES

PLA/CNC nanocomposites can be prepared using different processing methods, e.g., solvent casting, melt compounding and molding, and layer-by-layer coating technique. The final properties (e.g., mechanical, thermal, and optical) are governed by the dispersibility of CNC in the PLA matrix. Due to their tailorability, design flexibility, and processability, polymer-nanocellulose composites are extensively used in the automotive, packaging, electronics, and biomedical industries. However, some disadvantages of using nanocellulose as a reinforcing material are its high moisture absorption, poor wettability, incompatibility with most polymeric matrices, limitations in processing temperature, and self-agglomeration of CNC in the polymer matrix. Preventing the agglomeration of CNC remains a challenge in the field of polymer nanocomposites [29]. CNC tends to agglomerate due to the presence of the interacting surface hydroxyl groups. This agglomeration affects CNC nanoscale formation and can impede the mechanical reinforcement of the polymer matrix.

Surface modification and functionalization of CNC is one of the approaches to improve the dispersibility of CNC in the polymer matrix, which is attributed to the presence and abundance of hydroxyl groups on its surface. Some of the feasible functionalization strategies of CNC include etherification, oxidation, esterification, polymer grafting, and silylation. Several chemical modifications, like introducing stable negative or positive electrostatic charges on the nanocellulose surface, have been investigated to obtain better dispersion [30]. In the following section, two main techniques (i.e., solvent casting and melt mixing compounding) together with some surface functionalization strategies to produce PLA/CNC nanocomposites are discussed in order to improve the CNC dispersion.

2.4.1 SOLVENT CASTING

Solvent casting (solution casting) technology has been used for the production of engineering plastics, optical films, medical films, and sheet forming for electronic applications. The advantages of this technology include uniform thickness distribution, maximum optical purity, and extremely low haze. The cast film can be processed in-line with an optical coating design. The key elements of the solvent cast technology include the following: (1) The polymer must be soluble in a volatile solvent or water, (2) a stable solution with a reasonable minimum solid content and viscosity should be formed, and (3) the formation of a homogeneous film and its release from the casting support must be possible [31].

Both water-soluble polymers and polymer aqueous dispersions (e.g., latex) can be used to obtain nanocomposites with CNC. After mixing the CNC dispersion with the polymer solution/dispersion, a solid nanocomposite film can be obtained by simple casting and water evaporation. This mode of processing allows the preservation of the individualization state of the CNC nanoparticles resulting from their colloidal dispersion in water [32]. Attempts to avoid agglomerations are made to maintain a suspension of particles in an adequate solvent as long as possible. Suspension maintenance can continue into the processing stage by the use of a suitable water-soluble matrix or by performing solvent exchange from water to a convenient organic solvent [29].

Due to the high stability of CNC in water dispersion, the most preferred solvent is water and the difficulty of CNC redispersion after freeze dry can be eliminated. However, since PLA is hydrophobic, water is not a feasible option as solvent usage in the preparation of nanocomposites. Chloroform is normally used to produce PLA nanocomposites when solution casting method is considered. Figure 2.5 shows a simple laboratory-scale procedure of solution casting to prepare PLA/CNC nanocomposites. More details about the real production and manufacturing of solvent cast film (in general) have been documented by Siemann [31].

Liu et al. [33] prepared PLA/flax yarns of CNC (CNC loading: 2.5 and 5 wt%) by the solution casting of mixtures of PLA solution and CNC suspension in chloroform. The PLA/CNC film (thickness = 160 μm) showed good transparency. The tensile modulus and tensile strength of PLA were increased by 47% and 59%, respectively, by the addition of 5 wt% CNC. Panicker et al. [34] used the solution casting method to fabricate PLA bionanocomposite film reinforced with CNC obtained from sugarcane bagasse (SCB). The CNC was obtained by the dual acid hydrolysis of microcrystalline cellulose (MCC) using sulfuric acid and hydrochloric acid followed by ultrasonication. The resulting CNC exhibited excellent dispersion characteristics owing to the surface sulfate esters. Fortunati et al. [35] used a surfactant (acid phosphate ester of ethoxylated

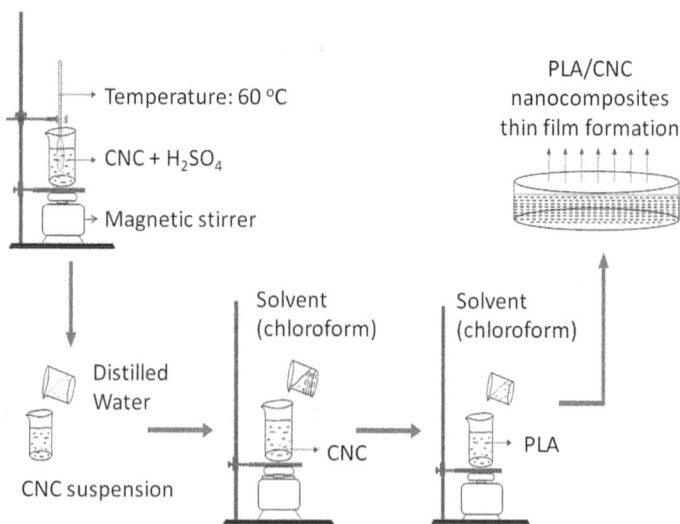

FIGURE 2.5 Laboratory-scale solvent casting of PLA/CNC nanocomposites.

nonylphenol, Beycostat A B09) to modify the surface of CNC and investigated its effects on PLA nanocomposites. The CNC was subjected to modification before freeze dry followed by solvent casting with PLA. It was found that dispersion of CNC is enhanced by the presence of surfactant as it helps in better interaction between CNC and PLA matrix. Reduction in 34% water permeability in the casted films displays good oxygen barrier properties, thus proving the effectiveness of the surface-modified CNC. Espino-Pérez et al. [36] investigated the effects of CNC surface grafting using *n*-octadecyl-isocyanate (ICN) on PLA solution-casted films. The CNC suspensions were solvent-exchanged to chloroform by centrifugation and then mixed with PLA/chloroform solution. The ICN-modified CNC exhibited better distribution in the PLA matrix compared to unmodified CNC. This can be justified by the higher tensile properties and better transparency of the PLA/ICN-modified CNC nanocomposites compared to those of PLA/unmodified CNC.

2.4.2 MELT COMPOUNDING AND MOLDING

Melt compounding and molding is one of the conventional ways to produce polymer blends and composite materials. Figure 2.6 shows common equipment used in melt compounding and molding. Melt processing is environmental-friendly (often solvent-free) and viable for commercial and industrial applications. Extrusion and injection molding is commonly used to process thermoplastic polymers. Extrusion temperature, time, and screw speed are usually optimized to avoid the agglomeration of fillers during the mixing process. In the case of CNC, strong hydrogen bonds are formed during drying process, which limits the particle dispersion and its reinforcement effects on the polymer. Sometimes, CNC may be prone to thermal degradation during the melt compounding if the processing temperature is too high. This is always related to the thermal stability of the respective CNC and the selection of the thermoplastic. In other words, the melting temperature of the thermoplastic should be tolerant to the thermal stability of the selected CNC. The melting compounding of PLA/CNC is still a feasible approach. Although there is not much problem in terms of processing temperature constraints, the CNC agglomeration issues remain challenging.

Oksman et al. [37] attempted to melt-mix PLA with CNC (extracted from MCC) using a corotating twin screw extruder. The extrusion was carried out in

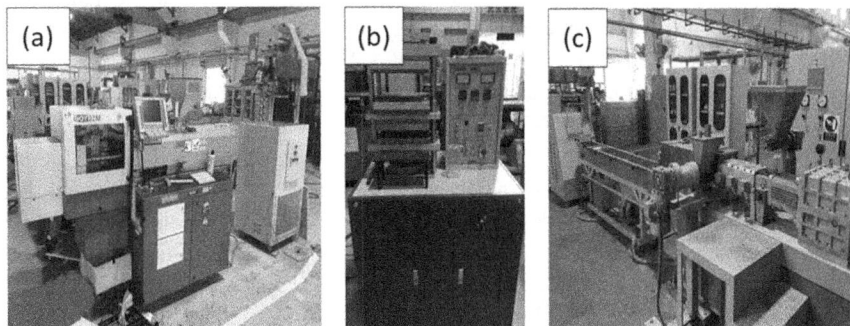

FIGURE 2.6   Melt processing technique: (a) twin screw extruder, (b) compression molding machine, and (c) injection molding machine.

the temperature range of 170°C–200°C, and the screw speed was held constant at 150 rpm. The total throughput was 5 kg/h, and the maximum capacity of this extruder was 50 kg/h. The MCC was treated with N,N-dimethylacetamide (DMAc) containing lithium chloride (LiCl) in order to swell the MCC and partly separate the CNC. The results showed that DMAc/LiCl could be used as a swelling/separation agent for MCC and pumping medium but caused the degradation of the composites during high-temperature processing. It was found that the addition of polyethylene glycol (PEG) interacts with CNCs, covers them, and improves the dispersion of the CNCs. The extrusion process using liquid feeding of the CNC suspension is a promising way to produce cellulose nanocomposites. It is practically possible to feed a liquid suspension of the CNC into the hot polymer melt; however, the concern is a proper venting system that should be designed for the removal of liquid during the extrusion.

A possible approach to prevent reaggregation of CNC is to add a water-soluble polymer into the CNC suspension. This water-soluble polymer is expected to encapsulate the single CNC and prevent reaggregation and formation of strong hydrogen bonds as the water is sublimated or evaporated, e.g., during freeze drying or direct pumping into a polymer melt during extrusion. Bondeson and Oksman [38] prepared PLA/CNC nanocomposites by corotating and fully intermeshing twin screw extruder. Polyvinyl alcohol (PVOH) was used to improve the dispersion of CNC in the PLA matrix. PVOH is used to encapsulate the single CNC and hinder reaggregation as the water is removed. The PLA/CNC nanocomposites are produced by extrusion, and two different feeding methods are used: dry feeding of freeze-dried PVOH/CNC pellets together with PLA and liquid feeding of PVOH/CNC suspension directly into molten PLA during extrusion. From an economical point of view, it could be beneficial with liquid feeding. Supplementary treatment of the suspension before extrusion, like freeze drying, will involve additional expenses such as investment costs for pieces of equipment and prolonged preparation times. Transmission electron microscopy (TEM) micrograph analysis showed that the CNCs were better dispersed in the PLA-PVOH/CNC nanocomposite produced using liquid feeding and that the CNCs were also partially dispersed in the PLA phase. However, it was found that the majority of the CNCs were located in the PVOH phase and only a negligible amount was located in the PLA phase, thus limiting the improvement in tensile properties. This should be taken into consideration if one needs to achieve enhancement in mechanical and thermal properties, since the properties of the PLA/CNC nanocomposites are governed by the CNC dispersion and localization in the PLA matrix.

Functionalization of cellulose such as grafting or silanization was carried out to improve the affinity between CNC and PLA [39]. Yin et al. [40] grafted poly(ethylene glycol) epoxide (PEG-EP) on CNC surfaces using γ-aminopropyltriethoxysilane (APS) as an intermediate, to improve the thermostability of CNC and compatibility of CNC and PLA matrix. APS was applied as intermediate to introduce reactive amido groups on CNC surfaces as initiating sites, which can nucleate the PEG-EP grafting of CNC. The surface-treated CNCs were then dry-mixed with PLA powder followed by the subsequent hot-pressing process (molding temperature = 170°C; molding time = 10 min; pressure = 40 MPa). Scanning electron microscopic (SEM) analysis indicated that the modified CNC showed better dispersion in PLA matrix.

2.5 MECHANICAL PROPERTIES OF PLA/CNC NANOCOMPOSITES

Use of CNC as a filler in nanocomposites has been actively studied due to its good mechanical properties such as high flexural strength and stiffness (e.g., Young's modulus ~150 GPa). Polymer/CNC nanocomposites often demonstrated good mechanical properties such as the aspect ratio, reinforcing ability, and interfacial interaction of CNC.

Since CNC suspensions have good stability in water, the preparation of nanocomposites becomes eco-friendly and advantageous if the process of dispersion of CNC into a polymer matrix (or into another polymer which is miscible or compatible with the matrix) is carried out in an aqueous medium in place of an organic solvent mixture. In the case of PLA/CNC composites, such a goal can be achieved by using an aqueous PVAc emulsion. Pracella et al. [41] investigated the reinforcement effects of glycidyl methacrylate (GMA)-functionalized CNC and predispersed CNC (in PVAc emulsion) on PLA nanocomposites. The tensile modulus and strength of PLA/PVAc/CNC nanocomposites are higher than those of pure PLA. The improvement in mechanical properties was attributable to homogenous dispersion of CNC in the PLA matrix, which is well correlated with the morphological observation of SEM.

Arias et al. [42] investigated the dispersion and properties of PLA/CNC nanocomposites prepared from a two-step process (i.e., solvent casting and melt mixing). Low molecular weight polyethylene oxide (PEO; M_w = 1,000 g/mol) and high molecular weight PEO (M_w = 5 × 10^6 g/mol) were used to prepare PEO/CNC blend before melt mixing with PLA.

When using the PEO (M_w = 1,000 g/mol) and higher PEO/CNC ratios (i.e., 1.25 and 12.5), finer dispersion of CNC was observed. PLA/PEO/CNC nanocomposites achieved a significant improvement in elongation at break. At PEO/CNC ratio of 12.5, the elongation at break was recorded at 80%, which is approximately 20 times compared to that of PLA.

Panicker et al. [34] investigated the mechanical properties of dual acid-hydrolyzed SCB CNC-reinforced PLA. PLA/CNC composite films were solution-casted with different compositions of CNC at 0, 0.2, 0.3, 0.4, and 0.5 wt%. The tensile modulus and strength of PLA/CNC were found to be increased by the incorporation of CNC, due to the reinforcing ability of CNC, increased crystallinity (CNC as a nucleating agent), and possible interfacial bonding between PLA and CNC. The tensile strength of PLA was increased by approximately 108% by the addition of 0.3 wt% CNC. Nevertheless, it should be mentioned that when the CNC loading exceeded 0.3 wt%, the tensile strength of the PLA/CNC nanocomposites was decreased gradually – this is always related to the agglomeration of CNC.

2.6 THERMAL PROPERTIES OF PLA/CNC NANOCOMPOSITE

Thermal properties of PLA such as melting temperature, co-crystallization temperature, glass transition temperature, and decomposition temperature are widely studied by researchers, because they are essential data for the PLA processing and application.

Sullivan et al. [43] investigated the thermomechanical effects of CNC on the properties of PLA films by dynamic mechanical analyzer (DMA). The storage modulus of PLA was found to be significantly increased from 1.9 to 2.9 GPa with the addition of 3 wt% CNC. This is attributed to the hindered polymer chain mobility and increased PLA crystallinity (X_c of PLA = 11.5%, whereas X_c of PLA/CNC 1 wt% = 29.7%). Figure 2.7 shows the storage modulus and the loss modulus curve as a function temperature for the PLA and PLA/CNC nanocomposites.

Khoo et al. [44] reported the thermal properties of PLA/CNC nanocomposites that have been prepared using the solution casting technique. Thermogravimetric analyzer (TGA) and differential scanning calorimeter (DSC) were used to characterize the thermal properties of the PLA/CNC nanocomposites. From the DSC results, the cold crystallization temperature (T_{cc}) of PLA and PLA/CNC nanocomposites is hardly observable. This indicates that the material is highly crystalline after the solution casting technique. The degree of crystallinity (X_c) of PLA was affected by the CNC loading. At 5 wt% CNC, higher crystallinity (34.5%) was obtained compared to the neat PLA (30.9%), which could be due to the presence of CNC that accelerated the crystallization of PLA. From TGA results, it was found that the decomposition temperature (T_d) of the PLA/CNC nanocomposites (CNC loading: 5 wt%) was recorded as 376°C, which is higher than that of pure PLA (T_d = 353°C). This indicates that the CNC enhanced the thermal stability of PLA.

Bagheriasl et al. [45] investigated the thermal properties of solvent-casted PLA/CNC nanocomposites using DMA and DSC. From the DMA results, the storage modulus of the PLA nanocomposites significantly increased with CNC content, particularly at low frequencies. DSC results demonstrated that the crystallinity of the PLA in the nanocomposites increased, which is associated with the nucleation effect of the CNC on the crystallization of PLA.

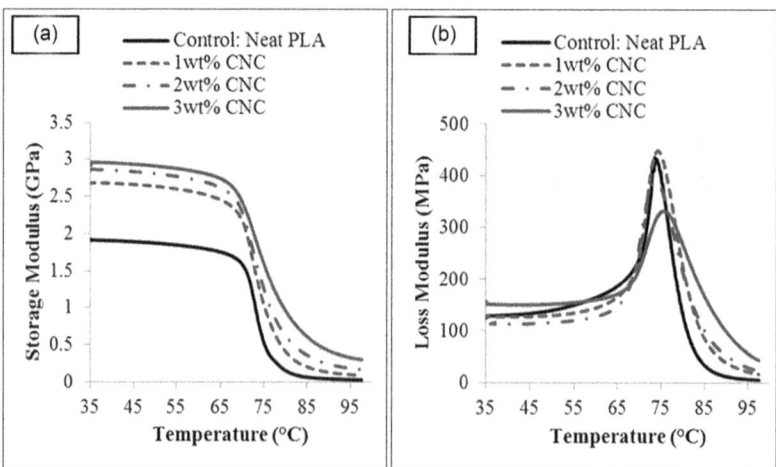

FIGURE 2.7 (a) Storage modulus and (b) loss modulus of the CNC/PLA films as a function of CNC content. (Reprinted with permission from Sullivan et al. [43].)

2.7   OPTICAL PROPERTIES OF PLA/CNC NANOCOMPOSITES

Optical transparency of the polymer matrix is an important characteristic to be preserved after the addition of filler materials. Optical properties of CNC-based polymer composites can be evaluated under polarized light to show the CNC flow birefringence and visual examination of the transparency of composites. UV-visible spectrometer can be used to investigate the optical properties of CNC-based nanocomposites at the wavelength between 200 and 1,000 nm. Effective cross-sectional area of CNC and its dispersion level in the host matrix are the fundamental aspects in determining the transparency of polymer nanocomposites. Higher transmittance values represent a good transparency as well as a homogenous dispersion of CNC in polymer matrix. Figure 2.8 shows the flow birefringence of CNC suspensions produced from medium-milled wood (MMW), acid sulfite-pretreated wood (SPW), and MCC [46]. CNC originated from MCC exhibits stronger birefringence owing to its well-organized particle sizes, compared to the other two materials. Birefringence of CNC is an evidence of the well-dispersed CNC suspension obtained from sulfuric acid hydrolysis [39].

Espino-Pérez et al. [36] produced PLA nanocomposite films reinforced with unmodified CNC and ICN-grafted CNC. CNCs were well dispersed in the polymer matrix, and the film transparency was similar to a neat PLA film. PLA films filled with ICN-grafted CNC were more transparent compared to unmodified CNC. At higher loading of CNC, granular structures can be observed in the PLA films due to the agglomeration of unmodified CNC, while ICN-grafted CNC only made the PLA films opaque and no aggregates were identified.

Orellana et al. [47] studied the optical properties of PLA films containing surfactant-modified CNCs. The decylamine (DA) surfactant at 1.0 wt% with respect to CNC loading was found to significantly enhance CNC compatibility and PLA property. Besides a good transparency, the PLA nanocomposites with high CNC concentration also exhibited interesting optical properties when observed between crossed polars. Colored phases appeared as the thickness was increased by stacking layers of the composite films (see Figure 2.9). This behavior is believed to be the result of the self-assembly of CNC within the films despite the agglomerations observed at high CNC concentrations.

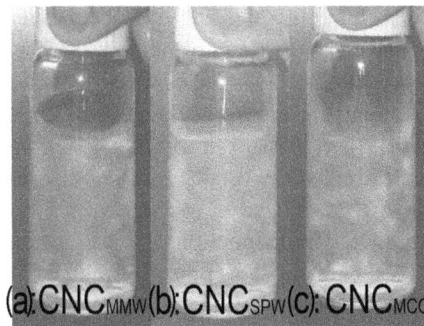

(a) CNC_{MMW} (b) CNC_{SPW} (c) CNC_{MCC}

FIGURE 2.8   Flow birefringence of CNC suspensions between two cross-polarizers (a) CNC_{MMW}, (b) CNC_{SPW}, and (c) CNC_{MCC}. (Reprinted with permission from Du et al. [46].)

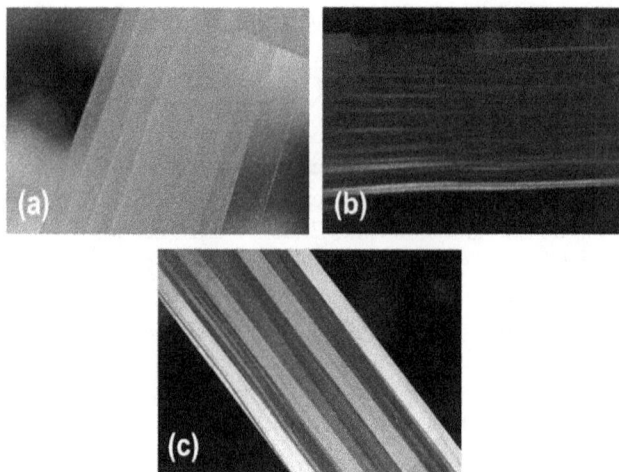

FIGURE 2.9 Photographs of four PLA-CNC (10%)–DA films that are stacked on top of one another with an offset such that the edges are one film thick and the center is four films thick. The stacked films are viewed (a) in unpolarized light, (b) between cross-polarizers at a 0° angle, and (c) between cross-polarizers at a 45° angle. (Reprinted with permission from Orellana et al. [47].)

2.8   ANTIMICROBIAL PROPERTIES OF PLA/CNC NANOCOMPOSITES

Food packaging has always been an important issue in producing waste materials as they were usually disposed into the environment after its use. Commercial packaging uses synthetic polymers such as polyethylene (PE) and polypropylene (PP) – these polymers are excellent in terms of barrier and thermomechanical properties but they are nonbiodegradable [48]. Hence, biopolymers like PLA were chosen as an alternative material due to its ease of processing, transparency, and biodegradability [49].

Bioactive packaging is the recent trend of food packaging products that fulfill consumers' demands of high quality and safe products besides providing additional functions like antimicrobial properties to prevent food spoilage. Incorporation of antimicrobial substances is an innovative way to synthesize active food packaging products to control the growth of microorganisms on the food surface [50]. Antimicrobial substances such as nisin [51], essential oils (EOs) [49,52], and silver nanoparticles [50] were used together with CNC to produce nanocomposites for food packaging applications.

Fortunati et al. [50] prepared multifunctional PLA nanocomposite films incorporating CNC and silver (Ag) nanoparticles into PLA matrix. The antibacterial properties of the ternary system were tested against microorganisms, i.e., *Escherichia coli RB* (*E. coli RB*) and *Staphylococcus aureus* 8325-4 (*S. aureus* 8325-4). PLA films reinforced with 5 wt% of CNC and 1 wt% Ag exhibited the best antibacterial activity on *E. coli* regardless of the temperature chosen (*T* = 4°C, 24°C and 37°C).

For *S. aureus*, PLA films containing 5 wt% of modified CNC and 1 wt% Ag showed the best antibacterial activity as it showed greatest percent of bacterial attachment. Overall, PLA samples carrying Ag nanoparticles have a better antibacterial effect on *E. coli* than on *S. aureus*. This may be due to a thicker cell wall of *S. aureus* that protects the cell from penetration of Ag ions into cytoplasm. The silver ions turn deoxyribonucleic acid (DNA) into a condensed form that will damage or cause destruction to microorganisms. Silver powder in nanometer scale also helps in inducing the release of high Ag+ ions, which is crucial to the antimicrobial activity in a system.

Salmieri et al. [51] prepared bioactive PLA/CNC films containing nisin to investigate their antimicrobial properties on cooked ham. Nisin is a type of bacteriocin, which is commonly used in bioactive food packaging. It is a small polypeptide (34 amino acids) produced from Gram-positive bacterium *Lactococcus lactis* subsp. *lactis*. Nisin is effective against a wide variety of foodborne pathogens such as *Listeria monocytogenes*, *S. aureus*, *Clostridium botulinum*, and its spore. PLA/CNC films were first compression-molded and placed in an aqueous solution of nisin for 24 h. The presence of CNC is critical as it is capable to enhance the nisin binding onto PLA/CNC chains via hydrogen bonding. *L. monocytogenes* was used in this study as this pathogen is commonly found on raw meat and cooked ham offers appropriate conditions for pathogen growth. PLA films that contain nisin showed a significant reduction of *L. monocytogenes* in ham from day 1 and a total inhibition from day 3. Besides, the percentage of nisin release also increased continuously from day 0 to day 14 (21% at day 14). These results confirm the antimicrobial effectiveness of PLA/CNC/nisin films in pathogen inhibition of meat products.

Salmieri et al. [52] demonstrated the antimicrobial capacity of oregano EO in PLA/CNC nanocomposite films. The phenolic-rich compounds in EO are effective against foodborne pathogens. Oregano EO is considered one of the best antimicrobial agents. PLA/CNC/oregano EO composite films were prepared by solution casting and tested against *L. monocytogenes* in mixed vegetables. It was found that the PLA/CNC/oregano EO films were capable of performing a quasi-total inhibition of bacteria in vegetables at day 14. This indicates that the films can be part of a potential bioactive packaging with strong antimicrobial capacity for vegetable products.

2.9 CONCLUSION REMARKS

PLA is one of the most popular and environmental-friendly biopolymers with huge prospects in various potential markets, such as food packaging, automotive, electronics, and biomedical devices. CNC offers a variety of enhancements to polymer matrix in terms of mechanical and thermal properties attributed to its renewability, biocompatibility, and surface tunable properties. Combination of PLA and CNC is a feasible approach to develop greener bionanocomposites. However, some issues should be taken into consideration, e.g., agglomeration of CNC and processability of PLA/CNC. Thus, more efforts should be made (1) to improve the dispersion of CNC (via surface modification and functionalization) and (2) to increase the processing feasibility and industrialization of PLA/CNC (via innovation in processing techniques that are suitable for mass production).

ACKNOWLEDGMENT

We gratefully acknowledge the financial support provided by Research University Grant (1001/PBAHAN/8014024) from Universiti Sains Malaysia.

REFERENCES

1. W. S. Chow and Z. A. Mohd Ishak, "Polyamide blend-based nanocomposites: a review," *Express Polym Lett,* vol. 9, pp. 211–232, 2015.
2. D. R. Paul and L. M. Robeson, "Polymer nanotechnology: nanocomposites," *Polymer,* vol. 49, pp. 3187–3204, 2008.
3. K. K. Maniar, "Polymeric nanocomposites: a review," *Polym-Plast Technol Eng,* vol. 43, pp. 427–443, 2004.
4. Q. Yuan and R. Misra, "Polymer nanocomposites: current understanding and issues," *Mater Sci Technol,* vol. 22, pp. 742–755, 2006.
5. J. M. Garcés, D. J. Moll, J. Bicerano, R. Fibiger, and D. G. McLeod, "Polymeric nanocomposites for automotive applications," *Adv Mater,* vol. 12, pp. 1835–1839, 2000.
6. C. Min, X. Shen, Z. Shi, L. Chen, and Z. Xu, "The electrical properties and conducting mechanisms of carbon nanotube/polymer nanocomposites: a review," *Polym-Plast Technol Eng,* vol. 49, pp. 1172–1181, 2010.
7. S. Li, M. Meng Lin, M. S. Toprak, D. K. Kim, and M. Muhammed, "Nanocomposites of polymer and inorganic nanoparticles for optical and magnetic applications," *Nano Rev,* vol. 1, p. 5214, 2010.
8. A. M. Youssef, "Polymer nanocomposites as a new trend for packaging applications," *Polym-Plast Technol Eng,* vol. 52, pp. 635–660, 2013.
9. N. Khan, A. Kausar, and A. U. Rahman, "Modern drifts in conjugated polymers and nanocomposites for organic solar cells: a review," *Polym-Plast Technol Eng,* vol. 54, pp. 140–154, 2015.
10. D. Feldman, "Polymer nanocomposites in medicine," *J Macromol Sci A,* vol. 53, pp. 55–62, 2016.
11. J. Silvestre, N. Silvestre, and J. De Brito, "Polymer nanocomposites for structural applications: recent trends and new perspectives," *Mech Adv Mater Struc,* vol. 23, pp. 1263–1277, 2016.
12. N. Pandey, S. Shukla, and N. Singh, "Water purification by polymer nanocomposites: an overview," *Nanocomposites,* vol. 3, pp. 47–66, 2017.
13. J.-W. Rhim, H.-M. Park, and C.-S. Ha, "Bio-nanocomposites for food packaging applications," *Prog Polym Sci,* vol. 38, pp. 1629–1652, 2013.
14. M. M. Reddy, S. Vivekanandhan, M. Misra, S. K. Bhatia, and A. K. Mohanty, "Biobased plastics and bionanocomposites: current status and future opportunities," *Prog Polym Sci,* vol. 38, pp. 1653–1689, 2013.
15. S. Mondal, "Review on nanocellulose polymer nanocomposites," *Polym-Plast Technol Eng,* vol. 57, pp. 1377–1391, 2018.
16. J.-M. Raquez, Y. Habibi, M. Murariu, and P. Dubois, "Polylactide (PLA)-based nanocomposites," *Prog Polym Sci,* vol. 38, pp. 1504–1542, 2013.
17. R. Z. Khoo, W. S. Chow, and H. Ismail, "Sugarcane bagasse fiber and its cellulose nanocrystals for polymer reinforcement and heavy metal adsorbent: a review," *Cellulose,* vol. 25, pp. 4303–4330, 2018.
18. I. Gan and W. Chow, "Antimicrobial poly (lactic acid)/cellulose bionanocomposite for food packaging application: a review," *Food Packaging Shelf,* vol. 17, pp. 150–161, 2018.
19. P. K. Bajpai, I. Singh, and J. Madaan, "Tribological behavior of natural fiber reinforced PLA composites," *Wear,* vol. 297, pp. 829–840, 2013.

20. D. Garlotta, "A literature review of poly (lactic acid)," *J Polym Env,* vol. 9, pp. 63–84, 2001.

21. A. J. Lasprilla, G. A. Martinez, B. H. Lunelli, A. L. Jardini, and R. M. Filho, "Poly-lactic acid synthesis for application in biomedical devices - a review," *Biotechnol Adv,* vol. 30, pp. 321–328, 2012.

22. T. Mukherjee and N. Kao, "PLA based biopolymer reinforced with natural fibre: a review," *J Polym Env,* vol. 19, p. 714, 2011.

23. W. S. Chow, E. L. Teoh, and J. Karger-Kocsis, "Flame retarded poly(lactic acid): a review," *Express Polym Lett,* vol. 12, pp. 396–417, 2018.

24. A. Khan, T. Huq, R. A. Khan, B. Riedl, and M. Lacroix, "Nanocellulose-based composites and bioactive agents for food packaging," *Crit rev in Food Sci Nutr,* vol. 54, pp. 163–174, 2014.

25. A. Kumar, Y. S. Negi, V. Choudhary, N. K. Bhardwaj, and S. S. Han, "Morphological, mechanical, and in vitro cytocompatibility analysis of poly(vinyl alcohol)–silica glass hybrid scaffolds reinforced with cellulose nanocrystals," *Int J Polym Anal Ch,* vol. 22, pp. 139–151, 2016.

26. M. A. Henrique, W. P. Flauzino Neto, H. A. Silvério, D. F. Martins, L. V. A. Gurgel, H. d. S. Barud, *et al.,* "Kinetic study of the thermal decomposition of cellulose nanocrystals with different polymorphs, cellulose I and II, extracted from different sources and using different types of acids," *Ind Crop Prod,* vol. 76, pp. 128–140, 2015.

27. M. A. Azizi Samir, F. Alloin, and A. Dufresne, "Review of recent research into cellulosic whiskers, their properties and their application in nanocomposite field," *Biomacromolecules,* vol. 6, pp. 612–626, 2005.

28. Z. Lu, L. Fan, H. Zheng, Q. Lu, Y. Liao, and B. Huang, "Preparation, characterization and optimization of nanocellulose whiskers by simultaneously ultrasonic wave and microwave assisted," *Bioresour Technol,* vol. 146, pp. 82–88, 2013.

29. H. Kargarzadeh, M. Mariano, J. Huang, N. Lin, I. Ahmad, A. Dufresne, *et al.,* "Recent developments on nanocellulose reinforced polymer nanocomposites: a review," *Polymer,* vol. 132, pp. 368–393, 2017.

30. E. Fortunati, W. Yang, F. Luzi, J. Kenny, L. Torre, and D. Puglia, "Lignocellulosic nanostructures as reinforcement in extruded and solvent casted polymeric nanocomposites: an overview," *Eur Polym J,* vol. 80, pp. 295–316, 2016.

31. U. Siemann, "Solvent cast technology–a versatile tool for thin film production," in Stribeck, N. and Smarsly, B. (eds), *Scattering methods and the properties of polymer materials,* ed. Springer, Berlin, Heidelberg, vol. 130, pp. 1–14, 2005.

32. F. Vilarinho, A. Sanches Silva, M. F. Vaz, and J. P. Farinha, "Nanocellulose in green food packaging," *Crit Rev Food Sci Nutr,* vol. 58, pp. 1526–1537, 2018.

33. D. Y. Liu, X. W. Yuan, D. Bhattacharyya, and A. J. Easteal, "Characterisation of solution cast cellulose nanofibre – reinforced poly(lactic acid)," *Express Polym Lett,* vol. 4, pp. 26–31, 2010.

34. A. M. Panicker, K. A. Rajesh, and T. O. Varghese, "Mixed morphology nanocrystalline cellulose from sugarcane bagasse fibers/poly(lactic acid) nanocomposite films: synthesis, fabrication and characterization," *Iran Polym J,* vol. 26, pp. 125–136, 2017.

35. E. Fortunati, M. Peltzer, I. Armentano, L. Torre, A. Jimenez, and J. M. Kenny, "Effects of modified cellulose nanocrystals on the barrier and migration properties of PLA nano-biocomposites," *Carbohydr Polym,* vol. 90, pp. 948–956, 2012.

36. E. Espino-Pérez, J. Bras, V. Ducruet, A. Guinault, A. Dufresne, and S. Domenek, "Influence of chemical surface modification of cellulose nanowhiskers on thermal, mechanical, and barrier properties of poly(lactide) based bionanocomposites," *Eur Polym J,* vol. 49, pp. 3144–3154, 2013.

37. K. Oksman, A. P. Mathew, D. Bondeson, and I. Kvien, "Manufacturing process of cellulose whiskers/polylactic acid nanocomposites," *Compos Sci Technol,* vol. 66, pp. 2776–2784, 2006.
38. D. Bondeson and K. Oksman, "Polylactic acid/cellulose whisker nanocomposites modified by polyvinyl alcohol," *Compos Part A: Appl Sci,* vol. 38, pp. 2486–2492, 2007.
39. M. Mariano, N. El Kissi, and A. Dufresne, "Cellulose nanocrystals and related nanocomposites: review of some properties and challenges," *J Polym Sci Polym Phys,* vol. 52, pp. 791–806, 2014.
40. Y. Yin, J. Ma, X. Tian, X. Jiang, H. Wang, and W. Gao, "Cellulose nanocrystals functionalized with amino-silane and epoxy-poly(ethylene glycol) for reinforcement and flexibilization of poly(lactic acid): material preparation and compatibility mechanism," *Cellulose,* vol. 25, pp. 6447–6463, 2018.
41. M. Pracella, M. M.-U. Haque, and D. Puglia, "Morphology and properties tuning of PLA/cellulose nanocrystals bio-nanocomposites by means of reactive functionalization and blending with PVAc," *Polymer,* vol. 55, pp. 3720–3728, 2014.
42. A. Arias, M.-C. Heuzey, M. A. Huneault, G. Ausias, and A. Bendahou, "Enhanced dispersion of cellulose nanocrystals in melt-processed polylactide-based nanocomposites," *Cellulose,* vol. 22, pp. 483–498, 2014.
43. E. Sullivan, R. Moon, and K. Kalaitzidou, "Processing and characterization of cellulose nanocrystals/polylactic acid nanocomposite films," *Materials,* vol. 8, pp. 8106–8116, 2015.
44. R. Z. Khoo and W. S. Chow, "Mechanical and thermal properties of poly(lactic acid)/sugarcane bagasse fiber green composites," *J Thermoplast Compos Mater,* vol. 30, pp. 091–1102, 2015.
45. D. Bagheriasl, P. J. Carreau, B. Riedl, and C. Dubois, "Enhanced properties of polylactide by incorporating cellulose nanocrystals," *Polym Compos,* vol. 39, pp. 2685–2694, 2018.
46. L. Du, J. Wang, Y. Zhang, C. Qi, M. P. Wolcott, and Z. Yu, "Preparation and characterization of cellulose nanocrystals from the bio-ethanol residuals," *Nanomaterials,* vol. 7(51), pp. 1–12, 2017.
47. J. L. Orellana, D. Wichhart, and C. L. Kitchens, "Mechanical and optical properties of polylactic acid films containing surfactant-modified cellulose nanocrystals," *J Nanomater,* vol. 2018, Article ID 7124260, 12 pages, 2018.
48. V. Guillard, M. Mauricio-Iglesias, and N. Gontard, "Effect of novel food processing methods on packaging: structure, composition, and migration properties," *Crit Rev Food Sci Nutr,* vol. 50, pp. 969–988, 2010.
49. M. Llana-Ruiz-Cabello, S. Pichardo, J. M. Bermudez, A. Banos, C. Nunez, E. Guillamon, *et al.,* "Development of PLA films containing oregano essential oil (*Origanum vulgare L. virens*) intended for use in food packaging," *Food Addit Contam Part A Chem Anal Control Expo Risk Assess,* vol. 33, pp. 1374–1386, 2016.
50. E. Fortunati, I. Armentano, Q. Zhou, A. Iannoni, E. Saino, L. Visai, *et al.,* "Multifunctional bionanocomposite films of poly(lactic acid), cellulose nanocrystals and silver nanoparticles," *Carbohydr Polym,* vol. 87, pp. 1596–1605, 2012.
51. S. Salmieri, F. Islam, R. A. Khan, F. M. Hossain, H. M. M. Ibrahim, C. Miao, *et al.,* "Antimicrobial nanocomposite films made of poly(lactic acid)-cellulose nanocrystals (PLA-CNC) in food applications: part A—effect of nisin release on the inactivation of *Listeria monocytogenes* in ham," *Cellulose,* vol. 21, pp. 1837–1850, 2014.
52. S. Salmieri, F. Islam, R. A. Khan, F. M. Hossain, H. M. M. Ibrahim, C. Miao, *et al.,* "Antimicrobial nanocomposite films made of poly(lactic acid)–cellulose nanocrystals (PLA–CNC) in food applications—part B: effect of oregano essential oil release on the inactivation of *Listeria monocytogenes* in mixed vegetables," *Cellulose,* vol. 21, pp. 4271–4285, 2014.

3 Preparation and Characterization of Poly(3-Hydroxybutyrate-*Co*-3-Hydroxyvalerate)/ Paddy Straw Powder Biocomposites

N.D. Yaacob
Universiti Malaysia Perlis

H. Ismail
Universiti Sains Malaysia

S.T. Sam
Universiti Malaysia Perlis

CONTENTS

3.1 INTRODUCTION

Over the past few years, there has been increasing interest in using natural fibers as a reinforcing agent in composite materials [1,2]. A combination of properties such as low cost, low density, nontoxicity, high specific properties, no abrasion during processing, and recyclability has attracted an increasing interest from the manufacturing industry.

Poly(β-hydroxybutyrate-*co*-β-hydroxyvalerate) (PHBV) is a member of the polyhydroxyalkanoate (PHA) family. PHAs are biodegradable polyesters that are synthesized and accumulated intracellularly as a carbon or energy storage material during unbalanced growth by a large variety of bacteria. Currently, more than 80 hydroxyalkanoates have been detected as constituents of PHAs, and more than 300 different microorganisms are known to synthesize and accumulate PHAs intracellularly. PHBV is an optically active thermoplastic aliphatic polyester with high stereoregularity [3]. Paddy straw is one of the lignocellulosic fillers, which is an agricultural coproduct abundantly found in Malaysia, since paddy (*Oryza sativa*) is the third important crop in terms of acreage, after oil palm and rubber. The lignocellulosic filler exhibits some excellent properties compared to mineral filler (e.g., calcium carbonate, kaolin, mica, and talc), such as low cost, renewable, high-specific strength-to-weight ratio, minimal health hazard, low density, less abrasion to machine, biodegradability, and environmental-friendly.

Recently, science and industry have focused on the development and application of biodegradable plastic materials. Biodegradable plastics are defined by the American Society for Testing and Materials as degradable plastics in which the degradation results from the action of naturally occurring microorganisms such as bacteria, fungi, and algae. Generally, biodegradation studies are carried out in soil and/or compost to enhance the biodegradation of these materials that occur in the presence of compost. This complex biological environment allows for a high microbial diversity, thereby resulting in an increase in the degradation potential for polymeric compounds [4,5]. In the near future, discarded biocomposite by-products will become biocomposite wastes, and these products will break down naturally by the air, moisture, climate, and soil, and disintegrate into the surrounding land. However, as more and more biodegradable materials pile up, there is an increased threat to the environment. To remedy this, during the processing of materials, additives can be included to enhance the rate of biodegradability of biocomposites, depending on the end uses of the products. Currently, there are a few studies that have reported positive findings when conducting tests on the biodegradability of biocomposites, especially on natural soil burial and weathering [6–8].

In view of this, it is of interest to conduct biodegradation studies of biocomposites of natural fiber-reinforced PHBV. The changes in the biocomposites after soil burial were studied using a tensile test machine and scanning electron microscopy (SEM).

3.2 PREPARATION OF POLY(3-HYDROXYBUTYRATE-CO-3-HYDROXYVALERATE)/PADDY STRAW POWDER BIOCOMPOSITES

3.2.1 MATERIALS

PHBV was purchased from Hasrat Bestari Sdn. Bhd., Penang. The paddy straw was obtained from the paddy field located at Perlis. The paddy straw was cut and ground into powder. The paddy straw powder (PSP) was dried at 80°C for 24 h. The average particle size of the PSP was 63 μm, which was measured using Malvern Particle size Analyzer Instrument (Italy).

3.2.2 SAMPLE PROCESSING

PHBV was dried in a vacuum oven for 24 h at 60°C, whereas PSP was dried for 3 h at 105°C prior to blending. PHBV and PSP were compounded in an internal mixer (Haake Rheomix Polydrive R 600/610) with corotating blades and a mixing head with a volumetric capacity of 69 cm^3. Compounding temperature and rotor speed were set to 170°C and 50 rpm, respectively. PHBV was fed into the mixing chamber and allowed to melt for 3 min followed by the sequential feeding of PSP which was done for 5 min. The materials were allowed to mix for 7 min, making the total mixing time of 15 min. Compounded materials were collected at the end of the mixing and kept in a sealed plastic bag for subsequent compression molding experiment.

Materials collected from the internal mixer were compression-molded to obtain test specimens. The compounded materials were placed into an appropriated steel mold covered by aluminum plates at both sides. The materials were pressed at 170°C into tensile specimens of 3 mm thickness, respectively, using Kao Tieh Go Tech compression molding machine. The molding cycle involved 3 min of preheating without pressure, 3 min of compression under 14 MPa pressure, and 5 min of cooling under 6.21 MPa pressure. The cooling process was done in an adjacent cool press equipped with tap water cooling.

3.3 PROCESSING CHARACTERISTICS AND CHARACTERIZATION OF POLY(3-HYDROXYBUTYRATE-CO-3-HYDROXYVALERATE)/PADDY STRAW POWDER BIOCOMPOSITES

3.3.1 PROCESSING CHARACTERISTICS

The melt processing characteristics of the PHBV/PSP biocomposites have been studied with processing torque–time curves. Figure 3.1 shows the compounding characterization of the melt PHBV with different PSP loadings. It can be observed that the initial torque sharply increased as PHBV was charged into the mixing chamber. Basically, the loading peak (maximum torque) is highly dependent on the charged amount of PHBV. Torque decreases with the increase in PSP loading due to the reduction in charge weight of PHBV significantly. After the PHBV was melted, the melt torque began to decrease gradually.

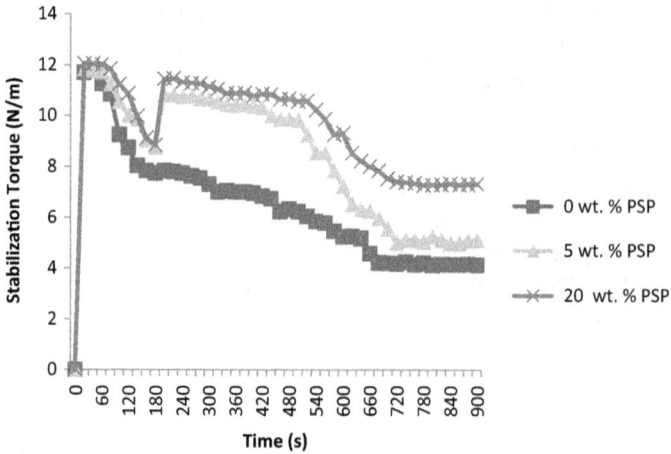

FIGURE 3.1 Processing torques of PHBV with different PSP loadings.

Under further compounding, PSP was dispersed into the molten PHBV by shear stress, which was significantly influenced by the loading amount of fiber. The inclusion of fillers caused a surface friction at the interface and thus increased the melt viscosity [9]. Thus, a sudden increase in torque was observed. The melt torque values rapidly increased with an increment of PSP loading from 10 to 20 wt%. Similar findings have been obtained by Ref. [10]. At this point, the increase in torque can be considered to be related to fiber that gives certain resistance to the deformability of matrix. With continuous compounding, the melt torque gradually decreased and stabilized.

The stabilization torques of composites at different amounts of PSP loading are shown in Figure 3.2. The stabilization torque increased up to 5.0 and 7.3 N/m as the composites were loaded with 10 and 20 wt% of PSP, respectively. The increase in stabilization torque prior to greater amount of PSP loaded in the biocomposites is attributable to higher surface friction generated by the total surface contact area of individual filler with matrix.

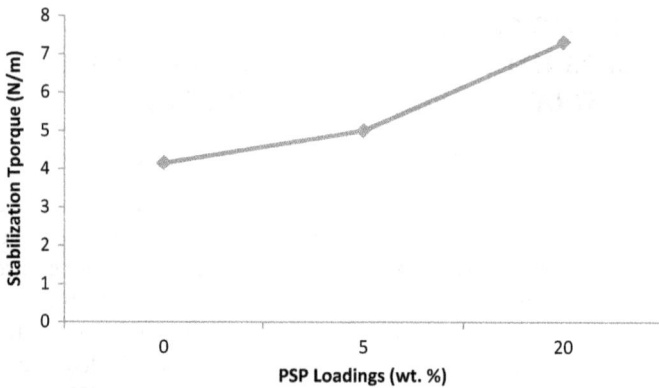

FIGURE 3.2 Stabilization torques of PHBV with different PSP loadings.

3.3.2 FOURIER TRANSFORM INFRA-RED (FTIR) ANALYSIS

IR spectra of pure PHBV and its biocomposites are shown in Figure 3.3. For pure PHBV, it can be detected that the pointed peak at $1,723\,cm^{-1}$ is allocated to C–O–C of the crystalline peaks and the bands at 1,211 and $1,125\,cm^{-1}$ are ascribed to the amorphous state of C–O--C stretching band. Zembouai et al. (2014) reported that the existence of absorption bands above $2,989\,cm^{-1}$ is generally related to the possibility of C–H···O hydrogen bond formation [11]. The peak at $971\,cm^{-1}$ is attributable to coupling of C–C backbone stretching with the CH_3 stretching vibration.

When 20 wt% of PSP is added to the biocomposites, it is characterized by broad absorption peak occurring at 1,260 to $900\,cm^{-1}$, which is attributed to the stretching of the C–O group (from a combination of hemicellulose, cellulose, and lignin) [12]. Apart from that, a broad absorption band in the region of $3,000$–$3,900\,cm^{-1}$ could also be observed in the IR spectra of PHBV/20PSP biocomposites. The characteristic peaks appeared due to the intermolecular hydrogen bond formation occurring in the biocomposites. The change in IR spectra indicates a distinct interaction between the chains of polymers. These results indicate that the PSP is distributed within PHBV matrix.

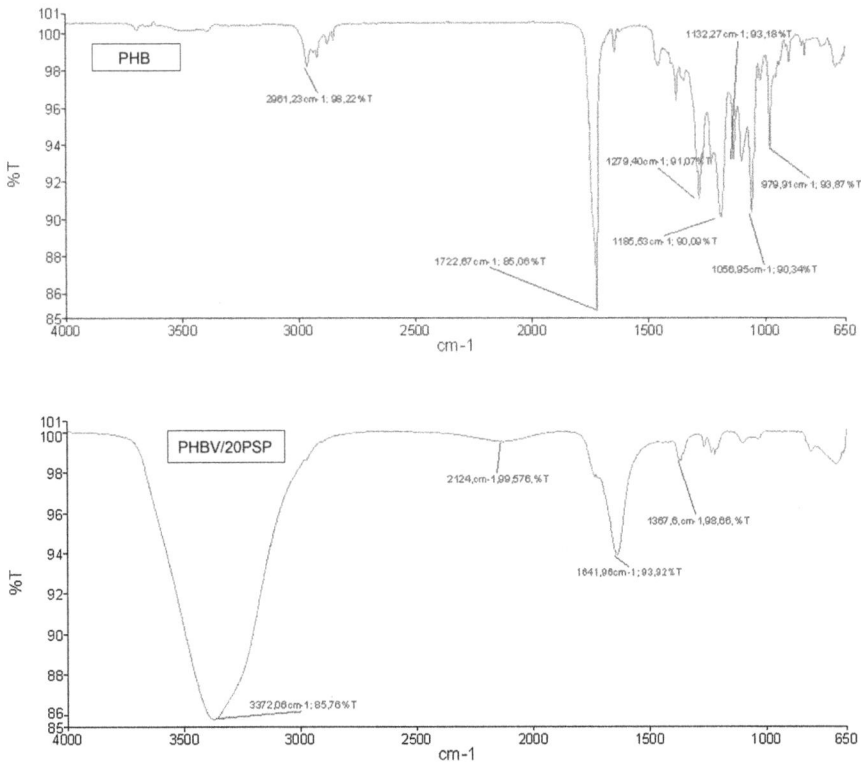

FIGURE 3.3   IR spectra of PHBV and PHBV/20PSP biocomposites.

3.3.3 Tensile Properties

Figures 3.4–3.6 show the plots of tensile strength, modulus of elasticity, and elongation at break (EB) of PHBV reinforced with different amounts of PSP. It can be seen that the incorporation of the PSP caused a reduction in tensile strength. The tensile strength was reduced by approximately 17.6%, 22.39%, 38.47%, and 41.06% at 5, 10, 15, and 20 wt% PSP loaded into biocomposites. Bledzki and Jaszkiewicz (2010) and Li et al. (2015) studied the tensile strength of the composites using PHBV matrix [2,13]. The decrement is most probably due to the excessive amount of PSP loaded in the system, which caused the discontinuity of the matrix. Indeed, the weight fraction of filler plays a significant role in the mechanical properties of composites. Usually, mechanical properties of biocomposites are found to be enhanced as the optimum amount of filler is loaded into the composites. Yet, beyond the optimum loading, a poor dispersion of filler within the matrix might occur [14]. Apart from that, the

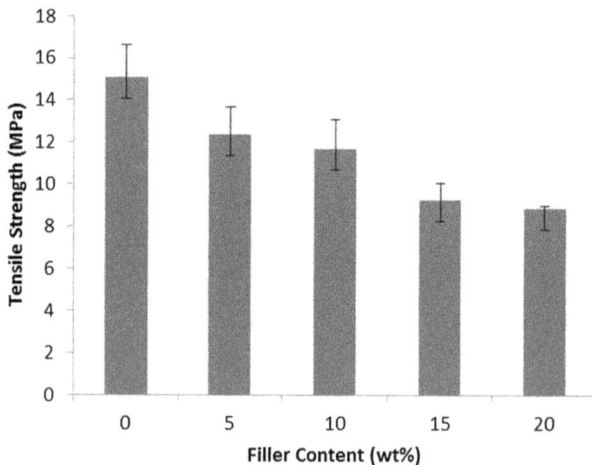

FIGURE 3.4 Tensile strength of PHBV with different PSP loadings.

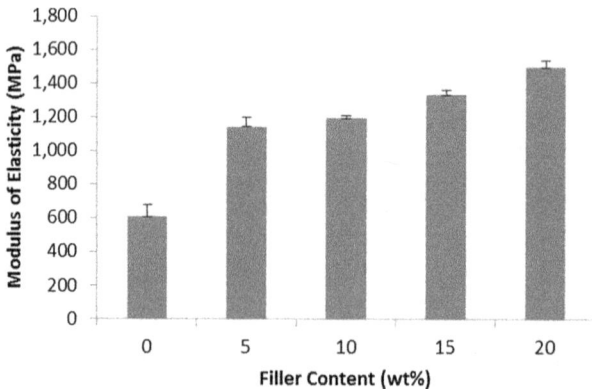

FIGURE 3.5 Modulus of elasticity of PHBV with different PSP loadings.

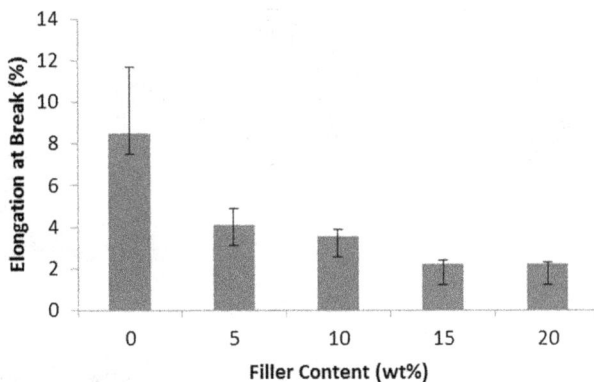

FIGURE 3.6   EB of PHBV with different PSP loadings.

reduction in strength might be due to incompatibility of highly hydrophilic PSP and nonpolar hydrophobic PHBV. Due to the different polarities, a weak interfacial adhesion between PSP and PHBV occurred, providing a site failure to initiate and propagate. Reduction in the effective surface area of the filler phase yielded inefficient stress transfer from the matrix. As a result, the tensile strength was not improved.

The drop in tensile strength becomes more drastic at higher PSP loading (20 wt%). This is due to the fact that the filler–filler interaction becomes more prominent than the filler–matrix interaction. Therefore, the applied force cannot be effectively transferred from the polymer matrix to the filler. This is justified by SEM micrograph, as shown in Figure 3.7.

As shown in Figure 3.5, the incorporation of PSP was found to increase the modulus of elasticity of biocomposites. The moduli were increased from 610.42 to 1,496.81 MPa when 5–20 wt% of PSP was incorporated in the biocomposite system. The addition of PSP was expected to increase the modulus resulting from the inclusion of rigid filler particles in the soft matrix [15]. Besides, this result suggests that the stiffening effect occurs because of the presence of fillers inside the biocomposites [16]. Tensile modulus specifies the stiffness of biocomposites. Hence, as the biocomposites become stiffer, higher tensile modulus is obtained. This finding is in agreement with the work reported by Abdul Khalil et al. (2001), whereby the incorporation of starch or cellulosic filler has improved the stiffness of the composite materials [17].

On the contrary, the addition of PSP brings about a drop in EB values (as shown in Figure 3.6). This is in agreement with other report elsewhere [2,13,18]. This effect is expected considering that the high rigidity of the cellulose fillers decreases the polymer chain mobility and deformability of an interface between filler and matrix, which leads to an inevitable diminution in the degree of ductility of the biocomposites.

3.3.4   MORPHOLOGICAL STUDIES

The SEM micrograph of the tensile fracture surface of PHBV and morphological effects of PSP incorporation is portrayed in Figure 3.7. As presented in Figure 3.7a, the surface of the pure PHBV is slightly smooth without any cracks.

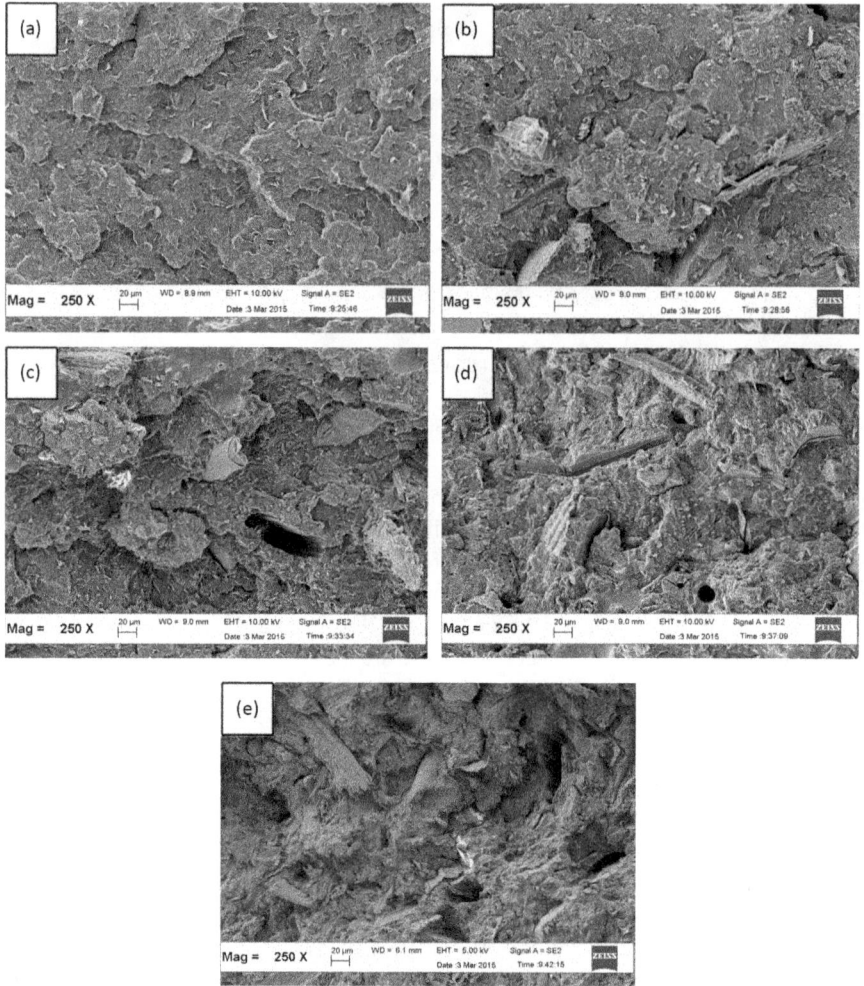

FIGURE 3.7 SEM morphology of PHBV with (a) 0 wt% PSP, (b) 5 wt% PSP, (c) 10 wt% PSP, (d) 15 wt% PSP, and (e) 20 wt% PSP.

Figure 3.7b illustrates the tensile fracture surface of PLA/5PSP. It can be observed that PSP embedded in the polymer and the filler surface is wetted by the PHBV. The fiber seems to be well dispersed in the matrix, and the fracture surface shows compensate mixing of filler within the matrix. The good dispersion corroborates the presence of chemical bonding between PSP and PHBV probably due to the formation of hydrogen bonding, good stress transfer from the matrix to the fiber, and reinforcing effect imparted by the filler [13].

On the other hand, Figure 3.7c illustrates the tensile fracture surface of biocomposites when PSP was loaded at 10 wt%. In this case, the presence of PSP was more visible. PSP appeared not wholly embedded in the matrix. Moreover, the fibers were slightly damaged and the some debonding phenomena indicating poor adhesion between the fiber and the matrix were found to be observed. And even there is the formation of voids as a result of fiber detachment in the matrix.

The fracture surface micrographs, shown in Figure 3.7d,e, showed poor wettability between the fiber and the matrix, and showed the most prominent fiber agglomeration present in the composites. The weaker bonding is illustrated by the greater fiber–fiber contact rather the than the fiber–matrix interaction. Deterioration in the aforementioned properties is likely a result of excessive content of PSP (15 and 20 wt%). The nonuniform distribution of fiber leads to the formation of stress concentration points. This situation is reflected in the mechanical properties. It is assumed that increase in fiber loading decreases the interfacial adhesion and homogeneity. The mechanical properties discussed earlier also support these findings (Section 3.3.3).

3.3.5 THERMOGRAVIMETRIC ANALYSIS

The weight loss of PHBV and its biocomposites as a function of temperature are shown in Figure 3.8. Figure 3.9 shows the derivative thermogravimetric (DTG) curves generated on the composites at various PSP loadings. As mentioned earlier, the composite samples undergo three steps of thermal degradation: the first region (250°C–280°C), the second region (335°C–400°C), and the third region (above 400°C). Table 3.1 summarizes the thermogravimetric analysis (TGA) results obtained from the graphical representation.

Referring to the table, it can be seen that the thermal stability of biocomposites was determined by the temperature at which 5% and 10% weight loss occurred. Obviously, PHBV decomposed at higher temperature when loaded with 20 wt% of PSP. With the incorporation of PSP in the PHBV, it was observed that the

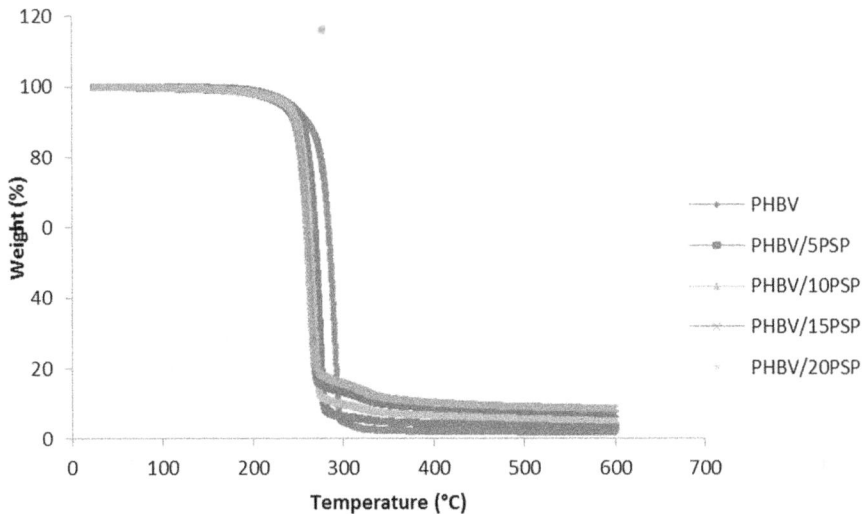

FIGURE 3.8   TGA thermogram of PHBV with different PSP loadings.

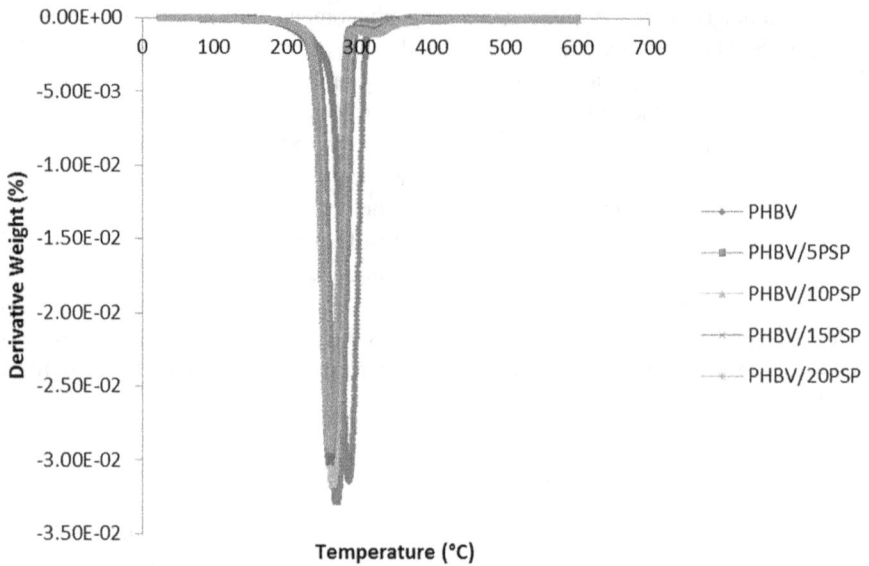

FIGURE 3.9   DTG thermogram of PHBV with different PSP loadings.

TABLE 3.1
TGA Results of PHBV with Different PSP Loadings

Sample	$T_{5\%}$ (°C)	$T_{10\%}$ (°C)	T_{max} (°C)	Residual Weight at T_{500}
PHBV	231	261	298	2.08
PHBV/5PSP	233	265	301	3.84
PHBV/10PSP	237	251	305	5.87
PHBV/15PSP	240	247	307	7.87
PHBV/20PSP	242	248	309	9.23

decomposition temperature of the composites slightly changed. Decomposition temperature of biocomposites at 5% and 10% weight loss increased by 0.86% and 1.53%, respectively.

In fact, comparing the maximum decomposition peak in DTG profile, the shift in the degradation temperature is noticeable for biocomposite samples. Accordingly, it seems that the T_{max} slightly rises up to 301°C as 5 wt% of PSP is loaded into the biocomposites. In fact, with higher content of PSP, the T_d relatively increases. The results show that increasing PSP content improves the thermal stability of biocomposites, which is attributable to the increase in the amount of cellulosic matter. It is possible to observe that the addition of PSP in PHBV increases the amount of residues after thermal degradation. The residue contains cellulose and lignin (PSP), which are higher in PHBV matrix.

Lignin is thermally stable in nature, therefore contributing to the greater percent of char residue (Table 3.1).

3.3.6 DIFFERENTIAL SCANNING CALORIMETRY (DSC)

Table 3.2 reports the melting temperature (T_m), crystallization temperature (T_c), enthalpy of melt (ΔH_m), and degree of crystallinity (X_c). It can be observed that the T_m is slightly increased from 165.50°C up to 167.38°C with the addition of the PSP (0–20 wt%). This indicates that the filler mat restricts the flow ability of PHBV molecules during melting.

On the other hand, the T_c seems shifted to lower temperature (124.47°C–117.41°C) as PSP content increases. This implies that the crystallization rate becomes slower. The presence of the PSP in the biocomposite system limits the mobility of the polymer, restricts its arrangement, and eventually weakens its ability to crystallize.

The addition of PSP to PHBV matrix seems to result in a decline in enthalpy of melt as well as crystallinity. This decline is associated with the decreased amount of PHBV in the biocomposites [13]. It is assumed that the presence of PSP will raise the crystallinity due to the nucleating effect. But the hindrance of polymer mobility by PSP appears to be overwhelmed rather than nucleation. Hence, the crystallinity dropped at approximately 31.30% and 48.10% when 5 and 20 wt% PSP were loaded in the biocomposites, respectively.

3.4 EFFECT OF SOIL BURIAL ON THE PROPERTIES OF PHBV/PSP BIOCOMPOSITES

3.4.1 VISUAL OBSERVATION

Figure 3.10 illustrates the physical appearance of PHBV/PSP biocomposites after soil burial test. It seems that the samples changed their colors from yellowish to dark when more amounts of PSP were loaded into the biocomposites. Also, the tiny deposits appeared on the surface of the biocomposites. Leaching of fiber can be clearly seen when the PSP contents are higher. This may be attributable to the climate change, which makes water to be immersed in the soil. Thus, the inner region of the biocomposites will undergo swelling process as loosely embedded fillers are leached out, thereby leaving pores and cracks.

TABLE 3.2
DSC Data for PHBV and PHBV/PSP Biocomposites

Samples	T_m (°C)	T_c (°C)	ΔH_m	Crystallinity (%)
Pure PHBV	165.50	124.47	64.66	44.28
PHBV/5PSP	167.01	122.35	44.41	30.42
PHBV/10PSP	167.07	119.78	43.75	29.96
PHBV/15PSP	167.34	118.94	42.64	29.20
PHBV/20PSP	167.38	117.41	33.56	22.98

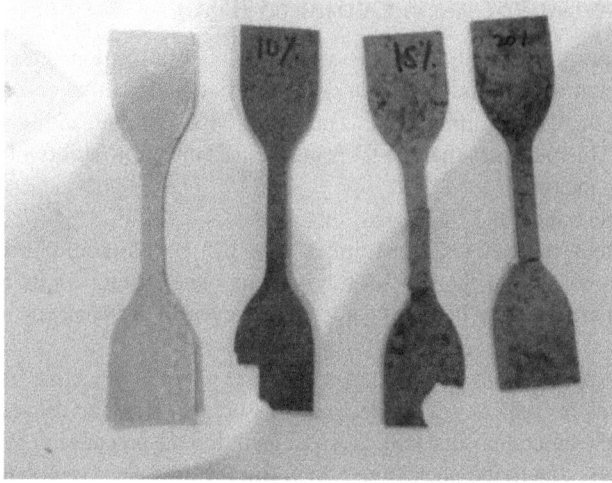

FIGURE 3.10 Visual observation of PHBV with different PSP loadings retrieved after being subjected to soil burial test.

3.4.2 TENSILE PROPERTIES

Figure 3.11 demonstrates the tensile strength properties of PHBV reinforced with PSP at different loading amounts (10–20 wt%) after being subjected to the soil burial test until 6 months. From the graph, it can be observed that by increasing the PSP amounts, a drastic drop in the tensile strength was observed. Prolonged burying time declined the tensile strength. After 6 months of the burial test, all the samples underwent fragmentation, except for pure PHBV. Biocomposite with 20 wt% of PSP content underwent fragmentation starting from 2 months. Table 3.3 shows the retention of tensile strength.

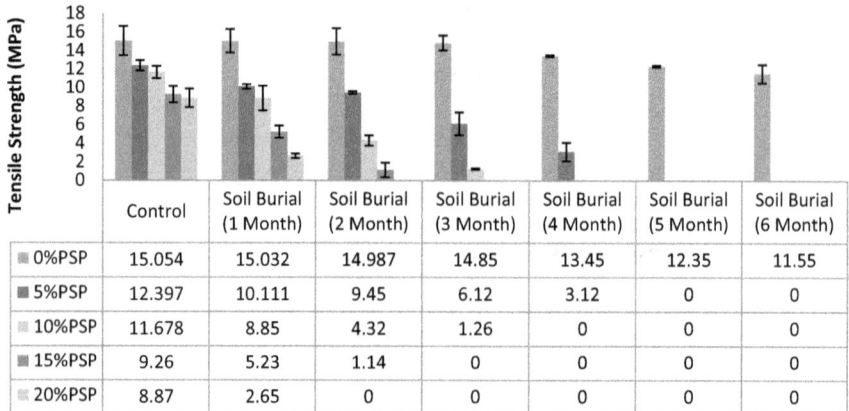

Tensile Strength (MPa)	Control	Soil Burial (1 Month)	Soil Burial (2 Month)	Soil Burial (3 Month)	Soil Burial (4 Month)	Soil Burial (5 Month)	Soil Burial (6 Month)
▬ 0%PSP	15.054	15.032	14.987	14.85	13.45	12.35	11.55
▬ 5%PSP	12.397	10.111	9.45	6.12	3.12	0	0
▬ 10%PSP	11.678	8.85	4.32	1.26	0	0	0
▬ 15%PSP	9.26	5.23	1.14	0	0	0	0
▬ 20%PSP	8.87	2.65	0	0	0	0	0

FIGURE 3.11 Tensile strength of PHBV with different PSP loadings retrieved after being subjected to soil burial test.

TABLE 3.3

Retention of Tensile Properties for PHBV/PSP Biocomposites after 6 Months of Soil Burial

Sample		Retention of Biocomposites (%)	
	Tensile Strength	Elongation at Break	Modulus of Elasticity
PHBV	76.72	66.86	415.49
PHBV/5 PSP	Fragmentation	Fragmentation	Fragmentation
PHBV/10PSP	Fragmentation	Fragmentation	Fragmentation
PHBV/15PSP	Fragmentation	Fragmentation	Fragmentation
PHBV/20PSP	Fragmentation	Fragmentation	Fragmentation

These results are explained by the combined effect of microbial attack and hydrolysis. During this test, the samples are greatly affected by the climate conditions. Due to sunny and rainy weather, there is an increased opportunity of water immersion in the soils, which eventually changes the structure and properties of filler as well as matrix, and also their interphase [17]. Paddy straw contains hydrophilic substances such as cellulose and hemicellulose that highly promote the water intake and favor the microbial growth [6]. Soil microflora constitutes a mixed microbial population (including bacteria, actinomycetes, and fungi), which may act synergistically during degradation and reproduce under naturally occurring conditions.

The formation of voids and gaps due to climate conditions will permit microbial attack on the filler. Generally, hemicellulose in the filler cell wall is prone to degradation by the action of particular enzyme system, which can hydrolyze the polymer to digestible units. It is also important to note that deeper microbial invasion might occur and assimilates humidity as well as carbon from the polymer over time. This elucidation is in agreement with the SEM morphology, as shown in Figure 3.14, whereby more filler content and prolonged burying time revealed further deterioration of biocomposites.

It is known that the presence of filler in the composite might decrease the polymer chain mobility and deformability of a rigid interface between filler and matrix, thereby restricting the biocomposites to elongate further upon tensile testing. All the samples show a decrease in EB as more filler is loaded into the matrix (Figure 3.12). Likewise, the buried samples exhibit the same trend. Yet, prolonged burying time results in a decrease in EB, as shown in Table 3.3. The decrement of EB value was attributable to the deterioration of the biocomposite structure.

Moduli of elasticity are also critically affected by the deterioration of the biocomposite samples. Referring to Figure 3.13, it can be clearly seen that the samples which have not been subjected to soil burial test result in an increase in modulus of elasticity when more PSP contents are added to the biocomposites. The presence of the filler in the biocomposites usually leads to the stiffening effect. Buried samples also show a similar trend.

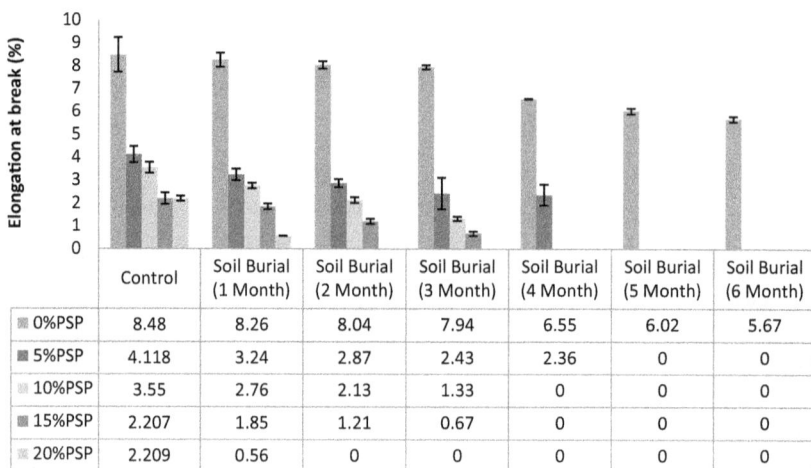

	Control	Soil Burial (1 Month)	Soil Burial (2 Month)	Soil Burial (3 Month)	Soil Burial (4 Month)	Soil Burial (5 Month)	Soil Burial (6 Month)
0%PSP	8.48	8.26	8.04	7.94	6.55	6.02	5.67
5%PSP	4.118	3.24	2.87	2.43	2.36	0	0
10%PSP	3.55	2.76	2.13	1.33	0	0	0
15%PSP	2.207	1.85	1.21	0.67	0	0	0
20%PSP	2.209	0.56	0	0	0	0	0

FIGURE 3.12 EB of PHBV with different PSP loadings retrieved after being subjected to soil burial test.

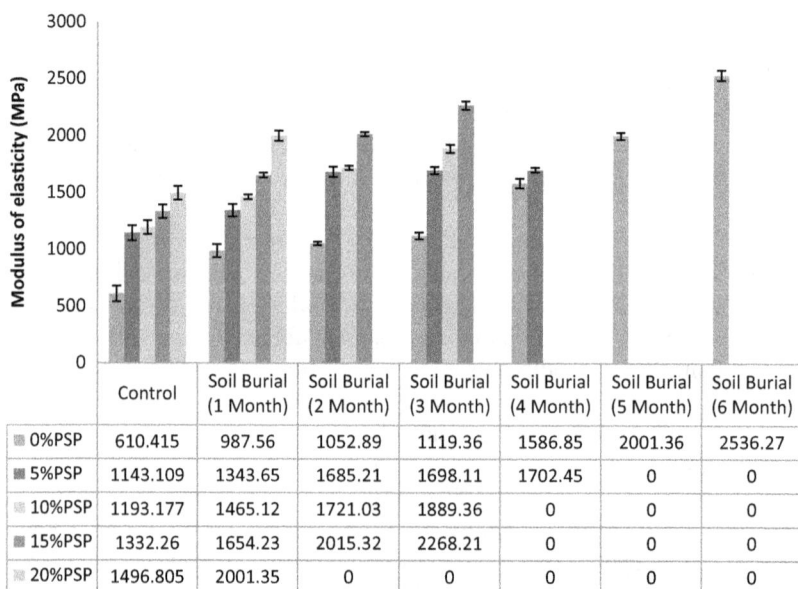

	Control	Soil Burial (1 Month)	Soil Burial (2 Month)	Soil Burial (3 Month)	Soil Burial (4 Month)	Soil Burial (5 Month)	Soil Burial (6 Month)
0%PSP	610.415	987.56	1052.89	1119.36	1586.85	2001.36	2536.27
5%PSP	1143.109	1343.65	1685.21	1698.11	1702.45	0	0
10%PSP	1193.177	1465.12	1721.03	1889.36	0	0	0
15%PSP	1332.26	1654.23	2015.32	2268.21	0	0	0
20%PSP	1496.805	2001.35	0	0	0	0	0

FIGURE 3.13 Modulus of elasticity of PHBV with different PSP loadings retrieved after being subjected to soil burial test.

3.4.3 MORPHOLOGICAL STUDIES

Figure 3.14 demonstrates a sequential morphology of PHBV-reinforced PSP biocomposites as a function of PSP loading subjected to the soil burial test for 3 and 6 months. After 3 months of the soil burial test, a rougher surface of PHBV can be observed (as shown in Figure 3.14a). In contrast, the biocomposites with the inclusion of 5 wt% of

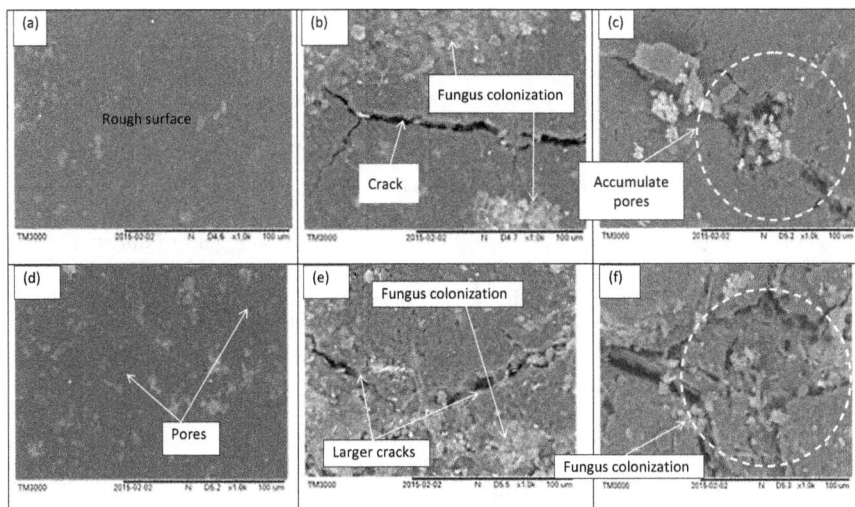

FIGURE 3.14 SEM morphology of PHBV with (a) 0 wt% PSP, (b) 10 wt% PSP, and (c) 20 wt% PSP after 3 months of soil burial and (d) 0 wt% PSP, (e) 10 wt% PSP, and (f) 20 wt% PSP after 6 months of soil burial (1K magnification).

PSP lead to the formation of crack and pores on the surface (Figure 3.14b). This may be due to the removal of the biodegradable component under soil environment [6]. Thus, the larger pores and more cracks are evident on the surface of the biocomposite with the inclusion of 20 wt% of PSP (Figure 3.14c).

After 6 months of soil exposure, the degradation of the samples becomes more obvious. As more amounts of PSP are loaded into the biocomposites, the accumulation of pores can be seen on the surface of buried samples (Figure 3.14e,f). As expected, by increasing the loading of PSP and prolonging burying time, more degradation can be observed. Filler (PSP) may act as a channel and enable the microbial attack in the biocomposites [19]. Eventually, within the created pores and cracks, filamentous fungi colonization can be seen (as shown in Figure 3.14f).

3.5 CONCLUSIONS AND REMARKS

Plastic utilization is increasing due to an increase in population, developmental activities, and industrialization, and consequently, the quantity of plastic waste contributes to the negative social and environmental impacts. It is believed that this problem can be overcome by developing the biocomposites. However, the main challenge of developing the biocomposites is the concern with cost of materials and processing methods.

The tensile strength and the EB of PHBV/PSP biocomposites decreased with increasing filler content. However, the modulus of elasticity of PHBV/PSP biocomposites increased with increasing filler content. The results from this work show that the degradation of PHBV in the soil can be accelerated by the addition of the lignocellulosic material, i.e., PSP. The tensile strength and the EB of the PHBV/PSP biocomposites decreased after composting in the soil, whereas the modulus of elasticity increased. These results were supported by observations from the SEM images.

REFERENCES

1. F. Bertini et al., "Effect of ligno-derivatives on thermal properties and degradation behavior of poly(3-hydroxybutyrate)-based biocomposites." *Polym. Degrad. Stab.*, vol. 97, no. 10, pp. 1979–1987. 2012.
2. L. Li et al., "Properties and structure of polylactide/poly (3-hydroxybutyrate-co-3-hydroxyvalerate) (PLA/PHBV) blend fibers." *Polymer*, vol. 68, pp. 183–194. 2015.
3. J. Tao et al., "Thermal properties and degradability of poly(propylene carbonate)/poly(β-hydroxybutyrate-co-β-hydroxyvalerate) (PPC/PHBV) blends." *Polym. Degrad. Stab.*, vol. 9, no. 4, pp. 575–583. 2009.
4. Y. Tokiwa & B.P. Calabia, "Biodegradability and biodegradation of poly(lactide)." *Appl. Microbiol. Biotechnol.*, vol. 72, no. 2, pp. 244–251. 2006.
5. S. Sukkhum & V. Kitpreechavanich, "New insight into biodegradation of poly (L-lactide), enzyme production and characterization." *Progr. Molecul. Environ. Bioeng.*, pp. 587–604. 2011.
6. V.A. Alvarez, R.A. Ruseckaite & A. Vazquez, "Degradation of sisal fibre/mater Bi-Y biocomposites buried in soil." *Polym. Degrad. Stab.*, vol. 91, no. 12, pp. 3156–3162. 2006.
7. H. Ismail, et al., "Effects of natural weathering on properties of recycled newspaper-filled polypropylene (PP)/natural rubber (NR) composites." *Polym. Plastics Technol. Eng.*, vol. 47, pp. 697–707. 2010.
8. S.T. Sam et al., "Soil burial of polyethylene-g-(maleic anhydride) compatibilised LLDPE/soya powder blends." *Polym. Plastics Technol. Eng.*, vol. 50, pp. 851–861. 2011.
9. S.Y. Chang, H. Ismail & Q. Ahsan, "Effect of maleic anhydride on kenaf dust filled polycaprolactone/thermoplastic sago starch composites." *BioResources*, vol. 7, no. 2, pp. 1594–1616. 2012.
10. R.A. Majid et al., "The effects of natural weathering on the properties of linear density polyethylene (LDPE)/thermoplastic sago starch (TPSS) blends." *Polym. Plastics Technol. Eng.*, vol. 49, pp. 1142–1149. 2010.
11. I. Zembouai et al., "Poly(3-hydroxybutyrate-co-3-hydroxyvalerate)/polylactide blends: thermal stability, flammability and thermo-mechanical behavior." *J.Polym. Environ.*, vol. 22, no. 1, pp. 131–139. 2014.
12. O. Faruk et al., "Biocomposites reinforced with natural fibers: 2000–2010." *Progr. Polym. Sci.*, vol. 37, no. 11, pp. 1552–1596. 2012.
13. A.K. Bledzki & A. Jaszkiewicz, "Mechanical performance of biocomposites based on PLA and PHBV reinforced with natural fibres – a comparative study to PP." *Compos. Sci. Technol.*, vol. 70, no. 12, pp. 1687–1696. 2010.
14. M.G. Lomelí-Ramírez et al., "Bio-composites of cassava starch-green coconut fiber: part II - structure and properties." *Carbohydr. Polym.*, vol. 102, no. 1, pp. 576–583. 2014.
15. X.V. Cao et al., "Mechanical properties and water absorption of kenaf powder filled recycled high density polyethylene/natural rubber biocomposites using mape as a compatibilizer." *BioResources*, vol. 6, no. 3, pp. 3260–3271. 2011.
16. W.V. Srubar et al., "Mechanisms and impact of fiber – matrix compatibilization techniques on the material characterization of PHBV/oak wood flour engineered biobased composites." *Compos. Sci. Technol.*, vol. 72, no. 6, pp. 708–715. 2012.
17. H.P.S. Abdul Khalil et al., "The effect of soil burial degradation of oil palm trunk fiber-filled recycled polypropylene composites." *J. Reinfor. Plastics Compos.*, vol. 29, no. 11, pp. 1653–1663. 2010.
18. H. Ismail, "Bamboo fibre filled natural rubber composites : the effects of filler loading and bonding agent." *Polym. Test.*, vol. 21, pp. 139–144. 2002.
19. Z.N. Azwa et al, "A review on the degradability of polymeric composites based on natural fibres". *Mater. Des.*, vol. 47, pp. 424–442. 2013.

4 Surface Modification of Kapok Husk on the Properties of Soy Protein Isolate Biocomposite Films Using Methyl Methacrylate

P.L. Teh and R.K.C. Pani Sellivam
Universiti Malaysia Perlis

CONTENTS

4.1 INTRODUCTION

Nowadays, biofilms based on agricultural materials have gained much consideration as packaging materials. To overcome such environmental problems rather than synthetic polymer packaging, biofilms are accepted to be a promising solution [1]. Additionally, biofilms are suitable to commercialize into various products such as garbage bags, grocery bags, composting yard waste bags, agriculture mulches, and agro bags [2]. Soy protein is naturally occurring edible material that has been broadly investigated due to its properties such as inexpensive nature, biodegradability, and bioavailability. From the comparisons between soy protein products, soy protein isolate (SPI) has higher protein content, which gives superior ability to form films [3]. From the investigation, researchers have been reported that SPI films have high barrier properties on both vapor and oxygen at low relative humidity, and good biodegradability [4].

Soy protein can be divided into three types, namely, SPI, soy protein concentrate, and soy flour. Among the three types, SPI was widely studied by researchers due to its high protein content and purity, low cost, biodegradability, and easy availability. SPI contains more than 90% of protein content compared to other soy protein products. To form SPI films, the protein structures of the native state would need to be denatured to reform new configurations via new linkages within the protein molecule. SPI-based biofilms have been extensively studied due to spur of potential industrial applications and fundamental research; however, there are two intrinsic issues that restrict their application: relatively high moisture sensitivity and poor mechanical properties [5].

Green products such as natural filler and biodegradable polymer are called biocomposite materials. Recently, natural fillers have received much attention because they have advantages such as low density, low costs, and high toughness with acceptable mechanical properties [6]. Biocomposites have been widely researched due to their enhanced material properties such as higher strength and modulus, improved barrier properties, and improved resistance to thermal decomposition as compared to the neat polymers [7]. Composite structures combine the strength and stiffness of fibers to reinforce dimensional instabilities inherent in the biopolymer matrix alone. The resulting effect is a material with new properties that could be designed for specific applications [8]. These novel materials also bring about economic and environmental benefits such as low cost, comparable specific strengths, carbon dioxide sequestration, and biodegradability. Some studies have used natural fillers such as jute, kenaf, hemp, flax, kapok, bamboo, cotton, *Nypa fruticans* husk, coir, and sisal to reinforce biopolymers at a macro level [9–11].

Kapok husk (KH) has engrossed increasing consideration from many researchers. The KH fiber has been used for the enrichment culture of lignocellulose-degrading bacteria [12] and also as oil absorbent [13]. KH is obtained from the seed pods of the kapok tree (*Ceiba pentandra*) from the Bombacaceae family [14]. The seeds are enclosed in capsules or pods that are picked and broken open with mallets. The KH is extremely light with circular cross-sectional thin walls and a spacious lumen. The KH is used as a natural filler in SPI matrix due to the low density, low cost, and easy availability, which gives the reinforcement effect on the biocomposite films [10].

All natural fillers are hydrophilic and show low moisture resistance, which results in a very poor interface between natural filler and hydrophobic matrix. However, the tensile strength is higher as the cellulose content of the natural filler increases [15]. To improve the interface, the filler surface can be treated by several methods. Different treatment methods will lead to the different characteristics of filler, such as changes in particle size and shape, surface area, and functional groups on the filler surface. Many studies have been undertaken to modify the performance of natural fillers. Different surface treatment methods such as alkali treatment [16], isocyanate treatment [17], acrylation [18], benzoylation [8], acetylation [19], silane [20], and peroxide treatment [21] have been applied on the filler to improve its strength, size, and shape and the filler–matrix adhesion. Those methods have their own merits and demerits: Alkali treatment improves the filler–polymer adhesion due to the removal of natural and artificial impurities, and changes the chemical composition of the filler by removing the cementing substances such as lignin and hemicelluloses. In this study, methyl methacrylate (MMA) was used to reduce the hydrophilicity of SPI/KH biocomposite films with different KH loadings at 10, 20, 30, and 40 wt% in order to enhance the tensile strength, morphology, thermogravimetric analysis (TGA), moisture content (MC), total soluble matter (TSM), gel fraction, and enzymatic biodegradation properties.

4.2 PREPARATION AND CHARACTERIZATION OF SPI/KH BIOCOMPOSITE FILMS

4.2.1 MATERIALS

SPI with 90% protein with an average particle size of 63 μm was supplied by Shandong Wonderful Industrial Group Co., Ltd., Dongying, China. KH was obtained from the rural area, Perlis, Malaysia. Kapok fiber was removed from kapok pod. Cleaned kapok pod was crushed and ground into powder. The average particle size of KH was 16 μm. MMA was supplied by Sigma-Aldrich, Malaysia.

4.2.2 PREPARATION OF BIOCOMPOSITE FILM

SPI/KH biocomposite films were prepared by the casting technique. The ratio of SPI to glycerol used was 2:1 for each blend biocomposite film. The KH powder was mixed using a mechanical stirrer into 3% (v/v) of chemical modification solution (MMA, sodium dodecyl sulfate, 2-ethyl hexylacrylate) in ethanol, respectively. To become homogeneous, the solution was stirred for 2 h and remained for 24 h. Then, the mixed solution was filtered and dried in the oven for 24 h at 80°C to fasten the evaporation process of ethanol. The treated SPI was dissolved in distilled water and stirred in a water bath at 90°C for 15 min. Then, KH was added to SPI, followed by the sequential addition of glycerol with constant stirring for another 15 min. The total mixing time was 30 min. Finally, the treated and untreated SPI/KH solutions were poured into the plastic mold and dried in the oven at 50°C for 24 h. The films were carefully peeled off from the plastic surface.

4.2.3 TESTING AND CHARACTERIZATION

Fourier transmission infrared (FTIR) spectra of the SPI/KH biocomposite films were analyzed on a Perkin Elmer Spectrometer 2000 FTIR for three times to identify the functional group after chemical modification. The scanned range was 650–4,000 cm^{-1} with a resolution of 4 cm^{-1}. Tensile test was performed by Instron Universal Machine, Model 5569, according to ASTM D 882 in order to determine the tensile properties such as tensile strength, modulus of elasticity, and elongation at break of the biocomposite films. Each biocomposite film specimen with the thickness of 0.2 ± 0.05 mm was cut into rectangular shape in dimensions of 100 mm × 15 mm. The test was performed at 25°C ± 3°C with a cross-head speed of 10 mm/min. Scanning electron microscopy (SEM), model JEOL JSM-6460-LA, was used to analyze the morphological characteristics of SPI/KH biocomposite film like tensile fracture surface. The tensile fracture surface of biocomposite film was mounted on aluminum stubs and coated with a thin layer of palladium to avoid electrostatic charging during SEM analysis. TGA was conducted using TGA Pyris Diamond, Perkin-Elmer. The samples were weighed at (7 ± 2) mg and were heated at the rate of 20°C/min from 30°C to 650°C. The nitrogen gas was used as purge gas at the flow rate of 50 mL/min. MC was determined according to a method described. Three samples for each ratio were weighed (W_o) into glass dishes, dried in an air-circulating oven at 105°C for 24 h, and weighed again (W_i). The MC for each film was determined by the following equation:

$$\%MC = \frac{W_o - W_i}{W_o} \times 100\% \tag{4.1}$$

where MC represents the moisture content, W_o represents the weight before drying, and W_i represents the weight after drying. TSM of films was determined using a method described by Gontard et al. (1992). Initial dry matter of film pieces (20 mm × 20 mm) was measured by drying in an air-circulating oven at 100°C for 24 h. These film pieces were placed in beakers containing 50 mL of distilled water and traces of sodium azide (0.02%, w/v) to prevent microbial growth and stored at room temperature for 24 h with occasional gentle stirring. Undissolved dry film matter was determined by taking the film pieces out of the beakers and drying them in an air-circulating oven (100°C for 24 h). The weight of solubilized dry matter was calculated by subtracting the weight of unsolubilized dry matter from the initial weight of dry matter, and its result was reported on an initial dry weight basis (Equation 4.2).

$$\%TSM = \frac{M_o - M_i}{M_o} \times 100\% \tag{4.2}$$

where TSM represents the total soluble matter, M_o represents the weight before drying, and M_i represents the weight after drying. The untreated and treated SPI/KH biocomposite films were allowed to swell in 1% of acetic acid for 24 h. The soluble fraction (sol) was dissolved in acetic solution; however, the cross-linked portion (gel) remained insoluble. The gel was filtered and dried in the oven at 50°C for 24 h. The percentage of the sol fraction was estimated by the following equation:

$$\%sol = \frac{N_o - N_i}{N_o} \times 100\% \tag{4.3}$$

where N_o is the weight of specimens before swelling and N_i the weight of the dried gel. The percentage of the gel fraction was calculated by the following equation:

$$\%gel = 100 - \%sol \tag{4.4}$$

For enzymatic degradation, a 50 mL buffer solution containing 10 mg of α-amylase with pH 7.3 was prepared by adding 4.8 mL of 0.2 M acetic acid solution to 45.2 mL of 0.2 M sodium acetate solution. The samples were immersed into the mixture for 14 days at 37°C. The SPI/KH biocomposite samples with a dimension of 2 cm × 3 cm were taken out every 2 days and rinsed with distilled water to remove the excess of α-amylase enzyme on the surface of samples. Then, the samples were transferred to an oven at 50°C for 24 h. The weight loss of enzymatic degradation can be evaluated by the following equation:

$$\text{Weight loss of enzymatic degradation } (\%) = \frac{Z_o - Z_i}{Z_o} \times 100\% \tag{4.5}$$

where Z_o is the initial weight loss of a specimen and Z_i the weight loss of a specimen after enzymatic degradation.

4.3 FOURIER TRANSFORM INFRARED SPECTROSCOPIC ANALYSIS

Figure 4.1 illustrates the FTIR spectra of untreated and treated SPI/KH biocomposite films with MMA. The main functional groups of KH are –OH stretch (3,306 cm⁻¹), –CH stretch (2,920 cm⁻¹), –NH bend (1,622 cm⁻¹), and C=O stretch (1,538 cm⁻¹). After treating KH with MMA, the hydrophilicity of the film was found to be significantly reduced because the absorption band of –OH groups decreased as compared to untreated biocomposite films. The ester bridge through transesterification reaction between MMA (ester group) and KH (hydroxyl group) was identified with a new peak at 1,736 cm⁻¹. This bonding further supported the absorption bands at 1,622 and 1,538 cm⁻¹ reduced to 1,615 and 1,537 cm⁻¹, indicating N–H bending vibration and amide I stretching vibration KH chain. Figure 4.2 demonstrates the schematic reaction between MMA and SPI/KH biocomposite film.

4.4 TENSILE STRENGTH

Figure 4.3 illustrates the tensile strength of untreated and treated SPI/KH biocomposite films with MMA. For both the untreated and treated SPI/KH biocomposite films, the addition of KH filler increased about 61.4% of tensile strength of

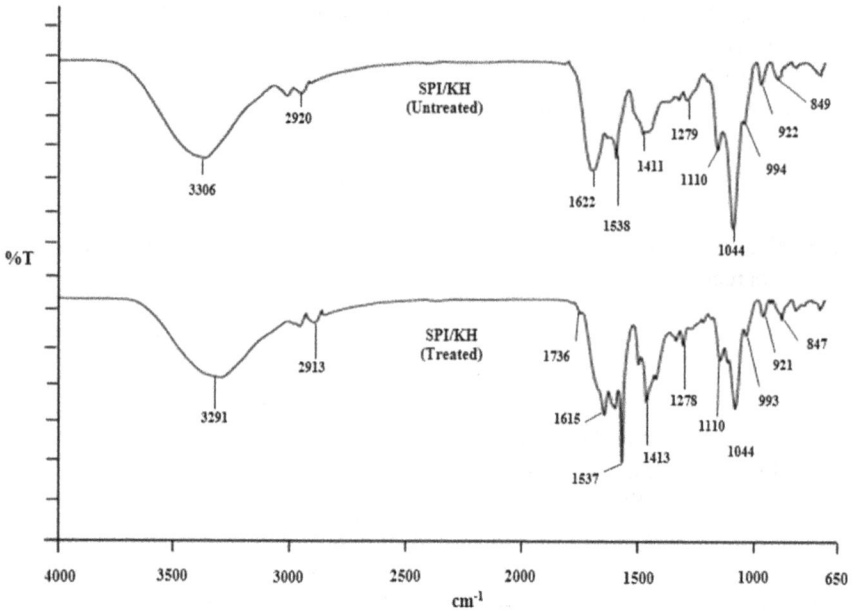

FIGURE 4.1   FTIR spectra of untreated and treated SPI/KH biocomposite films with MMA.

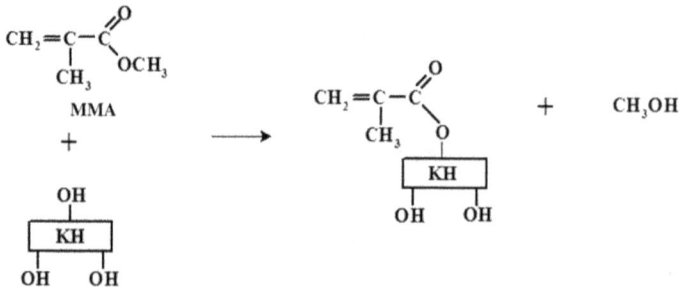

FIGURE 4.2   Schematic reaction between MMA and KH biocomposite film. (Adapted from Rasidi et al., 2014.)

the SPI/KH biocomposite films at 40 wt% of KH loading compared to the neat SPI film. The increase in tensile strength of the SPI/KH biocomposite films indicated that the KH filler serves as a reinforcing agent in the films. The presence of polar groups, for instance, hydroxyl, amine, and carboxyl, in SPI leads to strong chemical bonding with KH. Besides that, the increase in tensile strength of the biocomposite films can be further justified by the KH filler capability to support the stress transferred from the matrix (SPI) phase to the filler (KH) phase. However, the increase in the tensile strength is attributed to the formation of intermolecular hydrogen bonds between SPI matrix and KH filler, thus resulting in a more compact structure of biocomposite films.

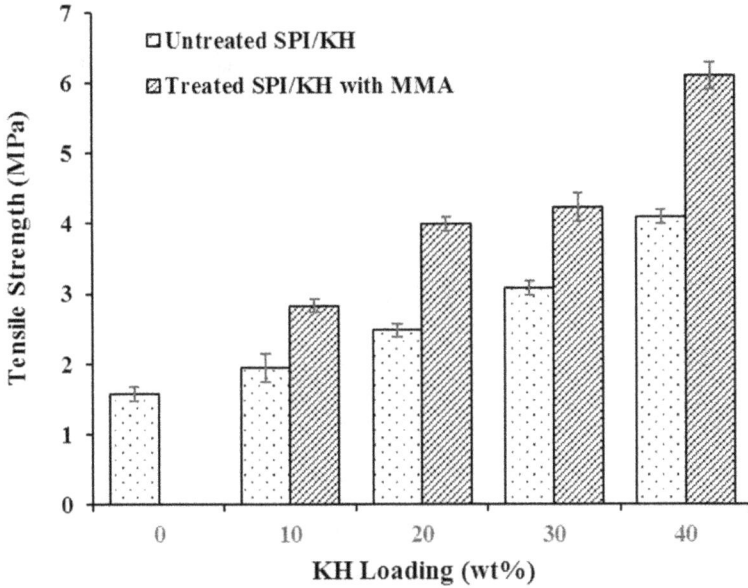

FIGURE 4.3   The effect of KH loading on tensile strength of untreated and treated SPI/KH biocomposite films with MMA.

It can be seen that the tensile strength of treated SPI/KH is higher than that of untreated SPI/KH biocomposite films. The tensile strength of treated SPI/KH biocomposite films with MMA showed an improvement with an average tensile strength of 35.4% as compared to untreated biocomposite films. This behavior can be attributed to the presence of strong interfacial adhesion and better dispersion between KH and SPI matrix with the addition of MMA. The MMA is bonded with the hydroxyl groups of KH, resulting in better filler–matrix interaction. It was believed that filler modification with MMA gave the better stress propagation of the film, which was caused by the improved compatibility and wetting to the matrix polymer.

4.5   ELONGATION AT BREAK

The elongation at break of untreated and treated SPI/KH biocomposite films is shown in Figure 4.4. The graph presents that the elongation at break decreased with increasing KH loading for both untreated and treated samples. The decrease in elongation at break is related to the increased stiffness of the films due to the incorporation of the KH filler. The compact matrix of the film indicates the reduction in chain mobility and the enhancement of rigidity in the biocomposite films, thus resulting in reduced elongation at break of biocomposite films. The similar trend in decreasing elongation at break is found by Lodha et al. [22] while studying the characterization of phytagel-modified soy protein resin and unidirectional flax yarn-reinforced green composites. Wang et al. [23] found that the addition of 0–30 wt% of cellulose whiskers reduced the elongation at break of soy protein/cellulose whisker films.

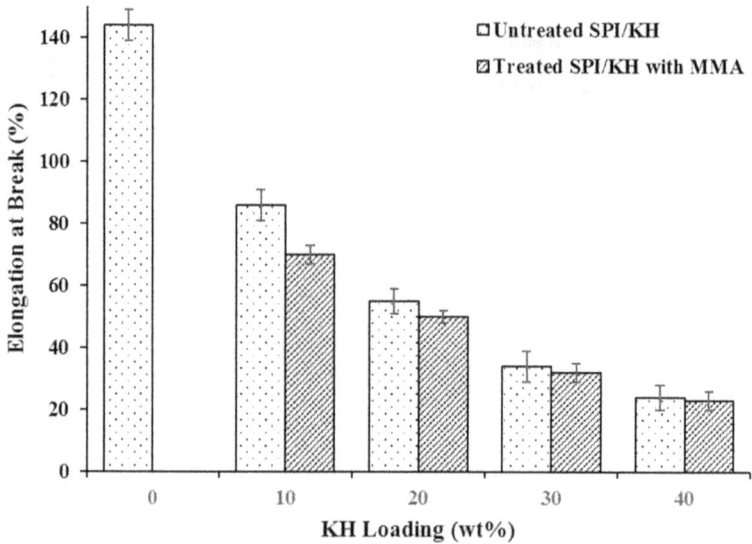

FIGURE 4.4 The effect of KH loading on elongation at break of untreated and treated SPI/KH biocomposite films with MMA.

At the similar KH loading, the elongation at break of the treated SPI/KH is lower compared to that of untreated SPI/KH biocomposite, which indicates the formation of ester linkages in SPI/KH biocomposite films that have increased rigidity and restricted chain mobility of KH. The presence of MMA in KH has also improved the adhesion at the interface between SPI and KH, which reduced the elongation at break of the treated biocomposite films. Moreover, the elongation at break of the treated SPI/KH with MMA was found to be decreased compared to that of untreated SPI/KH biocomposite films.

4.6 MODULUS OF ELASTICITY

The modulus of elasticity of both untreated and treated SPI/KH biocomposite films with MMA at different loading amounts of KH filler is illustrated in Figure 4.5. The results indicated that the modulus of elasticity of both untreated and treated SPI/ KH biocomposite films increased with increasing KH loading. The addition of KH reduced the chain mobility due to the increased stiffness of the SPI/KH biocomposite films. The modulus of elasticity is an indication of the relative stiffness of biocomposite films. The increase in the modulus of elasticity also depends on various factors such as the amount of filler used, the orientation of filler, the adhesion between the matrix and the filler, and the ratio of filler to matrix. The increase in modulus of elasticity was found because the amount of KH increased and the filler–filler interaction became more pronounced than the filler–matrix interaction. The increase in modulus of elasticity of blend films with the addition of cellulose derivatives/SPI was also reported by Zhou et al. [24]. The increase in modulus of elasticity due to natural

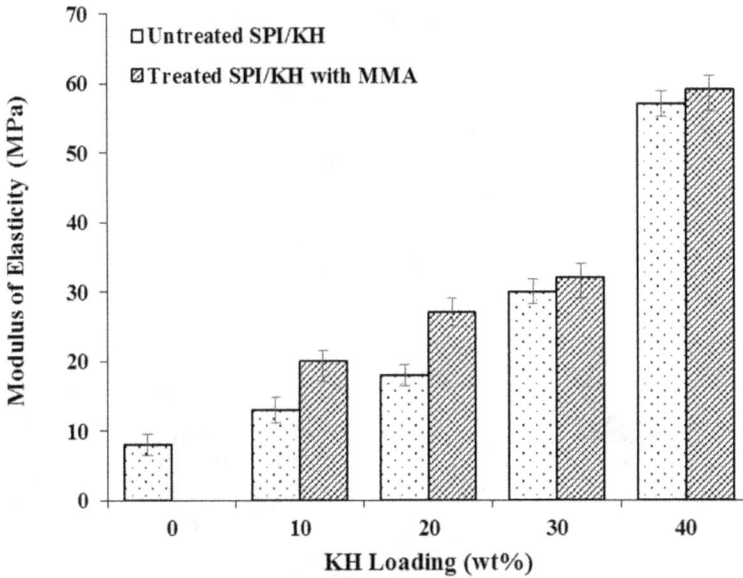

FIGURE 4.5   The effect of KH loading on modulus of elasticity of untreated and treated SPI/KH biocomposite films with MMA.

fillers is normally stiffer than the polymer matrix, thus increasing the stiffness of the composites.

In addition, the moduli of elasticity of the treated SPI/KH biocomposite films with MMA were higher than those of untreated biocomposite films. The treated SPI/KH biocomposite films with MMA exhibited higher average modulus of elasticity (26.7%) as compared to untreated biocomposite films. MMA was added to overcome the dispersion problem and enhance the modulus of elasticity of biocomposite films by improving the interfacial adhesion.

4.7   MORPHOLOGY

The SEM micrographs of untreated and treated SPI/KH biocomposite films with MMA at 20 and 40 wt% are shown in Figure 4.6a–e. The tensile fracture surface of the neat SPI (Figure 4.6a) exhibited a homogeneous, smooth, and continuous surface, demonstrating that SPI had good film-forming ability. The SEM micrograph of the untreated SPI/KH biocomposite film at 20 wt% KH is illustrated in Figure 4.6b. SEM shows there were no KH agglomerations that occurred in the matrix surface; it revealed that the dispersion of KH in the SPI matrix was oriented in a random manner, thus producing a reinforcement effect on the biocomposite films. Figure 4.6c presents the SEM micrograph of untreated SPI/KH biocomposite films at 40 wt% KH. It shows the KH is well embedded in the SPI matrix, signifying a good adhesion of the filler with the matrix. It can be assumed that the biocomposite film experienced good dispersion efficiency as there is less agglomeration of KH filler. This result supports

(a)

(b)

(c)

(d)

(e)

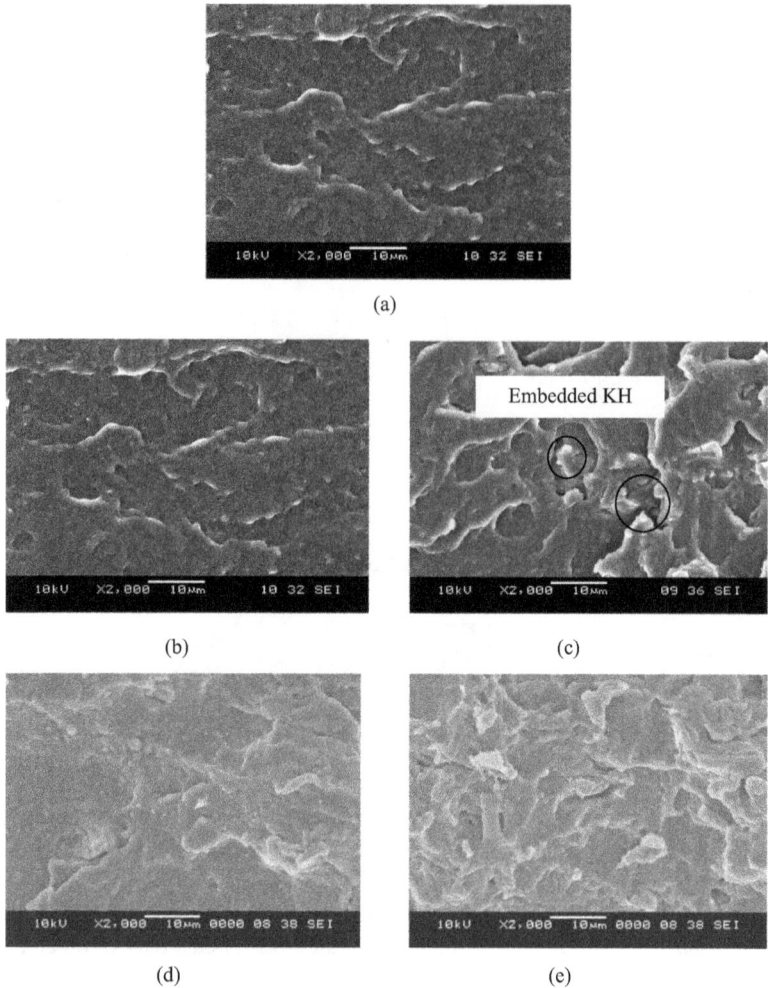

FIGURE 4.6 SEM micrograph of tensile fracture surface of untreated SPI/KH biocomposite films with MMA at (a) 0 wt%, (b) 20 wt%, and (c) 40 wt% of KH loading and treated SPI/KH biocomposite films with MMA at (d) 20 wt% and (e) 40 wt% of KH loading.

the increased tensile strength of the biocomposite film with increasing KH loading, thus proving that the stress is fairly distributed.

It was observed that the surface of both treated SPI/KH biocomposite films exhibited rough tearing with increased KH loading. After KH filler was treated with MMA, the fracture surface showed less KH pull-out and detachment from SPI matrix. From Figure 4.6d,e, it can be confirmed that the modification of KH filler by MMA has changed the morphology of the treated SPI/KH biocomposite films, i.e., films become brittle.

4.8   THERMAL STABILITY PROPERTIES

The TGA curves of untreated and treated SPI/KH biocomposite films with MMA are illustrated in Figure 4.7. Table 4.1 summarizes the TGA data of untreated and treated SPI/KH biocomposite films with MMA. It can be seen that the thermal degradation of the neat SPI film exhibits two stages of degradation of weight loss. The first stage (30°C–150°C) represents the loss of adsorption and bound moisture. The second stage represents degradation (150°C–650°C), which is attributable to the presence of glycerol in the SPI matrix. Furthermore, KH was degraded in two stages. The initial weight loss from room temperature to 150°C was attributed to moisture evaporation. The second-stage weight loss occurred between 150°C and 500°C, with maximum at 650°C, and referred to the thermo-oxidative reaction of the main organic compounds (decomposition of mainly cellulose, hemicellulose, and lignin).

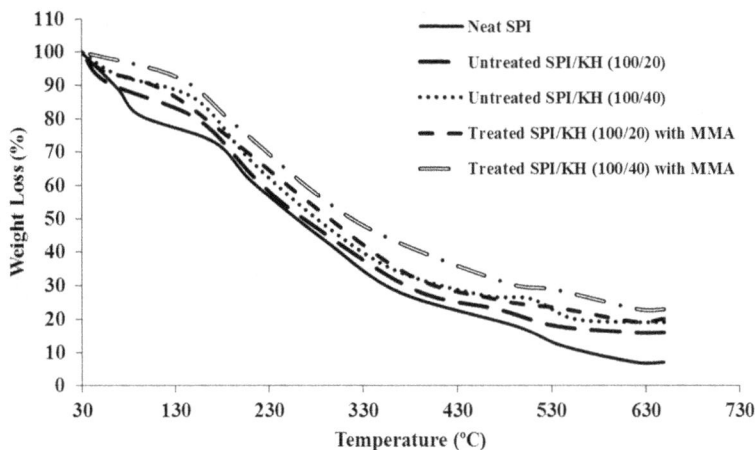

FIGURE 4.7   TGA curves of untreated and treated SPI/KH biocomposite films with MMA.

TABLE 4.1
The Percentage of Weight Loss of Untreated and Treated SPI/KH Biocomposite Films with MMA

Biocomposite Films	Weight Loss (%)			Residue Remaining at 650°C (%)
	$T_{200°C}$	$T_{300°C}$	$T_{400°C}$	
Neat SPI	34.58	59.88	74.78	7.05
Untreated SPI/KH (10/20)	33.76	56.79	72.25	16.33
Untreated SPI/KH (10/40)	28.85	51.16	68.55	19.12
Treated SPI/KH (10/20) with MMA	22.77	48.77	67.39	20.01
Treated SPI/KH (10/40) with MMA	19.99	41.79	57.85	23.41

The weight loss of both untreated and treated SPI/KH biocomposite films decreased with increasing KH loading. The untreated and treated SPI/KH biocomposite films also decomposed in the following three steps. The first step of thermal decomposition occurs between 30°C and 200°C, resulting in moisture loss from the biocomposite films. The second step of thermal decomposition begins from 200°C to 400°C, whereby degradation of SPI, glycerol, and cellulose from both biocomposite films occurs. The third step of thermal decomposition occurs between 400°C and 650°C, resulting in lignin decomposition and char residue. The weight loss of SPI/KH biocomposite films at T_{200} decreased with the increased KH loading, due to the reduction in MC. Thus, the weight loss at T_{300} and T_{400} decreased at highest KH content. The treated biocomposite films showed lower weight loss than untreated SPI/KH biocomposite films. This indicates that the formation of treated KH with MMA enhanced the thermal stability of biocomposite film with an average improvement of 18.4% compared to untreated SPI/KH biocomposite films. Moreover, the char residues at temperatures of 200°C, 300°C, and 400°C of SPI/KH biocomposite films reduced with increasing KH loading. This observation was an evidence of the better filler dispersion and filler–matrix interaction in SPI/KH biocomposite films.

4.9   MOISTURE CONTENT

The MC of untreated and treated SPI/KH biocomposite films with MMA is illustrated in Figure 4.8. The increased filler loading from 10 to 40 wt% for both untreated and treated SPI/KH biocomposite films decreased the percentage of MC. The MC

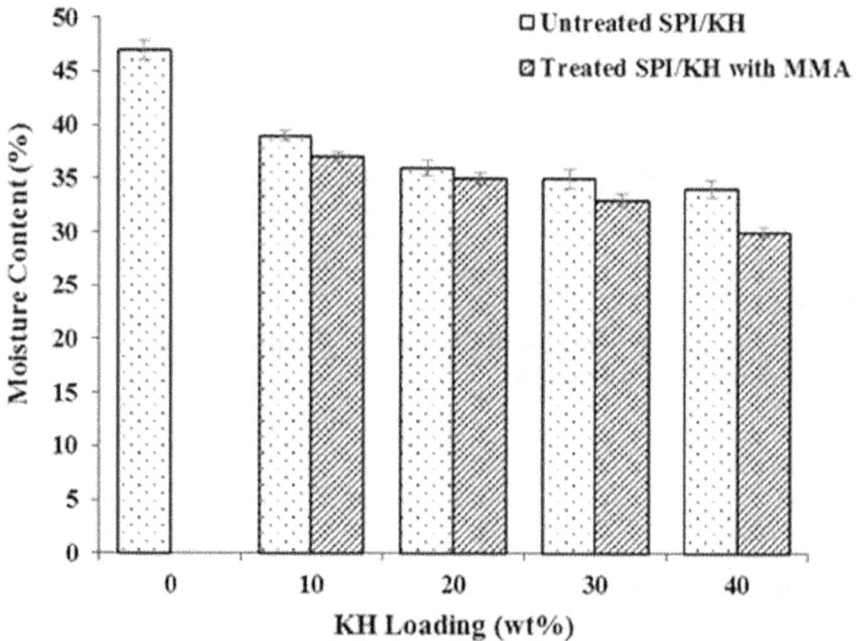

FIGURE 4.8   The MC of untreated and treated SPI/KH biocomposite films with MMA.

of SPI/KH biocomposite films decreased with increased KH loading from 10 to 40 wt%. The MC of untreated SPI/KH biocomposite films at 40 wt% of KH decreased around 28.4% compared to that of the neat SPI matrix. It is believed that the addition of KH fibers with denser network of cellulose might have formed a biocomposite structure that reduces the rate of moisture absorption through the SPI matrix. The MC decreased with increased KH loading due to the formation of a rigid hydrogen bonding network of cellulose in SPI/KH biocomposite films via percolation mechanism [25]. In addition, Lu et al. [26] reported that as the chitin of whiskers content increased in SPI matrix, it has the ability to enhance the water resistance property of SPI matrix, due to the formation of 3D networks of intermolecular hydrogen bonding interactions between "filler and filler" and "filler and matrix."

It was found that treated SPI/KH biocomposite films with MMA indicated lower MC absorption than untreated biocomposite films. However, at similar KH loading, the average moisture resistivity enhancement of treated SPI/KH biocomposite films was about 9.3% compared to that of untreated SPI/KH biocomposite films. This proved that the acetyl group ($-CH_3COO$) reacts with the hydrophilic hydroxyl groups of the fiber, thus removing the existing moisture [27]. Thus, the hydrophilicity nature of the KH was found to be decreased by improving the dimensional stability along with KH dispersion into SPI matrix.

4.10 TOTAL SOLUBLE MATTER

Figure 4.9 presents the comparison in terms of the TSM between untreated and treated SPI/KH biocomposite films with MMA at KH loading amounts of 0, 10, 20, 30, and 40 wt%. The increased filler loading for SPI/KH biocomposite films decreased the TSM. The SPI/KH biocomposite films tested did not break or dissolve

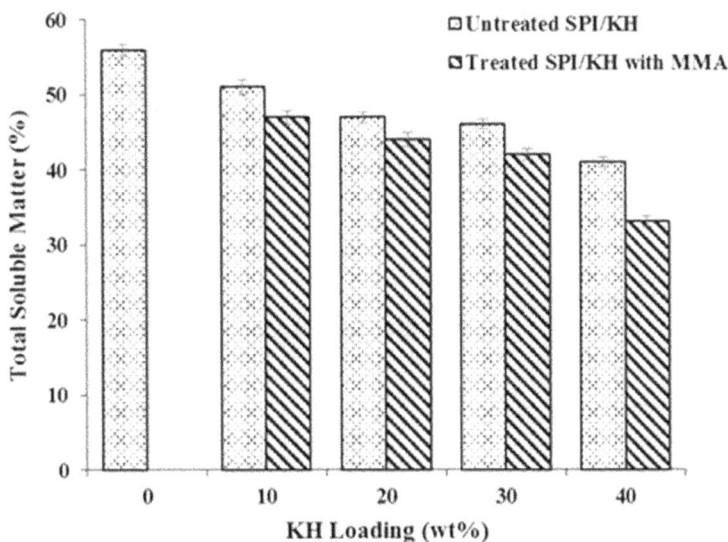

FIGURE 4.9   TSM of untreated and treated SPI/KH biocomposite films with MMA.

after 24 h of incubation. This has proven that the biocomposite film was stable and only some small molecules of peptides were soluble. The neat SPI film usually has the weaker bonding, which would be associated with the shorter chain length of protein molecules. This leads to a dipped interaction between the molecules, which resulted in a higher solubility of the resulting films. At 40 wt% of KH, the TSM of untreated SPI/KH biocomposite films decreased about 27.7% compared to that of the neat SPI film. This indicates that the network formed at higher KH loading hinders the solubility of the matrix into the water. In addition, the kapok pods were not washed before being crushed to KH; thus, the waxy material was not removed. Hence, the TSM was prevented to some extent. A similar trend of fiber loading was reported by Lomelí-Ramírez et al. [28] in the study of cassava starch/green coir fiber biocomposites. Such a trend could be illustrated due to the fact that higher loading of green coir fiber induced the cassava starch–fiber interaction, thus resulting in a decreased content of the TSM.

It was found that treated SPI/KH biocomposite films indicated lower TSM compared to untreated SPI/KH biocomposite films. The treated biocomposite films showed about 11.9% decrement of average TSM compared to untreated SPI/KH biocomposite films at similar KH loading. The surface modification of KH with MMA indicated that the MMA promotes hydrophobicity and acts as a barrier between KH surface and SPI matrix, thus increasing the resistivity towards film solubility in water.

4.11   GEL FRACTION

The effect of treated KH with MMA on gel fraction of SPI/KH biocomposite films is shown in Table 4.2. From this table, it can be seen that the gel fraction of both untreated and treated biocomposite films increased slightly with KH loading. The result shows that the gel fraction of the neat SPI was lower than that of SPI/KH biocomposite films, because KH is a natural filler that cannot be dissolved in acetic acid solvent.

However, the gel fraction of the treated SPI/KH biocomposite films was higher than that of untreated biocomposite films. Yet, this testing did not perform as expected due to less cross-linking reactions that occur between SPI matrix and treated KH with MMA. At similar KH loading, about 3.7% of the average gel fraction increased compared to untreated SPI/KH biocomposite films.

TABLE 4.2
Gel Fraction of Untreated and Treated SPI/KH Biocomposite Films with MMA

Biocomposite Films	Gel Fraction (%)
Neat SPI (untreated)	25.78
SPI/KH:100/20 (untreated)	32.21
SPI/KH:100/40 (untreated)	41.10
SPI/KH:100/20 (treated with MMA)	33.74
SPI/KH:100/40 (treated with MMA)	42.34

4.12   ENZYMATIC DEGRADATION

Figure 4.10 shows the effect of enzymatic degradation on weight loss of untreated and treated SPI/KH biocomposite films with MMA in α-amylase buffer solution. Table 4.3 represents the summary of weight loss on enzymatic degradation of untreated and treated SPI/KH biocomposite films with MMA after 14 days. Apparently, the weight loss of both untreated and treated SPI/KH biocomposite films on enzymatic degradation decreased with KH loading. The untreated SPI/KH biocomposite film at 40 wt% of KH improved the weight loss of biodegradation percentage at about 38.5% compared to the neat SPI film, because α-amylase penetrated into SPI/KH biocomposite films and attacked the amino acids of SPI matrix first followed by cellulose and hemicellulose of KH filler. Besides that, this might be due to the good dispersion of KH filler into the SPI matrix and chemical bonding between them.

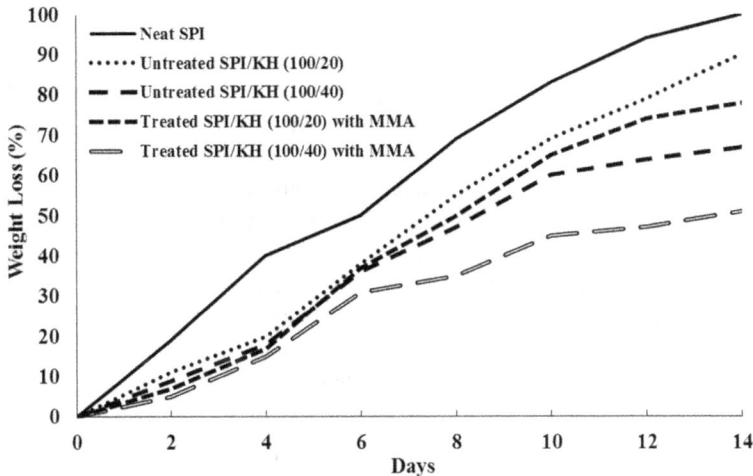

FIGURE 4.10   The weight loss of untreated and treated SPI/KH biocomposite films with MMA on enzymatic degradation.

TABLE 4.3
Weight Loss of Enzymatic Degradation of Untreated and Treated SPI/KH Biocomposite Films with MMA after 14 Days

Biocomposite Films	Weight Loss of SPI/KH Biocomposite Films on Enzymatic Degradation after 14 Days (%)
Neat SPI	99 ± 0.7
Untreated SPI/KH (100/20)	90 ± 1.1
Untreated SPI/KH (100/40)	61 ± 0.3
Treated SPI/KH (100/20) with MMA	78 ± 1.1
Treated SPI/KH (100/40) with MMA	51 ± 0.4

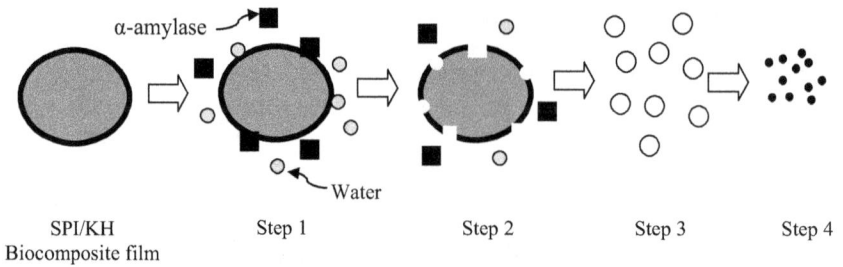

FIGURE 4.11　Proposed mechanism of enzymatic degradation of untreated and treated SPI/KH biocomposite films.

The enzymatic hydrolysis of biocomposite materials is a heterogeneous process that is affected by the interaction between the enzymes and polymeric chains (Yeng et al., 2015). Thus, this mechanism involves four steps: (1) diffusion of α-amylase and water from the bulk solution to the untreated and treated SPI/KH biocomposite film surface; (2) adsorption of the α-amylase and water on the substrate, resulting in the formation of the enzyme–substrate complex; (3) the continuous degradation of SPI and KH and the occurrence of weight loss; and (4) diffusion of the soluble degradation from the SPI/KH substrate to the solution. Figure 4.11 illustrates the proposed steps of enzymatic degradation of SPI/KH biocomposite films.

The chemically treated SPI/KH with MMA showed lower percentage weight loss compared to untreated SPI/KH biocomposite films. The average weight loss of enzymatic degradation after treated KH with MMA reduced about 17.9% compared to untreated SPI/KH biocomposite films. The ether bonds from treated SPI/KH biocomposite films restrict the diffusion of α-amylase enzyme molecules into SPI/KH biocomposite films, thus lowering the degradation rate.

4.13　CONCLUSION

The utilization of KH in SPI film-forming solution showed a positive effect on the properties of biocomposite films. Moreover, the KH loading at 40 wt% showed the highest improvement in tensile properties, thermal properties, morphological characteristics, moisture resistivity, TSM, gel fraction, and enzymatic degradation. Chemical modification using MMA on the surface of the filler showed better results. This behavior is attributed to the presence of better dispersion and strong interfacial adhesion between KH filler and SPI matrix as the modification created more active sites for hydrogen bonding and reduced the moisture absorption through the ester bridge between MMA and KH. The tensile strength and modulus of elasticity of treated SPI/KH biocomposite films increased; however, the elongation at break decreased with increasing KH loading amount. The SEM micrograph showed the better interfacial adhesion between SPI matrix and KH filler in treated SPI/KH biocomposite films. The addition of KH also increased the thermal stability of SPI/KH biocomposite films. Both the MC and the TSM of the treated SPI/KH biocomposite films decreased. Furthermore, the gel fraction of the treated SPI/KH biocomposite films increased. The biodegradability for the treated SPI/KH biocomposite films in enzyme buffer solution also decreased.

ACKNOWLEDGMENT

The financial support of Fundamental Research Grant Scheme (FRGS) Grant No: FRGS/1/2018/TK05/UNIMAP/02/13 is gratefully acknowledged.

REFERENCES

1. B. Poyraz, "Enzyme treated CNF biofilms: characterization", *International Journal of Biological Macromolecules*, vol. 117, pp. 713–720, 2018.
2. R.R. Ali, W.W.A Rahman, N.B. Ibrahim, and R.M. Kasmani, "Starch-based biofilms for green packaging", In *Developments in Sustainable Chemical and Bioprocess Technology*, Edited by R. Pogaku, B. Awang, C. Chu. Springer, Boston, MA, pp. 347–354, 2013.
3. M. Wihodo and C.I. Moraru, "Physical and chemical methods used to enhance the structure and mechanical properties of protein films: a review", *Journal of Food Engineering*, vol. 114(3), pp. 292–302, 2013.
4. D. Carpine, J.L.A. Dagostin, L.C. Bertan, and M.R. Mafra, "Development and characterization of soy protein isolate emulsion-based edible films with added coconut oil for olive oil packaging: barrier, mechanical, and thermal properties", *Food and Bioprocess Technology*, vol. 8(8), pp. 1811–1823, 2015.
5. L. Fernandez-Espada, C. Bengoechea, F. Cordobes, and A. Guerrero, "Protein/glycerol blends and injection-molded bioplastic matrices: soybean versus egg albumen", *Journal of Applied Polymer Science*, vol. 133(6), pp. 42980, 2016.
6. D.R. Kumar and P. Mohanraj, "Review on natural fiber in various pretreatment conditions for preparing perfect fiber", *Asian Journal of Applied Science and Technology (AJAST)*, vol. 1(2), pp. 66–78, 2017.
7. R.K. Morchhale, M.D. Goel, P. Patel, and S. Murali, "An experimental investigation on the development of saw dust polymer composite for door shutter application as a substitute to natural wood", *Asia Journal of Civil Engineering (BHRC)*, vol. 17(3), pp. 335–346, 2016.
8. A.K. Mohanty, M. Misra, and G. Hinrichsen, "Biofibres, biodegradable polymers and biocomposites: an overview", *Macromolecular Materials and Engineering*, vol. 276(1), pp. 1–24, 2000.
9. M. Ramesh, "Kenaf (Hibiscus cannabinus L.) fibre based bio-materials: a review on processing and properties", *Progress in Materials Science*, vol. 78, pp. 1–92, 2016.
10. S. Husseinsyah, M.Y. Chan, A.R. Kassim, M.M. Zakaria, and H. Ismail, "Kapok husk-reinforced soy protein isolate biofilms: tensile properties and enzymatic hydrolysis", *BioResources*, vol. 9(3), pp. 5636–5651, 2014.
11. V. Govindan, S. Husseinsyah, and P.L. Teh, "Treated Nypa fruticans Husk - filled regenerated cellulose biocomposite films", *BioResources*, vol. 11(4), pp. 8739–8755, 2016.
12. Y. Zheng, J. Wang, Y. Zhu, and A. Wang, "Research and application of kapok fiber as an absorbing material: a mini review", *Journal of Environmental Sciences*, vol. 27, pp. 21–32, 2015.
13. R.S. Rengasamy, D. Das, and C. Praba Karan, "Study of oil sorption behavior of filled and structured fiber assemblies made from polypropylene, kapok and milkweed fibers", *Journal of Hazardous Materials*, vol. 186(1), pp. 526–532, 2011.
14. S.Y. Yoon, D.J. Kim, Y.J. Sung, S.H. Han, N.S. Aggangan, and S.J. Shin, "Enhancement of enzymatic hydrolysis of kapok [Ceiba pentandra (L.) Gaertn.] seed fibers with potassium hydroxide pretreatment", *Asia Life Sciences*, vol. 25(1), pp. 17–29, 2016.
15. F. Ahmad, H.S. Choi, and M.K. Park, "A review: natural fiber composites selection in view of mechanical, light weight, and economic properties", *Macromolecular Materials and Engineering*, vol. 300(1), pp. 10–24, 2015.

16. V. Fiore, G. Di Bella, and A. Valenza, "The effect of alkaline treatment on mechanical properties of kenaf fibers and their epoxy composites", *Composites Part B: Engineering*, vol. 68, pp. 14–21, 2015.

17. G.H.D. Tonoli, R.F. Mendes, G. Siqueira, J. Bras, M.N. Belgacem, and H. Savastano, "Isocyanate-treated cellulose pulp and its effect on the alkali resistance and performance of fiber cement composites", *Holzforschung*, vol. 67(8), pp. 853–861, 2013.

18. M.S. Huda, L.T. Drzal, A.K. Mohanty, and M. Misra, "Effect of fiber surface-treatments on the properties of laminated biocomposites from poly (lactic acid)(PLA) and kenaf fibers", *Composites Science and Technology*, vol. 68(2), pp. 424–432, 2008.

19. P. Larsson-Brelid, M.E.P. Wålinder, M. Westin, and R. Rowell, "Ecobuild a center for development of fully biobased material systems and furniture applications.", *Molecular Crystals and Liquid Crystals*, vol. 484(1), pp. 257–623, 2008.

20. M.K. Thakur, R.K. Gupta, and V.K Thakur, "Surface modification of cellulose using silane coupling agent", *Carbohydrate Polymers*, vol. 111, pp. 849–855, 2014.

21. N.I.A. Razak, N.A. Ibrahim, N. Zainuddin, M. Rayung, and W.Z. Saad, "The influence of chemical surface modification of kenaf fiber using hydrogen peroxide on the mechanical properties of biodegradable kenaf fiber/poly (lactic acid) composites", *Molecules*, vol. 19(3), pp. 2957–2968, 2014.

22. P. Lodha and A.N. Netravali, "Characterization of Phytagel® modified soy protein isolate resin and unidirectional flax yarn reinforced "green" composites", *Polymer Composites*, vol. 26(5), pp. 647–659, 2005.

23. Y. Wang, X. Cao, and L. Zhang, "Effects of cellulose whiskers on properties of soy protein thermoplastics", *Macromolecular Bioscience*, vol. 6(7), pp. 524–531, 2006.

24. Z. Zhou, H. Zheng, M. Wei, J. Huang, and Y. Chen, "Structure and mechanical properties of cellulose derivatives/soy protein isolate blends", *Journal Of Applied Polymer Science*, vol. 107(5), pp. 3267–3274, 2008.

25. Y. Pu, J. Zhang, T. Elder, Y. Deng, P. Gatenholm, and A.J. Ragauskas, "Investigation into nanocellulosics versus acacia reinforced acrylic films", *Composites Part B: Engineering*, vol. 38(3), pp. 360–366, 2007.

26. Y. Lu, L. Weng, and L. Zhang, "Morphology and properties of soy protein isolate thermoplastics reinforced with chitin whiskers", *Biomacromolecules*, vol. 5(3), pp. 1046–1051, 2004.

27. N. Gontard, S. Guilbert, and J.L. Cuq, "Edible wheat gluten films: influence of the main process variables on film properties using response surface methodology", *Journal of Food Science*, vol. 57(1), pp. 190–195, 1992.

28. M.G. Lomelí-Ramírez, S.G. Kestur, R. Manríquez-González, S. Iwakiri, G.B. de Muniz, and T.S. Flores-Sahagun, "Bio-composites of cassava starch-green coconut fiber: part II—structure and properties", *Carbohydrate Polymers*, vol. 102, pp. 576–583, 2014.

29. M.S.M. Rasidi, S. Husseinsyah, P.L. Teh, "Chemical modification of Nypa fruticans filled polylactic Acid/recycled low-density polyethylene biocomposites", *Bioresources*, vol. 9(2), pp. 2033–2050, 2014.

30. C.M. Yeng, S. Husseinsyah, S.T. Sam, "A comparative study of different crosslinking agent-modified chitosan/corn cob biocomposite films", *Polymer Bulletin*, vol. 72, pp. 791–808, 2015.

5 Protein-Based Fillers in Biodegradable Polymer Composites

K.I. Ku Marsilla, A. Rusli, and Z.A.A. Hamid
Universiti Sains Malaysia

CONTENTS

5.1 INTRODUCTION: BIODEGRADABLE POLYMER COMPOSITES

In recent years, polymer scientist has been focusing on the development of new biodegradable polymer composite materials from renewable resources due to the disposal of synthetic material. The biodegradable plastics market is projected to grow from $3.02 billion in 2018 to $6.12 billion by 2023 [1]. The term "polymer composites" refers to polymers that consist of two or more components/phases that include inorganic or organic additives, while biocomposites/biodegradable composites are derived from natural resources.

Generally, biodegradable polymer composites can be classified into three categories according to their synthesis process: natural polymers, synthetic polymers from natural sources, and polymers from microbial fermentation. Table 5.1 lists the classification of the main biodegradable polymers. Natural polymers such as cellulose, fiber, starches, polysaccharide, and protein can be extracted from nature. Among all, starch

TABLE 5.1

Classification of Main Biodegradable Polymers

Classification of Biodegradable Polymers	Example of Polymers
Natural polymers	Cellulose
	Natural fiber
	Starches
	Polysaccharides
	Protein
Synthetic polymers from natural sources	PLA
	PBS
	PGA
	PVA
Polymers from microbial fermentation	PHB
	PHAs
	PCL

has been widely used in producing thermoplastic blends or composites. For example, biodegradable foam tray is developed from cassava starch with 30% kraft fiber and 4% chitosan, which has properties similar to those of polystyrene (PS) foam tray [2].

On the other hand, synthetic polymers from natural resources are the polymers that are produced by polycondensation or ring-opening polymerization of biologically derived monomers such as polylactic acid (PLA) and polybutylene succinate (PBS), while polymers from microbial fermentation are produced by microorganisms. PLA and PBS show a good tensile strength but are slow-degrading polymers due to their high crystallization rate [3]. Therefore, numerous approaches such as blending and formation of composites have been used, particularly to improve the physical properties.

Polymers from microbial fermentation such as PHA (poly(hydroxyalkanoates)), PHB (poly-3-hydroxybutyrate), and PCL (polycaprolactone) have a promising potential in many applications, especially in food packaging. To date, over 90 different types of PHA were reported. PHA has a wide range of properties comparable to flexible plastics like polyethylene (PE) and elastomeric materials. However, due to the production cost, only a few suppliers exist in the market.

5.2 FILLERS IN BIODEGRADABLE POLYMER COMPOSITE

Recent development of filler usage in blends and composites has been driven by a desire to enhance the properties of the materials. Global polymer filler market is expected to reach an estimated $49.1 billion by 2021 and is forecast to grow at a compound annual growth rate (CAGR) of 6.3% from 2016 to 2021 [4]. Typically, filler is used to reduce the formulation or manufacturing cost, but current development is focusing on improving the physical and mechanical properties. Fillers are used for distinctly different purposes when incorporated into polymers, such as increased stiffness, heat distortion temperature, and thermal stability or improved processing. The purpose of such incorporation depends on the types of fillers.

Fillers have been classified in various ways, depending on their shapes (fibers, flakes, spheres, particulate) and specific properties (fire retardants, electrical and magnetic modifiers, processing aids, surface property modifiers). Generally, fillers can be grouped into two types: (1) extender filler and (2) reinforcement filler. The main usage of extender fillers is to occupy the volume of composite and reduce the cost of expensive binder material. An ideal extender filler usually has a broad range of particle sizes for particle packing in composite. There is no chemical reactivity occurred with the polymer or additives. Reinforcement fillers, on the other hand, have a specific and vital role in improving the composite interphase through chemical and/or physical bonding.

Inorganic fillers hold the largest global demand and accounted for 78.9% of the total market volume in 2015. However, organic fillers are expected to witness a brisk growth of 5.6% over the forecast period [5]. The availability of sources and environmental concerns have attracted more industries to utilize organic fillers rather than traditional inorganic fillers. Table 5.2 provides the summary of inorganic and organic fillers used in most applications, especially plastics and composites. Calcium carbonate holds the largest global demand, while carbon black is the second largest market with a demand of 11.7 million tonnes [6]. It is expected that organic fillers such as fibers, cellulose, and starch will have a positive impact on the market growth.

5.3   ORGANIC FILLERS IN POLY LACTIC ACID (PLA)

PLA is a biodegradable plastic that is produced from the ring-opening polymerization of lactide, a dimer of lactic acid. Lactic acid (2-hydroxy propionic acid) is produced via fermentation. Currently, corn is used as a raw material for lactic acid production, but other potential starting materials such as corn stalks, wheat bran, cassava, cellulose, barley, starch, potato starch, beet molasses, rye flour, and carrot processing waste are also considered [7].

PLA presents great potential for industrial and commodity applications due to its good mechanical properties, biodegradability, transparency, and sustainable biomass resources. PLA has packaging and automotive applications, similar to polyethylene terephthalate (PET), PS, and polycarbonate (PC). However, the use of PLA for the cases may be limited because of its high T_g (60°C), high brittleness, and poor barrier properties [8]. Therefore, the addition of reinforcing fillers and additives into PLA matrix is one powerful way to alter specific end-use characteristics and properties.

Table 5.3 lists the recent trends in the field of composites based on PLA. Reinforcement using fibers is usually much stronger than that using polymer, and

TABLE 5.2
Types of Inorganic and Organic Fillers

Inorganic Fillers	Organic Fillers
Oxides – glass, Al_2O_3	Carbon, graphite
Hydroxides – $Al(OH)_3$	Natural polymers – cellulose, protein, starch
Salts – $CaCO_3$, $BaSO_4$, phosphates	Synthetic polymers – polyamide, polyester
Silicates – talc, mica, kaolin, clays	

TABLE 5.3
Recent Trends in the Field of Composites Based on PLA

Main Component		Fillers/Additives	References
PLA	Natural fibers	Kenaf	[12]
		Hemp	[9–11]
		Flax	[13,14]
PLA	Cellulosic	Cellulose nanowhiskers	[15,16]
		Nanocellulose	[17–20]
		Nanofiber cellulose	[3,21–23]
PLA	Starches	Tapioca	[24,25]
		Corn	[26,27]
		Cassava	[28]
		Maize	[29]
PLA	Protein	Soy protein	[30–34]
		Eggshell membrane	[35,36]
		Blood meal	[37,38]
		Fish	[39]

increases its modulus and strength of composites. For composites with long fibers with certain arrangement and orientation, the fibers become the major component of the composite, while when using short fibers or flakes, the content usually does not exceed 30%–40% by volume [9]. However, other than compositions and orientation of fibers, the processing method also plays a major role. Alignment of discontinuously treated fibers of harakeke and hemp was made using dynamic sheet former (DSF), and the tensile strength was 90% and 60%, respectively, which was higher than PLA. Comparison of mechanical property between DSF and injection-molded sample showed that composite produced using DSF was better due to the enhanced reinforcement between fibers and PLA matrix [10]. In other research, using PLA/hemp-co-wrapped hybrid yarns by compression molding, the mechanical property was found to be two times higher compared to the neat PLA [11].

Recently, composite using cellulosic materials at nanoscales has attracted a lot of attention due to high potential in improving the interfacial adhesion between matrix and fillers. By using chemical or mechanical treatment, the cellulose material can be converted into nanocellulose. Nanoparticles offer homogenous dispersion, have larger surface, and possess greater aspect ratio (ratio of the largest to the smallest dimension) in the production of nanocomposites. It can improve different properties of the materials incorporated, such as mechanical, electrical, fire retardant, or thermal properties. Nanocellulose filler in its different forms such as whiskers [15,16], particles [17–20], or nanofibers [3,21–23] shows an increasing demand in various applications, such as packaging, food industry, cosmetics, and medical.

Starch-based materials are among the most widely used due to their low cost and renewable properties. Starch can be extracted from different raw materials such as tapioca, corn, cassava, and maize. The strength of starch materials was tuned using the incorporation of different additives such as plasticizers and cross-linkers. In addition, starch was produced as copolymer where PLA was grafted on the surface of starch granules to improve the interfacial adhesion between starch granules and PLA matrix in starch/PLA blends [25].

In a blend of soy protein isolate (SPI) and PLA at 80/20 composition, with 10 phr triacetin (TA), a highly ordered porous matrix of SPI and homogenously dispersed PLA domains are formed, resulting in a complex coarsened phase structure for both SPI and PLA domains [31]. In other research of PLA and soluble eggshell membrane protein (SEP), PLA is in a semicrystalline state when SEP is added as a filler and its morphology changes to amorphous PLA when SEP content is increased. Increasing SEP also improves the biocompatibility of the composites [35,36].

5.4 PROTEIN FILLERS

Natural protein is linear, unbranched, and have a precise length consisting of up to 20 amino acids joined by peptide bonds, forming a polypeptide chain. Different amino acids interact differently with their environment, and the physical properties of the protein are determined by the sequence of amino acid groups. The most reactive amino acids are cysteine and lysine where at different pH values, they react to give stable derivatives. There are four target functional groups for the majority of cross-linking or chemical interactions as listed in Table 5.4. Manipulation of these target functional groups has attracted many developments to produce complex functional polymeric materials. On top of that, a mixture of hydrophilic and hydrophobic groups that can unfold in different environments (denaturant) offers researchers to design random configuration of protein for diverse applications.

5.4.1 Sources of Protein

Numerous sources of protein (from vegetable and animals) have been explored as biodegradable fillers in composites. Figure 5.1 shows the schematic presentation of bio-based polymers from sources of protein based on their origin.

TABLE 5.4
Four Chemical Targets Account for the Major Cross-linking and Chemical [40]

Target Side		Explanation
Primary amines	$(-NH_2)$	This group exists at the N-terminus of each polypeptide chain and in the side chain of lysine (Lys) residues
Carboxyl	$(-COOH)$	This group exists in the C-terminus of each polypeptide chain and in the side chains of aspartic acid (Asp) and glutamic acid (Glu)
Sulfhydryl	$(-SH)$	This group exists in the side chain of cysteine (Cys). Often, as part of a protein's secondary or tertiary structure, cysteine is joined together between their side chains via disulfide bonds
Carbonyls	(RCHO)	These aldehyde groups can be created by oxidizing carbohydrate groups on glycoprotein

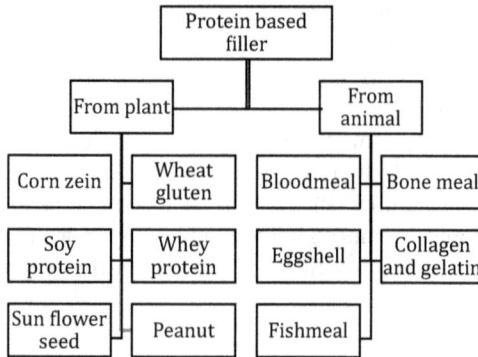

FIGURE 5.1   Schematic presentation of protein-based fillers based on their origin.

5.4.1.1   Protein from Plant

Plant protein is one of the major biopolymers obtained from agricultural feedstocks. Table 5.5 lists the source, composition, and major protein group from different plants. The United States is one of the major producers of soy beans in the world. It is commercially available in three different grades from soy bean-processing plants: SPI (90% protein), soy protein concentrate (SPC) (65%–72% protein), and soy flour (SF) (54% protein). Soy protein (SP) is hydrophilic due to the presence of a high proportion of glutamic acid and aspartic acid compared to other proteins [41].

SP consists of both polar and nonpolar side chains. Between these chains, there are strong intra- and intermolecular interactions, such as hydrogen bonding, dipole–dipole, charge–charge, and hydrophobic interactions. The strong charge and polar interactions between side chains of SP molecules restrict segment rotation and molecular mobility, which increases stiffness, yield point, and tensile strength of SP films [16]. Molecular conformation can be altered by physical [42,43], chemical [44, 45], or enzymatic agents [46].

Whey protein is the by-product of cheese and casein manufacturing that contains approximately 7% dry matter. In general, the dry matter includes 13% protein, 75% lactose, 8% minerals, approximately 3% organic acids, and less than 1% fat [47]. Whey protein concentrate (WPC, protein concentration 65%–80% in dry matter) or whey protein isolate (WPI, protein concentration over 90% in dry matter) can be obtained through a membrane filtration process followed by spray drying.

TABLE 5.5
Protein from Plants

Sources	Protein Content (%)	Amino Acid Residues	References
Gluten	80	Glutenin	[50]
Corn zein	65	Glutamine, proline, alanine, leucine	[51]
SP	90	Glutamic acid, aspartic acid, leucine, arginine	[52]
Sunflower seed	30–50	Lysine, isoleucine, arginine, histidine	[49]
Peanut	85	Arginine, histidine	[53]

Zein extracted from maize gluten meal (corn) has hydrophobic properties, which are characterized by high percentage of proline. The most popular application of zein is in the textile fiber market [48]. Sunflower meal is a by-product obtained after oil extraction from sunflower seeds. The high amount of histidine and arginine in sunflower meal increases its nutritive value, and thus, the sunflower meal can be potentially used as an additive in the food industry [49]. Peanut protein is extracted from defatted flour, which provides the food industry with a new high protein food ingredient for product formulation.

5.4.1.2 Protein from Animals

Proteins from animal by-products such as bovine blood meal, fish, gelatin, and collagen can be processed to form thermoplastics, fibers, films, composites, and coatings with interesting properties. Gelatin, e.g., has been produced on a large scale and widely used in the pharmaceutical and food industries. Gelatin film from mammals has the highest mechanical properties compared to gelatin film from fish [54]. Meat bone meal containing 50% protein, 9.5% fat, 10.1% calcium, and 4.8% phosphorus is considered as an adhesive for polywood industry [55] and sand replacement in cement-based materials [56]. In other research, milk protein is effective in delaying browning reactions on the surface of sliced fruits [57]. Dragline silk is one of the strongest fibers produced by spiders. The fiber is stronger, is one-tenth the weight of the high tensile steel, and consists of mainly hydrophobic amine groups such as glycine and alanine [58]. However, the mass production of the silk is too expensive and time-consuming, and limits its widespread applications.

5.5 MISCIBILITY AND COMPATIBILIZATION OF FILLER AND POLYMER/COMPOSITES

The mechanical properties of continuous phase and filler as dispersed phase usually in polymer blends and composites depend on the compositions of the components and the interactions between polymer and fillers. The definition of polymer blends and composites is often confounding. Polymer blend is a combination of two and more polymers to achieve miscibility or single-phase system. However, the definition of miscibility may be rather ambiguous. In many instances, it is desirable to have two phases present, as long as the multicomponent systems can be manipulated for structure, polymer interactions, and phase domain sizes.

On the other hand, polymer composites are the compound made up of two or more elements which finally form a multiphase system where each element reflects its identities and properties. Both mixtures can form different morphologies, such as compatible and incompatible blends. Figure 5.2 shows the terminology used that describes the differences between miscible, immiscible, compatible, and incompatible blends.

Towards achieving miscibility, the interfacial adhesion increases, resulting in a single-phase component. Immiscible blends usually form aggregates that are totally phase-separated. In compatible blends, the mixing between two polymers has high interfacial tension, giving a rough structure and poor adhesion. The interfacial tension is much higher with increasing particle size with uneven distribution in incompatible blends.

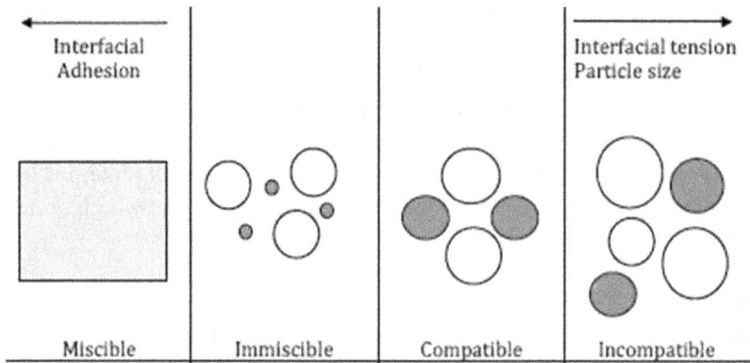

FIGURE 5.2 Blends terminology.

5.6 FACTORS AFFECTING COMPATIBILITY

5.6.1 COMPOSITION OF FILLERS

Changes in morphology may affect mechanical properties and can be related to a phase inversion with changes in composition. A phase inversion occurs when the blend's structure changes from a matrix of polymer A containing a dispersed polymer B to a dispersed polymer A within matrix B. For example, the morphology of SPI/PCL blend changes from smooth surface to a rough and heterogeneous fracture surface as the PCL content increases and shows a bicontinuous phase at 60 SPI/40 PCL accompanied by an increase in toughness [31].

It was observed that as the amount of protein increased in the modified polyester blends, more interactions between polyester and reactive groups in the protein occurred, resulting in an increase in tensile strength. However, blends containing more than 65% protein were difficult to be injection-molded [59]. The T_g of SPI in SPI/PCL blends decreased as PCL content increased (50% PCL) accompanied by decreasing crystallinity (X_c), indicating that the crystallization of PCL in blends was difficult [31]. In a blend of SPI/PLA (80/20) with 10 phr TA, a highly ordered porous matrix of SPI and homogenously dispersed PLA domains were involved in the formation of a complex coarsened phase structure for both SPI and PLA domains at 60/40 [60].

It is known that SPI is rich in protein than SPC, which also affects the properties of the blends. For example, SPC/PLA blends show a fine co-continuous phase structure, whereas SPI/PLA blends present severe phase coarsening. Due to higher protein content in SPI and its highly polar and hydrophilic nature compared to PLA which is hydrophobic, poor adhesion was observed. A high-viscosity disparity between SPI and PLA further exaggerated this [32].

5.6.2 COUPLING AGENT

In practice, polymer composites are said to be compatible if they exhibit two phases on a microscopic level but the interactions between polymer groups might be reasonable in a manner that provides useful properties of the multicomponent system. One of the strategies to improve compatibilization involves the addition of at least one substance with a highly reactive group that can interact with more than one component of the composites [61]. There are two methods for blend compatibilization (Figure 5.3): first the addition of a third component and second, reactive compatibilization.

5.6.3 ADDITION OF A THIRD COMPONENT

Introducing a third component in blends reduces the interfacial tension between components, improving the dispersion and enhancing the adhesion between phases. The coupling agent should be miscible and have a high affinity for both phases (Figure 5.3). Polymeric methylene diphenyl diisocyanate (pMDI) and poly-2-ethyl-2-oxazoline (PEOX) are the most commonly used coupling agent in protein composites. pMDI is the polymeric form of methylene diphenyl diisocyanate (MDI), and distillation of MDI yields pMDI. pMDI is highly reactive with hydroxyl

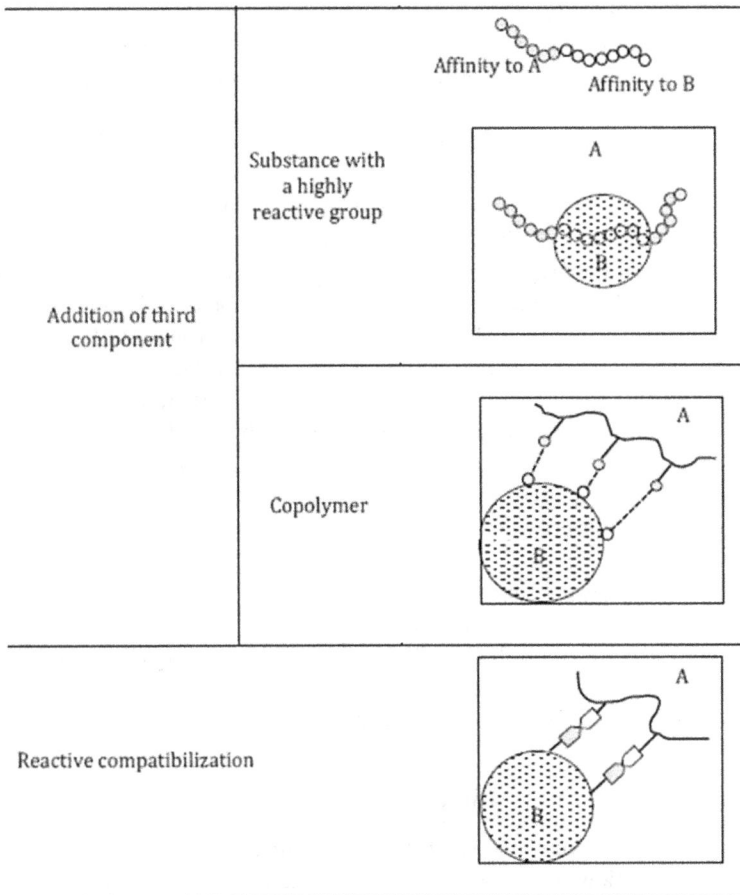

FIGURE 5.3 Methods for blends compatibilization.

functional groups to form urethane linkages [62], and the residues of untreated MDI are not expected in the blends due to high reactivity of the isocyanate groups [63]. Compatibility between SPI and PCL was improved with the addition of MDI in blends containing 50% of each component [31].

Another coupling agent often used in protein blends is poly(2-oxazoline), the properties of which can be tuned by changing the side chain of its monomer. In recent years, the use of poly (2-oxazoline) in biomedical applications has evolved as its biocompatibility and biodistribution are similar to those of polyethylene oxide (PEO) [64]. PEO has been widely used in protein adsorption, protein conjugates, and drug carriers. In PLA/SPC composites, hydrogen bonding between carbonyl groups in PEOX and hydroxyl and carboxylic end-groups in PLA reduced the inclusion size of SPC and increased interfacial bonding between PLA and SPC [33]. The carboxyl groups and/or amino group functionalities present in proteins are capable of reacting with the oxazoline functional group. The reaction between the carboxylic and oxazoline groups will result in an ester–amide linkage and is fast.

Maleic anhydride (MA) and its isostructural derivatives such as fumaric, citraconic, and itaconic acids are being widely used as a copolymer through the graft modification of various thermoplastic polymers. Graft modification onto polyolefins, especially polypropylene (PP), started back in 1969s when MA was grafted onto isotactic PP below its melting point [65]. Since then, much effort has been put to study the mechanism of grafting polyolefins under a variety of conditions [66].

These MA-grafted copolymers have been used as compatibilizers to improve the interfacial adhesion between two polymers. The compatibility between synthetic polyolefin-MA group with SP and starch was improved, which was evident from their morphological characteristics, mechanical properties, and thermal properties [59,67]. PLA-*g*-MA has been successfully used in starch and PLA blends. The maleation of PLA proved to be very effective in promoting strong interfacial adhesion with the hydroxyl group on starch [68]. Only recently, PLA-*g*-MA with a degree of grafting between 0.25% and 0.9% was used as compatibilizer in PLA/SP composites that improved the tensile strength and elongation at break of SP/PLA composites. A finer domain size of SPC phase was observed, suggesting improved dispersion [34]. L.A. Utracki [61] listed the three main factors that need to be considered in designing copolymers as compatibilizers:

1. To maximize miscibility of the appropriate part of its macromolecule with the specific polymeric component of the blends;
2. To minimize its molecular weight to just above the entanglement molecular weight for each interaction segment;
3. To minimize its concentration in the blend.

5.7 REACTIVE COMPATIBILIZATION

The second method is known as reactive compatibilization. From an economic point of view, this technique is more attractive than the addition of a third component. The concept involves in situ formation of a block or graft copolymer at the interface between the phases of a polymer blend during melt mixing. Reactive extrusion has been shown to be a cost-effective processing method as it is a continuous process that involves introducing reagents at optimum points in the reaction sequence. During extrusion, the reagents are homogenized at a longer residence time for a high conversion. Batch mixers are also used and have been found to be effective as they are easy to control processing parameters such as temperature, mixing time, and mixing intensity via the rotation speed of motors.

Reactive extrusion is an effective method for the chemical modification of polyesters along with the production of compatibilized blends between biodegradable polyesters and fillers [69]. During the chemical modification of polyesters like maleation, the reaction kinetics are dependent on the concentration of initiator and monomer, while a suitable screw configuration is required to meet residence time and mixing requirements to reach a high conversion. Also, polymerization needs to be fast enough if processing is in a continuous one-stage process.

Another advantage of using reactive extrusion is the formation of additional interactions between two phases at the interphase. During polymerization in an extruder, besides the production of graft polymers, there are unwanted reactions such as chain scission, cross-linking, and homopolymerization. The initiator radicals attack the polymer to generate macroradicals that might form cross-linking, which could produce a higher molecular weight polymer. Other advantages of cross-linking are that it can improve the properties of the original polymer, such as increased thermal stability, solvent resistance, and mechanical properties. Using reactive extrusion, cross-linking is also possible through ionic bonds and physical cross-linking, i.e., van der Waals or hydrogen bonds. In PLA, the cross-linked structure of triallyl isocyanurate (TAIC) promotes crystallization. Although this was achieved using a cross-linking agent, with further increase in cross-linking (from 12.1% to 41.2%), the onset crystallization temperature increased, resulting in the perfection of crystal lamellas [70].

5.8 CONCLUSION

Organic fillers in biodegradable polymer composites appear to have a very promising future for a wide range of applications and will have a positive impact on the market growth. Protein-based filler, on the other hand, may soon be competitive with existing fillers. Nevertheless, the main drawback for the commercialization of new protein-based filler lies in the miscibility and compatibilization issue, which affected the processing condition. Although reactive functional groups in protein for the majority of cross-linking or chemical interactions offer researchers to design a random configuration of protein, it is a trade-off between economic of scale in production and cost of product.

REFERENCES

1. MarketsandMarkets. *Biodegradable Plastics Market Worth $6.12 Billion by 2023*. Available: https://www.marketsandmarkets.com/PressReleases/biodegradable-plastics.asp
2. N. Kaisangsri, O. Kerdchoechuen, and N. Laohakunjit, "Biodegradable foam tray from cassava starch blended with natural fiber and chitosan," *Industrial Crops and Products*, vol. 37, pp. 542–546, 2012.

3. X. Zhang, J. Shi, H. Ye, Y. Dong, and Q. Zhou, "Combined effect of cellulose nano-crystals and poly(butylene succinate) on poly(lactic acid) crystallization: the role of interfacial affinity," *Carbohydrate Polymers*, vol. 179, pp. 79–85, 2018.

4. Ceresana. *Market Study: Fillers* (5th Edition) [Online].

5. Grand View Research, Inc. *Polymer Filler Market Size to Reach $62.54 Billion by 2024: Grand View Research, Inc.* Available: https://www.prnewswire.com/news-releases/polymer-filler-market-size-to-reach-6254-billion-by-2024-grand-view-research-inc-580617591.html

6. Grand View Research, Inc. (2018, 21 February 2019). *Calcium Carbonate Market Worth $34.28 Billion By 2025 | CAGR: 5.7%*. Available: https://www.grandviewre-search.com/press-release/global-calcium-carbonate-market

7. G. Reddy, M. Altaf, B. Naveena, M. Venkateshwar, and E. V. Kumar, "Amylolytic bacterial lactic acid fermentation—a review," *Biotechnology Advances*, vol. 26, pp. 22–34, 2008.

8. K. M. Nampoothiri, N. R. Nair, and R. P. John, "An overview of the recent develop-ments in polylactide (PLA) research," *Bioresource technology*, vol. 101, pp. 8493–8501, 2010.

9. M. A. Sawpan, K. L. Pickering, and A. Fernyhough, "Improvement of mechanical per-formance of industrial hemp fibre reinforced polylactide biocomposites," *Composites Part A: Applied Science and Manufacturing*, vol. 42, pp. 310–319, 2011.

10. K. L. Pickering and M. G. Aruan Efendy, "Preparation and mechanical properties of novel bio-composite made of dynamically sheet formed discontinuous harakeke and hemp fibre mat reinforced PLA composites for structural applications," *Industrial Crops and Products*, vol. 84, pp. 139–150, 2016.

11. B. Baghaei, M. Skrifvars, and L. Berglin, "Manufacture and characterisation of thermoplastic composites made from PLA/hemp co-wrapped hybrid yarn prepregs," *Composites Part A: Applied Science and Manufacturing*, vol. 50, pp. 93–101, 2013.

12. S. H. Kamarudin, L. C. Abdullah, M. M. Aung, C. T. Ratnam, and E. R. Jusoh, "A study of mechanical and morphological properties of PLA based biocomposites prepared with EJO vegetable oil based plasticiser and kenaf fibres, *Materials Research Express*, vol 5, p. 085314, 2018.

13. T. Gobi Kannan, C. M. Wu, and K. B. Cheng, "Effect of different knitted structure on the mechanical properties and damage behavior of Flax/PLA (poly lactic acid) double covered uncommingled yarn composites," *Composites Part B: Engineering*, vol. 43, pp. 2836–2842, 2012.

14. M. Bulota and T. Budtova, "Highly porous and light-weight flax/PLA composites," *Industrial Crops and Products*, vol. 74, pp. 132–138, 2015.

15. S. Qian, H. Zhang, W. Yao, and K. Sheng, "Effects of bamboo cellulose nanowhisker content on the morphology, crystallization, mechanical, and thermal properties of PLA matrix biocomposites," *Composites Part B: Engineering*, vol. 133, pp. 203–209, 2018.

16. S. Qian and K. Sheng, "PLA toughened by bamboo cellulose nanowhiskers: role of silane compatibilization on the PLA bionanocomposite properties," *Composites Science and Technology*, vol. 148, pp. 59–69, 2017.

17. E. Robles, I. Urruzola, J. Labidi, and L. Serrano, "Surface-modified nano-cellulose as reinforcement in poly(lactic acid) to conform new composites," *Industrial Crops and Products*, vol. 71, pp. 44–53, 2015.

18. M. R. Kamal and V. Khoshkava, "Effect of cellulose nanocrystals (CNC) on rheologi-cal and mechanical properties and crystallization behavior of PLA/CNC nanocompos-ites," *Carbohydrate Polymers*, vol. 123, pp. 105–114, 2015.

19. E. Lizundia, E. Fortunati, F. Dominici, J. L. Vilas, L. M. León, I. Armentano, L. Torre, and J. M. Kenny, "PLLA-grafted cellulose nanocrystals: role of the CNC content and grafting on the PLA bionanocomposite film properties," *Carbohydrate Polymers*, vol. 142, pp. 105–113, 2016.

20. N. Bitinis, E. Fortunati, R. Verdejo, J. Bras, J. M. Kenny, L. Torre, and M. A. López-Manchado, "Poly(lactic acid)/natural rubber/cellulose nanocrystal bionanocomposites. Part II: properties evaluation," *Carbohydrate Polymers*, vol. 96, pp. 621–627, 2013.

21. A. N. Frone, S. Berlioz, J.-F. Chailan, and D. M. Panaitescu, "Morphology and thermal properties of PLA–cellulose nanofibers composites," *Carbohydrate Polymers*, vol. 91, pp. 377–384, 2013.

22. A. Abdulkhani, J. Hosseinzadeh, A. Ashori, S. Dadashi, and Z. Takzare, "Preparation and characterization of modified cellulose nanofibers reinforced polylactic acid nanocomposite," *Polymer Testing*, vol. 35, pp. 73–79, 2014.

23. S. Spinella, G. Lo Re, B. Liu, J. Dorgan, Y. Habibi, P. Leclère, J.-M. Raquez, P. Dubois, and R. A. Gross, "Polylactide/cellulose nanocrystal nanocomposites: efficient routes for nanofiber modification and effects of nanofiber chemistry on PLA reinforcement," *Polymer*, vol. 65, pp. 9–17, 2015.

24. D. Wu and M. Hakkarainen, "Recycling PLA to multifunctional oligomeric compatibilizers for PLA/starch composites," *European Polymer Journal*, vol. 64, pp. 126–137, 2015.

25. X. Yang, A. Finne-Wistrand, and M. Hakkarainen, "Improved dispersion of grafted starch granules leads to lower water resistance for starch-g-PLA/PLA composites," *Composites Science and Technology*, vol. 86, pp. 149–156, 2013.

26. P. Müller, J. Bere, E. Fekete, J. Móczó, B. Nagy, M. Kállay, B. Gyarmati, and B. Pukánszky, "Interactions, structure and properties in PLA/plasticized starch blends," *Polymer*, vol. 103, pp. 9–18, 2016.

27. M. Akrami, I. Ghasemi, H. Azizi, M. Karrabi, and M. Seyedabadi, "A new approach in compatibilization of the poly(lactic acid)/thermoplastic starch (PLA/TPS) blends," *Carbohydrate Polymers*, vol. 144, pp. 254–262, 2016.

28. J. Muller, C. González-Martínez, and A. Chiralt, "Poly(lactic) acid (PLA) and starch bilayer films, containing cinnamaldehyde, obtained by compression moulding," *European Polymer Journal*, vol. 95, pp. 56–70, 2017.

29. J. M. Ferri, D. Garcia-Garcia, L. Sánchez-Nacher, O. Fenollar, and R. Balart, "The effect of maleinized linseed oil (MLO) on mechanical performance of poly(lactic acid)-thermoplastic starch (PLA-TPS) blends," *Carbohydrate Polymers*, vol. 147, pp. 60–68, 2016.

30. A. González and C. I. Alvarez Igarzabal, "Soy protein – poly (lactic acid) bilayer films as biodegradable material for active food packaging," *Food Hydrocolloids*, vol. 33, pp. 289–296, 2013.

31. L. Calabria, N. Vieceli, O. Bianchi, R. V. Boff de Oliveira, I. do Nascimento Filho, and V. Schmidt, "Soy protein isolate/poly(lactic acid) injection-molded biodegradable blends for slow release of fertilizers," *Industrial Crops and Products*, vol. 36, pp. 41–46, 2012.

32. J. Zhang, L. Jiang, L. Zhu, J.-l. Jane, and P. Mungara, "Morphology and properties of soy protein and polylactide blends," *Biomacromolecules*, vol. 7, pp. 1551–1561, 2006.

33. B. Liu, L. Jiang, H. Liu, and J. Zhang, "Synergetic effect of dual compatibilizers on in situ formed poly (lactic acid)/soy protein composites," *Industrial & Engineering Chemistry Research*, vol. 49, pp. 6399–6406, 2010.

34. R. Zhu, H. Liu, and J. Zhang, "Compatibilizing effects of maleated poly (lactic acid) (PLA) on properties of PLA/Soy protein composites," *Industrial & Engineering Chemistry Research*, vol. 51, pp. 7786–7792, 2012.

35. X. Xiong, Q. Li, J.-W. Lu, Z.-X. Guo, and J. Yu, "Poly(lactic acid)/soluble eggshell membrane protein blend films: preparation and characterization," *Journal of Applied Polymer Science*, vol. 117, pp. 1955–1959, 2010.

36. B. Ashok, S. Naresh, K. O. Reddy, K. Madhukar, J. Cai, L. Zhang, and A. V. Rajulu, "Tensile and thermal properties of poly(lactic acid)/eggshell powder composite films," *International Journal of Polymer Analysis and Characterization*, vol. 19, pp. 245–255, 2014.

37. A. Walallavita, C. Verbeek, and M. Lay, "Morphology and mechanical properties of itaconic anhydride grafted poly (lactic acid) and thermoplastic protein blends," *International Polymer Processing*, vol. 33, pp. 153–163, 2018.
38. S. C. Izuchukwu, C. J. Verbeek, and J. M. Bier, "Decoloured Novatein® and PLA blends compatibilized with itaconic anhydride," *Applied Mechanics and Materials*, pp. 3–13, 2018.
39. R. Saiwaew, P. Suppakul, W. Boonsupthip, and C. Pechyen, "Development and characterization of poly (lactic acid)/fish water soluble protein composite sheets: a potential approach for biodegradable packaging," *Energy Procedia*, vol. 56, pp. 280–288, 2014.
40. P. Gupta and K. K. Nayak, "Characteristics of protein-based biopolymer and its application," *Polymer Engineering & Science*, vol. 55, pp. 485–498, 2015.
41. P. Lodha and A. N. Netravali, "Thermal and mechanical properties of environment-friendly 'green' plastics from stearic acid modified-soy protein isolate," *Industrial Crops and Products*, vol. 21, pp. 49–64, 2005.
42. U. Kalapathy, N. S. Hettiarachchy, D. Myers, and K. C. Rhee, "Alkali-modified soy proteins: effect of salts and disulfide bond cleavage on adhesion and viscosity," *Journal of the American Oil Chemists' Society*, vol. 73, pp. 1063–1066, 1996.
43. W. Huang and X. Sun, "Adhesive properties of soy proteins modified by urea and guanidine hydrochloride," *Journal of the American Oil Chemists' Society*, vol. 77, pp. 101–104, 2000.
44. Y. Wang, X. Mo, X. S. Sun, and D. Wang, "Soy protein adhesion enhanced by glutaraldehyde crosslink," *Journal of Applied Polymer Science*, vol. 104, pp. 130–136, 2007.
45. K. L. Franzen and J. E. Kinsella, "Functional properties of succinylated and acetylated soy protein," *Journal of Agricultural and Food Chemistry*, vol. 24, pp. 788–795, 1976.
46. Y. Liu, J. Li, L. Yang, and Z. Shi, "Graft copolymerization of methyl methacrylate onto casein initiated by potassium ditelluratocuprate(III)," *Journal of Macromolecular Science, Part A*, vol. 41, pp. 305–316, 2004.
47. E. Bugnicourt, M. Schmid, O. M. Nerney, J. Wildner, L. Smykala, A. Lazzeri, and P. Cinelli, "Processing and validation of whey-protein-coated films and laminates at semi-industrial scale as novel recyclable food packaging materials with excellent barrier properties," *Advances in Materials Science and Engineering*, vol. 2013, 2013.
48. J. W. Lawton, "Zein: a history of processing and use," *Cereal Chemistry*, vol. 79, pp. 1–18, 2002.
49. P. Ivanova, V. Chalova, L. Koleva, and I. Pishtiyski, "Amino acid composition and solubility of proteins isolated from sunflower meal produced in Bulgaria," *International Food Research Journal*, vol. 20, p. 2995, 2013.
50. B. P. Mooney, "The second green revolution? Production of plant-based biodegradable plastics," *Biochemical Journal*, vol. 418, pp. 219–232, 2009.
51. R. Garratt, G. Oliva, I. Caracelli, A. Leite, and P. Arruda, "Studies of the zein-like α-prolamins based on an analysis of amino acid sequences: implications for their evolution and three-dimensional structure," *Proteins: Structure, Function, and Bioinformatics*, vol. 15, pp. 88–99, 1993.
52. V. Poysa, L. Woodrow, and K. Yu, "Effect of soy protein subunit composition on tofu quality," *Food Research International*, vol. 39, pp. 309–317, 2006.
53. T. Kholief, "Chemical composition and protein properties of peanuts," *Zeitschrift für Ernährungswissenschaft*, vol. 26, pp. 56–61, 1987.
54. R. Avena-Bustillos, B.-s. Chiou, C. Olsen, P. Bechtel, D. Olson, and T. McHugh, "Gelation, oxygen permeability, and mechanical properties of mammalian and fish gelatin films," *Journal of food science*, vol. 76, pp. E519–E524, 2011.
55. S. K. Park, D. Bae, and N. Hettiarachchy, "Protein concentrate and adhesives from meat and bone meal," *Journal of the American Oil Chemists' Society*, vol. 77, pp. 1223–1227, 2000.

56. M. Cyr and C. Ludmann, "Low risk meat and bone meal (MBM) bottom ash in mortars as sand replacement," *Cement and Concrete Research*, vol. 36, pp. 469–480, 2006.

57. C. Le Tien, C. Vachon, M. A. Mateescu, and M. Lacroix, "Milk protein coatings prevent oxidative browning of apples and potatoes," *Journal of Food Science*, vol. 66, pp. 512–516, 2001.

58. J. Gosline, P. Guerette, C. Ortlepp, and K. Savage, "The mechanical design of spider silks: from fibroin sequence to mechanical function," *Journal of Experimental Biology*, vol. 202, pp. 3295–3303, 1999.

59. J. John and M. Bhattacharya, "Properties of reactively blended soy protein and modified polyesters," *Polymer International*, vol. 48, pp. 1165–1172, 1999.

60. L. Calabria, N. Vieceli, O. Bianchi, R. V. B. De Oliveira, I. do Nascimento Filho, and V. Schmidt, "Soy protein isolate/poly (lactic acid) injection-molded biodegradable blends for slow release of fertilizers," *Industrial Crops and Products*, vol. 36, pp. 41–46, 2012.

61. L. A. Utracki, "Compatibilization of polymer blends," *The Canadian Journal of Chemical Engineering*, vol. 80, pp. 1008–1016, 2002.

62. G. Oertel, "Polyurethane handbook," *Carl Hanser Verlag*, p. 626, 1985.

63. H. Wang, X. Sun, and P. Seib, "Strengthening blends of poly (lactic acid) and starch with methylenediphenyl diisocyanate," *Journal of Applied Polymer Science*, vol. 82, pp. 1761–1767, 2001.

64. R. Hoogenboom, "Poly (2-oxazoline) s: a polymer class with numerous potential applications," *Angewandte Chemie International Edition*, vol. 48, pp. 7978–7994, 2009.

65. Y. Minoura, M. Ueda, S. Mizunuma, and M. Oba, "The reaction of polypropylene with maleic anhydride," *Journal of Applied Polymer Science*, vol. 13, pp. 1625–1640, 1969.

66. D. Shi, J. Yang, Z. Yao, Y. Wang, H. Huang, W. Jing, J. Yin, and G. Costa, "Functionalization of isotactic polypropylene with maleic anhydride by reactive extrusion: mechanism of melt grafting," *Polymer*, vol. 42, pp. 5549–5557, 2001.

67. U. R. Vaidya and M. Bhattacharya, "Properties of blends of starch and synthetic polymers containing anhydride groups," *Journal of Applied Polymer Science*, vol. 52, pp. 617–628, 1994.

68. D. Carlson, L. Nie, R. Narayan, and P. Dubois, "Maleation of polylactide (PLA) by reactive extrusion," *Journal of Applied Polymer Science*, vol. 72, pp. 477–485, 1999.

69. J.-M. Raquez, P. Degée, Y. Nabar, R. Narayan, and P. Dubois, "Biodegradable materials by reactive extrusion: from catalyzed polymerization to functionalization and blend compatibilization," *Comptes Rendus Chimie*, vol. 9, pp. 1370–1379, 2006.

70. Y. Zhang, C. Wang, H. Du, X. Li, D. Mi, X. Zhang, T. Wang, and J. Zhang, "Promoting crystallization of polylactide by the formation of crosslinking bundles," *Materials Letters*, vol. 117, pp. 171–174, 2014.

6 Study of Thermoplastic Starch Incorporation on Polylactic Acid/Natural Rubber Blends via Dynamic Vulcanization Approach

A.W.M. Kahar and K.Y. Low
Universiti Malaysia Perlis

H. Ismail
Universiti Sains Malaysia

CONTENTS

6.1 INTRODUCTION: BACKGROUND AND DRIVING FORCES

The use of plastics has replaced paper, glass, metal, and many other materials as they are superior in terms of lightness, stability, strength, ease of processing, durability, and impermeability properties [1]. As a consequence of these properties of plastic products, large volumes of plastic materials are used in daily life year by year. Meanwhile, these plastics that have the difficulties in decomposition have created a great continuing concern over environmental issues [2]. The improper handling of plastic materials, for instance, plastic littering, will affect the aquatic environments, which endangers aquatic lives and affects the biosystem [3]. Another problem encountered would be the chemical modification in the plastic materials, thus leading to the difficulty of reusing and recycling. Conventional plastics such as polyethylene terephthalate (PET) and polyvinyl chloride (PVC) may require decades or even longer period of time to degrade, and during the degradation process, they will also release additives and by-products, which would pose threats and endanger the organisms' state of health and also the environment to a certain extent that the plastics would be declared as hazardous materials [4,5].

Currently, starch is widely used because of its biodegradability and low cost in order to fabricate biodegradable products. Starch is a polysaccharide consisting of two microstructures, namely, amylose (a linear chain of glucose) and amylopectin (a branched chain of glucose) [6,7]. Amylose is a linear polysaccharide poly(α-1,4-glucopyronosyl) formed from α-1,4-linked glucose units bonded together by the $\alpha(1\rightarrow4)$ glycosidic bonds. Amylopectin, on the other hand, has higher molecular weight and therefore lower chain mobility, and there is a lower possibility to pack closely to allow hydrogen bonding formation. Amylopectin is a highly branched poly(α-1,4-glucopyronosyl) unit with 95% α-1,4 chains, and the remnants will be the 5% branches containing α-1,6 bonds [8]. The major starch in Southeast Asia is processed from the tapioca root. The tapioca root consists of moisture (70%), starch (24%), fiber (2%), protein (1%), and other substances, including minerals (3%). Because of its low cost, biodegradability, and easy to undergo modifications for changes in physical, chemical, and rheological properties, it can function as biopolymer having applications in the food industries, paint, textiles, paper, and detergent soap, and it is also used as adhesive and glue [7].

According to da Matta et al. [9], the starch film moisture can be affected by the incorporation of plasticizer, e.g., glycerol, due to its hygroscopic character. The glycerol molecules can act as the plasticizer by interacting and spacing between the starch polymer chains apart. Using high temperature, shear force, and pressure, plasticizer molecules can embed and penetrate into the starch granules in order to demolish the interior hydrogen bonds of starch. Because of the effect of plasticizer, there will be interactions between starch and plasticizer to form starch–plasticizer bonding, which destructs and supplants the starch–starch interactions. The plasticizer molecule is more mobile and smaller than the starch molecule, and therefore, the starch network can undergo deformation easily without rupture. The process of starch destructurization is known as gelatinization, in which the structure of the starch granules undergoes destruction in order to produce a homogeneous polymer melt known as thermoplastic starch (TPS) [10,11]. Kahar et al. [12] showed in their research that TPS was plasticized by glycerol with different molecular weight through the calendering process. The plasticizer used had significantly reduced the WVP (water vapor permeability) as a result of the TPS hygroscopic properties. Rosa et al. [13] found out that different types of plasticizer used will affect both thermal and mechanical properties in the PP/TPS blend. The TPS was prepared by 80% of tapioca starch and 20% of plasticizer. It was found that the mechanical properties of the blend was found to be reduced as a result of the TPS addition. Carvalho et al. [14] had also done research by melt processing the TPS with the use of glycerol, carboxylic acid, and water. It was found out that the viscosity of TPS was reduced when the carboxylic acids were used to demolish the macromolecules present in the starch structure. This could control the molecular weight of the starch. However, starch is hydrophilic and it will lead to moisture absorption, leading to poor mechanical properties and limited performance in TPS-based products. To improve its weakness, it is advisable to blend the TPS with another biodegradable polymer found to be more suitable to be used in packaging and industrial applications [15,16]. One of the appropriate polymers would be the polylactic acid (PLA), which is also derived from a renewable resource with inherently high strength and stiffness properties.

PLA is a class of an aliphatic polyester, which is used for bio-based products as it has good biodegradability and biocompatibility properties, making it as an ideal candidate as an alternative to petroleum-based materials without creating environmental problems and at the same time providing great mechanical properties as compared to conventional polymers [17]. PLA is very versatile so it is capable of producing many products in replacement such as plastic cups, straws, plastic plates, plastic packaging films, and plastic containers for foods and drinks instead of using traditional plastics, and that is why, it can be applicable in many fields such as textile, medicine, and agriculture [18]. By blending the TPS and PLA, the biodegradation rate increases and so does the rigidity while lowering the material cost [16]. However, the blending between the two will lead to a reduction in its impact resistance as well as elongation at break. The downgrading on its mechanical properties in the PLA/TPS blend is due to the inherent brittleness of the PLA with low toughness and low tensile strain at break [19]. Another reason is the incompatibility of hydrophobic PLA and hydrophilic starch, which greatly influences the interfacial adhesion in its morphology, leading to a poor barrier and mechanical properties of the blend [16].

To improve the drawbacks from the brittleness of PLA, one of the economical and practical resorts will be blending the PLA with natural rubber (NR). The NR has great ductility and elasticity properties that toughen the brittle PLA despite being high strength and modulus and a T_g (ranging from 55°C to 65°C), which is too brittle for room temperature applications [20]. The NR is an ideal candidate, which is biocompatible in nature and can be easily obtained from natural resources [21]. Being the stress concentrators, the elastomer particles are acting as the impact modifiers to absorb the fracture energy of the brittle PLA to make it to be a toughened material [17]. That is why, the PLA/NR/TPS is produced as each of the components compensates the weakness of each other and this will turn out to be a very interesting research. However, as mentioned beforehand, the different nature of hydrophilic properties of starch and hydrophobic properties of PLA will lead to an incompatibility problem, and another notable point is that the nonpolarity of NR makes PLA/NR blends to be incompatible. To improve the phase compatibility and interaction of the blend, other than doing surface treatment, using reinforcement fillers, coupling agent or even addition of compatibilizer as the third macromolecule component can be done to reduce the interfacial tension by manipulating the interfacial properties and to ensure good adhesion between phases [22].

NR is widely available and easily to be supplied from the *Hevea brasiliensis* tree in the latex form. NR mainly consists of *cis*-1,4-polyisoprene repeating isoprene units. According to Yuan et al. [23], NR, also known as green elastomer, is widely available from natural resources like *Hevea brasiliensis* rubber tree. Its ductility, flexibility, and elasticity properties make it an ideal choice to compensate for the breakability of PLA while blending together. By internal mixer and twin screw extruder, Bitinis et al. [24] also reported a study on PLA/NR blend. They found out that the blend ratio of PLA to NR of 90:10 would be the optimal ratio in order to obtain the best performance whereby the rubber phase was finely dispersed with droplet size being controlled in PLA continuous phase. The use of NR component also increased the crystallization rate of the blend. Last, the ductility and elongation at break have been improved greatly from 5% of pure PLA to 200% at this ratio. The PLA/NR blends can be improved via dynamic vulcanization (DV) technique, whereby it involves a selective cross-linking of the rubber component and a fine dispersion in molten thermoplastic continuous phase through intensive mixing to produce thermoplastic vulcanized (TPV). DV is an ideal technique to improve the TPV blend properties. The mechanical properties of the TPV can be greatly enhanced when the elastomer particles are being vulcanized in the unvulcanized thermoplastic continuous phase imparting better mechanical properties in its resultant blend. The intention is to improve the strength and stability in the NR phase by having the intramolecular or intermolecular cross-linking in elastomer chains. Other than mechanical properties, DV technique also improves the permanent set, thermal resistance, chemical resistance by fluids, and better phase stability morphology in melt as compared to unvulcanized system [24,25]. The NR in the PLA/NR blends was dynamically vulcanized through three different curing systems, namely, sulfur, peroxide, and phenolic resin system. It was found out that the cross-linked elastomer phase was in the continuous structure in all curing systems. After doing the testing and characterizations, they compared the results and data obtained between the three curing systems and they concluded that the PLA/NR with a blend ratio of 60/40 through the peroxide curing system has the greatest compressive strength followed by the sulfur curing system and last the phenolic resin system. Both sulfur and peroxide showed a small effect on PLA degradation. The notched impact strengths of peroxide and sulfur systems were 42.5 and 25.75 kJ/m^2, respectively, which have been improved as much as 16 times and 10 times better than that of the neat PLA. However, it showed a serious degradation of PLA chains in the phenolic resin system, and that is why the mechanical properties and thermal stability became worse. Huang et al. [26] pointed out the incompatibility problem between the PLA and NR, and found that melt blending the PLA/NR with the use of an internal mixer and dicumyl peroxide (DCP) as the cross-linking agent was used to enhance the properties of the blend. The impact strength and the elongation at break were 1.8 and 2.5 times better than those of the neat PLA when 2 and 0.2 wt% of DCP were used in DV of PLA/NR. Chen et al. [27] blended the PLA/NR by using the internal mixer and DCP as the curing agent via DV process. They found out that the impact of fracture surface of PLA/NR with 65/35 showed a good adhesion between PLA/NR phases, using SEM (scanning electron microscopy). As compared with the neat PLA, the PLA/NR with 65/35 gave an impact resistance as high as 58.3 kJ/m^2, which was 21 times better. Through swelling test, it could be observed that PLA was susceptible to dichloromethane and soluble into it instantly as it took around 10 min for complete dissolution, whereas the NR swelled slightly even after immersion for 20 min. For the PLA/NR with 65/35, it was crimped for the first 3 min as the PLA was dissolved into the solvent leaving the soft rubber phase, and it was swollen slightly after 10 min, meaning that the addition of NR into the blend had somehow improved the solvent resistance in the blend.

In previous work done by Kahar et al. [28], HDPE and NR were used as a matrix and blending with TPS was done by DV technique by using an internal mixer. The blend ratio of HDPE to NR was fixed at 70:30 for determining the various amounts of TPS. Through their experiment, HVA-2 (N, N'-m-phenylene bis-maleimide) cross-linker was used as the free radical cross-linking agent in the DV process, which certainly enhanced the mechanical properties of the HDPE/NR/TPS blend. Since the PLA and NR are also bio-based natural materials, the incorporation of starch is expected to improve the biodegradability. To investigate the effect of the starch on the PLA/NR blend, it was not only focused on the aspect of the resultant mechanical and morphological properties, but also focused on the biodegradability of the blend, which would also be another interesting case study. In this study, the different DV systems such as HVA-2-based and DCP-based systems are also evaluated in order to determine the best system in the aspect of their properties and performance. DV technique is a promising method to improve the TPV blends. The NR particles were cross-linked through an intramolecular or intermolecular cross-linking in elastomer chains in the unvulcanized thermoplastic. The intention is to improve the strength and stability of the NR phase. The combination between the two has the elasticity performance from the cross-linked elastomer phase, and at the same time, it has the thermoplastic reprocessing behavior.

6.2 PREPARATION OF THERMOPLASTIC STARCH TO REINFORCE THE PLA/NR BLENDS SYSTEM

6.2.1 MATERIALS

PLA is a bio-based material, which is used as the matrix. It was purchased from the Titan Chemicals Corp, Pasir Gudang, Johor, Malaysia. Standard Malaysia Rubber (SMR-L) plays a role as the dispersed phase in the PLA matrix. The SMR-L was purchased from the Lembaga Getah Malaysia (LGM), Selangor, Malaysia. Cassava starch, which is in white powder form, was obtained from Thye Huat Chan Sdn. Bhd. The tapioca starch is used as the biodegradable agent, and it is mixed with the glycerol to produce the TPS in order to blend into the PLA/NR. The reagent-grade glycerol was supplied from HmbG Chemicals with a molar mass of 92.10 g/mol. The glycerol is used as the plasticizer and mixed with the tapioca starch to make it into TPS. HVA-2 and dicumyl peroxide function as free radical cross-linking agents in the PLA/NR vulcanization system, and they were supplied from Aldrich Chemical Company, St. Louis, USA.

6.2.2 SAMPLE FABRICATION OF THERMOPLASTIC STARCH

Prior to the production of the TPS, 65 wt% of tapioca starch and 35 wt% of glycerol were mixed in a large container manually with a surgical glove until a homogeneous mixture was obtained [29]. It was then stored in a dry place for 24 h to allow the glycerol molecules to imbed well within the tapioca starch in order to stabilize it prior to use. After storing overnight, the heated two roll mills were used and set at the temperature of 150°C with 10 rpm for 10 min to process the mixture. The scrapper was used to scrap the film on the heated two roll mills. The product thus obtained was the TPS.

6.2.3 SAMPLE FABRICATION OF PLA/NR/TPS BLEND VIA DIFFERENT DYNAMIC VULCANIZATION APPROACH

Using the heated two roll mills, PLA/NR ratio (65/35) was chosen to ensure that the additives added would be homogeneously dispersed in the matrix phase while mixing. The total weight of the compounding for PLA/NR/TPS was 100 wt%. First, the NR was masticated by the two roll mills in order to transform NR into a sheet to make it easier to be cut into smaller pieces according to the weight percentage, and was kept inside a sealing bag. The weight percentage for each PLA and also TPS was also determined, and they were collected into a sealing bag, respectively. Table 6.1 shows the weight proportion of PLA/NR with varying contents of TPS in two different vulcanization systems. For HVA-2 and DCP vulcanization systems, the PLA/NR/TPS could then be compounded with the additives according to its vulcanization system recipes, respectively, as shown in Table 6.2.

TABLE 6.1
The Weight Proportion of PLA/NR with Varying Contents of TPS in Two Different Vulcanization Systems

Type of System	Sample Name	Weight Proportion (wt%) PLA/NR (65/35)	TPS
Unvulcanized system PLA/NR/TPS	Controlled system 0% TPS (CONTROL-0%)	100	0
	Controlled system 5% TPS (CONTROL-5%)	95	5
	Controlled system 10% TPS (CONTROL-10%)	90	10
	Controlled system 20% TPS (CONTROL-20%)	80	20
HVA-2 dynamic-vulcanized system PLA/NR/TPS	HVA-2-vulcanized 0% TPS (HVA-0%)	100	0
	HVA-2-vulcanized 5% TPS (HVA-5%)	95	5
	HVA-2-vulcanized 10% TPS (HVA-10%)	90	10
	HVA-2-vulcanized 20% TPS (HVA-20%)	80	20
DCP dynamic-vulcanized system PLA/NR/TPS	DCP-vulcanized 0% TPS (DCP-0%)	100	0
	DCP-vulcanized 5% TPS (DCP-5%)	95	5
	DCP-vulcanized 10% TPS (DCP-10%)	90	10
	DCP-vulcanized 20% TPS (DCP-20%)	80	20

TABLE 6.2
Compounding Recipe for Each of the Vulcanization System (Yahya et al., 2011)

Vulcanization System	Ingredients	Composition (phr)
Unvulcanized system	PLA/NR/TPS	100
HVA-2 dynamic-vulcanized system	PLA/NR/TPS	100
	HVA-2	3
DCP dynamic-vulcanized system	PLA/NR/TPS	100
	DCP	3

6.2.4 Testing and Characterizations

The samples were first cut into a 2 cm × 2 cm dimension. Each of the samples was then immersed in toluene in a glass bottle for 2 h. All the samples were then taken out, filtered, and weighed. The weights of these remaining insoluble and swollen parts were known as swollen mass (M_s). The insoluble part was then rinsed with ethanol to remove the toluene solvent. It was then allowed to pat-dry, dried inside the oven at 80°C for 5 h, and then weighed again [30]. The dried mass was then recorded and referred to as M_d. The cross-linking density could then be determined after knowing the values for normalized gel mass (NGM) and normalized swelling degree (NSD) by the following formulas:

$$NGM = M_d / A \qquad (6.1)$$

$$NSD = (M_s - M_d)/M_d * A \qquad (6.2)$$

where M_s represents the swollen and insoluble mass; M_d represents the dried mass; and A represents the surface area of the sample immersed in the solvent.

The Instron universal testing machine (Model 3366) was used to examine the tensile properties of each PLA/NR/TPS blend by setting the crosshead speed of 10 mm/min under an ambient temperature. An average value was taken for at least five samples in order to get an accurate value of tensile strength, modulus of elasticity, and elongation at break for each blend formulation. The Perkin Elmer Spectrum RX1 FTIR equipped with PIKE Miracle TM Single Reflection Horizontal ATR accessory was utilized to determine the compositions with absorption peaks in order to examine the functional group compositions inside the PLA/NR/TPS blend. Overall scan was operated at four scans with transmittance spectral regions between 4,000 and 800 cm^{-1} with a resolution of 4 cm^{-1}. The morphological studies were carried out on the tensile fracture surface which was the broken sample as a result from the tensile testing for each PLA/NR/TPS formulation sample by using the analytical SEM (Model JEOL JSM-6460 LA) with the supply of accelerating voltage at 10 kV. The purpose of this testing was to analyze the topographic imaging of samples such as the fracture surfaces, homogeneity and interphase of rubber in the PLA/NR matrix, and the distribution and dispersion of TPS embedded in a vulcanized PLA/NR matrix. To prevent the charging effect on the surface morphology that would affect the testing results, the fracture samples were required to be coated with palladium or gold layer of about 1.5–3.0 nm thickness with the aid of utilizing the Auto Fine Coater (Model JEOL JFC 1600).

The biodegradability of PLA/NR/TPS was determined by soil burial tests under the laboratory scale. Rectangular-shaped samples with a dimension of 2 cm × 2 cm were prepared from each type of blend. The soil was filled up a container, and it was exposed to the environmental condition during the whole testing duration. The samples were then buried into the soil with 10 cm depth from the soil surface to allow the action of the microorganisms for 60 days. About 100 mL of water was sprayed into the container every day to keep the soil moist. For every 10 days, the samples were taken out, cleaned with water, and put into the oven for drying before weighing. After 60 days, they were then taken out and cleaned with water before taking for surface analysis using an optical microscope. The model used for the optical microscope was the National Optical 131. The optical microscope was capable of producing quality images and also formed primary images up to micro size. The complex multi-lens assembly focused the light waves originating from the specimen which were then magnified by the eyepieces over a range of 380 times. In this case, it was used to detect the presence of microorganisms on the surfaces before and after the degradation through the soil burial test.

6.3 RESULTS AND DISCUSSION

The reason for using TPS despite deteriorating the properties of the PLA/NR matrix was its biodegradability properties. Therefore, in this study, the effect of the TPS on the PLA/NR blend to produce the partial bio product could be studied by comparing the properties of the blend as the weight percentage of TPS increased from 5, 10, and 20 wt%. According to Thongpin et al. [31], the weakness and deterioration of the properties from the TPS could be compensated by the DV approach. To study the effect of DV of PLA/NR/TPS blends, different types of vulcanization systems were designed by different types of curing agents. In this section, testing and characterizations of the PLA/NR/TPS blends were done in order to study the mechanical, structural, morphological and biodegradability properties depending on the interesting field of study and the research objectives being set.

6.3.1 Cross-linking Density of DCP- and HVA-2-Vulcanized PLA/NR/TPS Blends

The cross-linking density was related to the amount of molecules involved inside the cross-linked network formation. To determine the cross-linking density, two important pieces of information that could be derived in this testing were (1) the surface area NGM and (2) NSD, which were correlated with the cross-linking density. Generally, the higher the cross-linking density, the lower the NSD and the higher the NGM [32]. Based on the NGM as shown in Figure 6.1, it is observed that the PLA/NR/TPS with curing agents (HVA-2 and DCP) had higher values than unvulcanized compounds (controlled samples). After immersing into the polar solvent (toluene), most of the polar PLA agents were dissolved in the polar solvent, leaving the cross-linked rubber and cross-linked thermoplastics as the gel. The higher gel mass signifies that there was higher cross-linking formation found in the rubber phase, and also strong

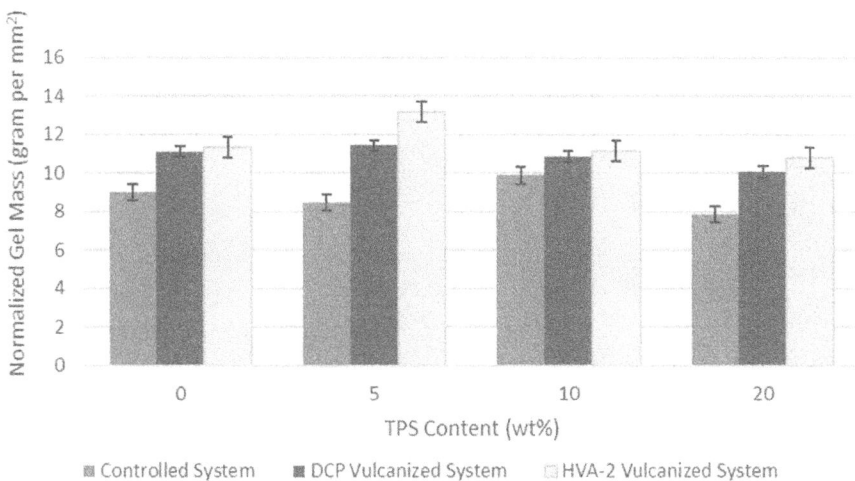

FIGURE 6.1 NGM determination.

H-bonding between TPS gives a compact packing structure [30]. For the case in between the HVA-0% and HVA-5%, the HVA-5% had higher NGM and lower NSM than the HVA-0%, signifying that the optimum ratio of HVA-5% to HVA-0% had the highest cross-linking density. This could be explained by the fact that the extra 5 wt% of the TPS, which constituted a large number of hydroxyl groups in its structure, remained associated with one another by inter- and intramolecular H-bonding in the blend. The cross-linking agents might also react with the –OH groups present in starch. However, excessive TPS content with 10 and 20 wt% would give an inferior result due to hindrance effect from the excessive large TPS particles in an efficient cross-linking formation [30]. While comparing between the two vulcanization systems, the HVA-2-vulcanized compounds had slightly higher NGM than DCP-vulcanized compounds. Dluzneski [33] stated that the DCP tended to undergo decomposition upon processing at high temperature and so the polymer chains would undergo chain scission into shorter chains, and this made the DCP blend system to be inferior to that of HVA-2 system [34]. That is why based on the result obtained, the NGM of HVA-5% was as high as 13.2 g/mm^2, whereas the DCP-5% was 11.1 g/mm^2.

With respect to NSD, generally the compounds with higher NGM would have higher cross-linking density and this would contribute to a lower value in NSD. This could be proven as shown in Figure 6.2. For unvulcanized compounds, the solvent would penetrate easily into the chain as compared to vulcanized compounds [30]. As a result, unvulcanized compounds showed higher swollen degree. The HVA-5% had the highest value of NGM of 13.2 g/mm^2 so the toluene solvent molecules found it difficult to penetrate into the closely packed chains in the HVA-5% with its NSD as low as merely 13.8 mm^2 due to the high cross-linking density and optimum amount of TPS with inter- and intramolecular bonding resulting in the lowest NSD [34]. The control-20% with the lowest NGM value, on the other hand, was found to have the highest NSD value as much as 29.32 mm^{-2}.

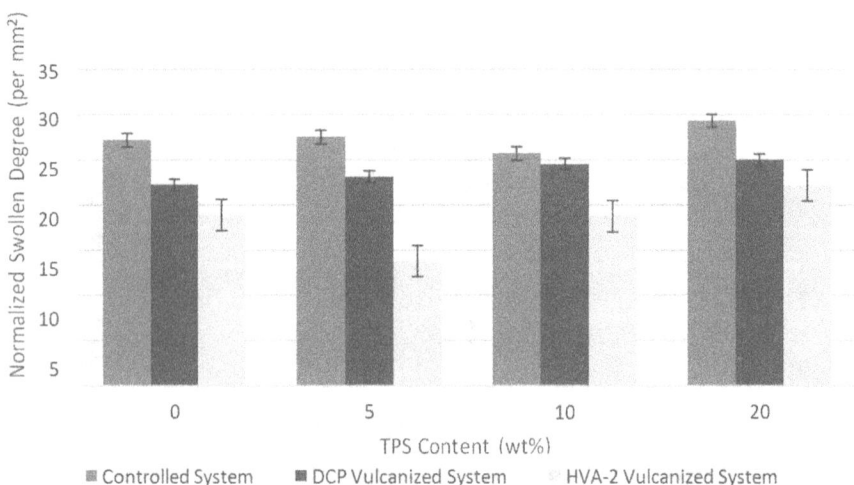

FIGURE 6.2 NSD determination.

6.3.2   TENSILE TEST OF DCP- AND HVA-2-VULCANIZED PLA/NR/TPS BLENDS

Tensile test was conducted to determine the physical and mechanical properties on tensile strength, Young's modulus, and elongation at break of different vulcanization systems of PLA/NR/TPS blends, as shown in Figure 6.3. The result showed that the increasing TPS content could reduce the tensile strength of the PLA/NR blends. The formation of agglomeration of TPS acted as the stress concentration point and caused the TPE to break easily when the force was applied [35]. The incorporation of TPS also led to poor adhesion of PLA/NR blend. The effect of these problems would cause the blend to not support the high stress and not to transfer the stress efficiently between the phases in PLA/NR/TPS blends. The distribution and dispersion of the TPS particles were the main factors of adhesion and tensile properties of the blend. Therefore, the higher the TPS content, the higher the deterioration of tensile properties, which resulted in the breaking of blend easily when the external force was applied [36]. As shown from the graph, the tensile strength of the CONTROL-0% was 12 MPa and it reduced to 11.2, 10.9, and 10.6 MPa, respectively, when the TPS was added at 5, 10, and 20 wt%. The deterioration of mechanical properties had shown the same trend when the HVA-2-vulcanized system and DCP-vulcanized system were used and when the TPS content was increased from none and slowly up to 20 wt%. However, there was an interesting finding that the tensile strength of dynamic-vulcanized systems of PLA/NR/TPS blending with DCP and HVA-2 was higher than that of unvulcanized systems. The improvement in tensile strength was a result of the enhancement in the adhesion between the matrix blends by the formation of cross-linking in the NR phase. The reduction in tensile strength as a result of increasing the TPS content was also less significant in the vulcanized system. It was because the TPS particles were becoming smaller and distributing better well in the PLA/NR matrix so the formation of weak concentration points would not be so prominent. The vulcanized rubber phase could prevent the chain slippage and provided better dimensional stability to the blends [37]. With respect to

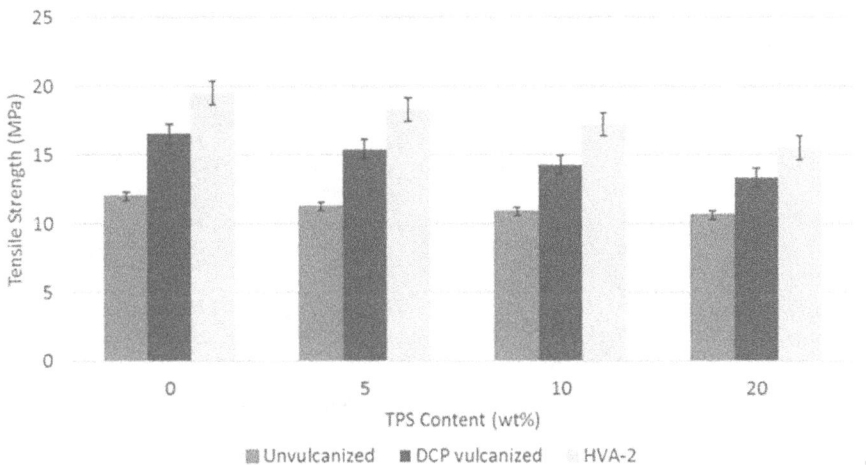

FIGURE 6.3   Tensile strength of different vulcanization systems.

the comparison between HVA-2- and DCP-vulcanized systems, the tensile strength of the HVA-2 blend systems was better than that of the peroxide-vulcanized system. As shown in Figure 6.3, the tensile strength of HVA-0% was 19.5 MPa, whereas the tensile strength of DCP-0% was merely 16.5 MPa. This could be explained from the results obtained from the cross-linking density test. The cross-linking density was more significant in the blend with the use of HVA-2. The better in strength was attributed to the extensive cross-linking formation in the HVA-2-vulcanized blends. This indicated that the HVA-2 was a more efficient cross-linker to be used in this vulcanization system as compared to DCP. The percentage of improvement in tensile strength by using HVA-2 with PLA/NR without TPS was 62.5% (from 12 to 19.5 MPa), while the percentage of improvement from DCP with PLA/NR without TPS was merely 37.5% (from 12 to 16.5 MPa).

The same case happened when it came to the modulus of elasticity of the PLA/NR/TPS blends as shown in Figure 6.4, whereby the increase in TPS content would reduce the Young's modulus of the blends. The modulus of elasticity of CONTROL-0%, DCP-0%-, and HVA-0%-vulcanized blends without TPS content was 774, 902.6, and 1,037.8 MPa and reduced to as low as 529.6, 698.3, and 729.4 MPa, respectively, when the TPS content was increased up to 20 wt%. The incorporation of TPS would soften the PLA/NR/TPS blends due to the presence of glycerol in TPS and the natural properties of TPS, which were naturally poor in mechanical properties and susceptible to environmental factors [38]. In other words, it was shown that the higher the TPS content, there would be more reduction in mechanical properties. However, the application of DV approach compensated for the loss in the mechanical properties as a result of the TPS incorporation. The percentage of improvement in modulus of elasticity by using HVA-2 with PLA/NR without TPS was 34.1% (from 774 to 1,037.8 MPa), while the percentage of improvement from DCP with PLA/NR without TPS was 16.6% (from 77 to 902.6 MPa). Other than the reason for the difference in cross-linking density

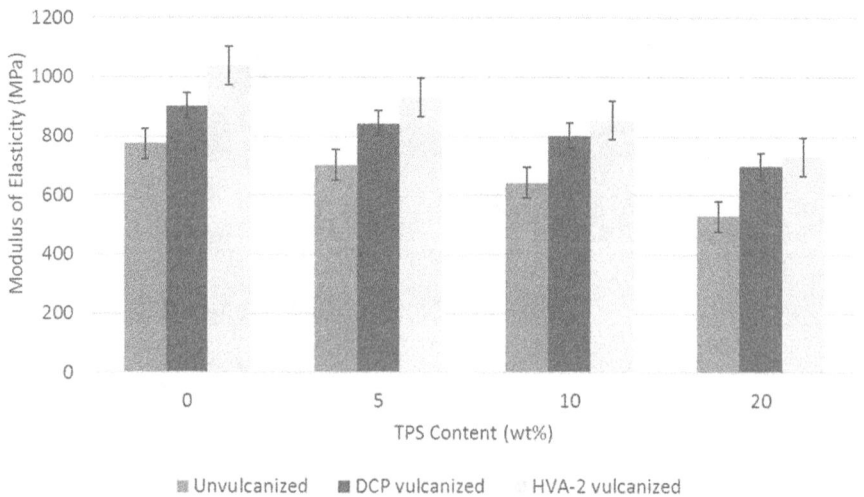

FIGURE 6.4 Modulus of elasticity of different vulcanization systems.

between the HVA-2 and DCP that leaded to a difference in properties, the presence of DCP in the blends whereby it produced reactive radicals upon decomposition at elevated temperature via exothermic reaction would induce chain scission towards the polymer and would also induce a rapid cure under the vulcanization temperatures. This instant curing would limit the processing time because it would be not so ideal for the case of blending process whereby a homogeneous blending was always required [33,34]. Overall, the vulcanized rubber phase could prevent the chain slippage and provide better dimensional stability to the blends.

Figure 6.5 shows the effect of TPS content on the elongation at break for the unvulcanized, and DCP- and HVA-2-vulcanized systems in PLA/NR blends. The similar observation was reported by Ren et al. [39], whereby they found that the percentage of elongation at break would decrease with increasing TPS content. The deterioration of the properties was attributable to the poor dispersion and distribution of the TPS within the PLA/NR matrix. The higher amount of TPS content would lead to weaker stress concentration points so it broke easier when the tensile force was being applied [16]. It could further be based on the data obtained from the elongation at break of CONTROL-0%, and DCP-0%- and HVA-0%-vulcanized blends without TPS content, which were 7.8%, 8.2%, and 11.6% and reduced to as low as 4.3%, 4.6%, and 7.7%, respectively, when the TPS content was increased up to 20 wt%. However, there were some improvements whereby all the dynamic-vulcanized blends showed better elongation at break than unvulcanized blends. The percentage of improvement in elongation at break by using HVA-2 with PLA/NR without TPS was 48.7% (from 7.8% to 11.6%), while the percentage of improvement from DCP with PLA/NR without TPS was just slightly 5.1% (from 7.8% to 8.2%). The contribution was a result of the cross-linked NR in the rubber phase, thereby improving the flexibility of the blends and restricting the deformability of the PLA/NR phase.

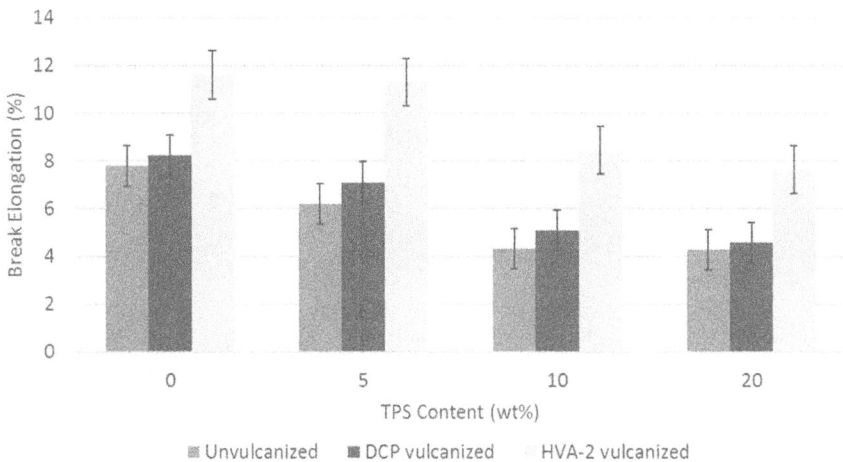

FIGURE 6.5 Elongation at break of different vulcanization systems.

6.3.3   Structural Analysis of DCP- and HVA-2-Vulcanized PLA/NR/TPS Blends

In this research study, the Fourier transform infrared (FTIR) spectroscopy was carried out to analyze the modifications of the blend components. IR is a useful technique to determine the possible reaction between active groups of PLA/NR blend with starch molecules for both unvulcanized and vulcanized blend systems. To achieve the objectives as listed in 1.3(a) section, the FTIR test was carried out onto unvulcanized PLA/NR blend (CONTROL-0%) and unvulcanized PLA/NR blend with 20 wt% (CONTROL-20%) of TPS to study the effect of incorporation of TPS into the blend, as shown in Figure 6.6. The interested peaks were broad peaks found at 3200, 3400, 2935, 1750, 1446, 1193, and 854 cm^{-1}. The most prominent peak to differentiate the CONTROL-0% and CONTROL-20% was observed in the band range occurring at 3,200–3,400 cm^{-1} corresponding to the hydroxyl groups which were mainly from the TPS phase and a small contribution from PLA. It could be observed that the CONTROL-20% had a lower peak intensity at 3,260 cm^{-1} than the CONTROL-0% at 3,283 cm^{-1}, signifying more TPS content was present in the CONTROL-20% blend. The peak at 2,935 cm^{-1} for CONTROL-20% and that at 2,938 cm^{-1} for CONTROL-0% represent C–H stretch and symmetry stretching vibrations of CH$_2$ groups, respectively. The chemical structure of the TPS consists of a lot of –CH$_2$ substituents leaving a lower peak intensity. The CONTROL-0% and

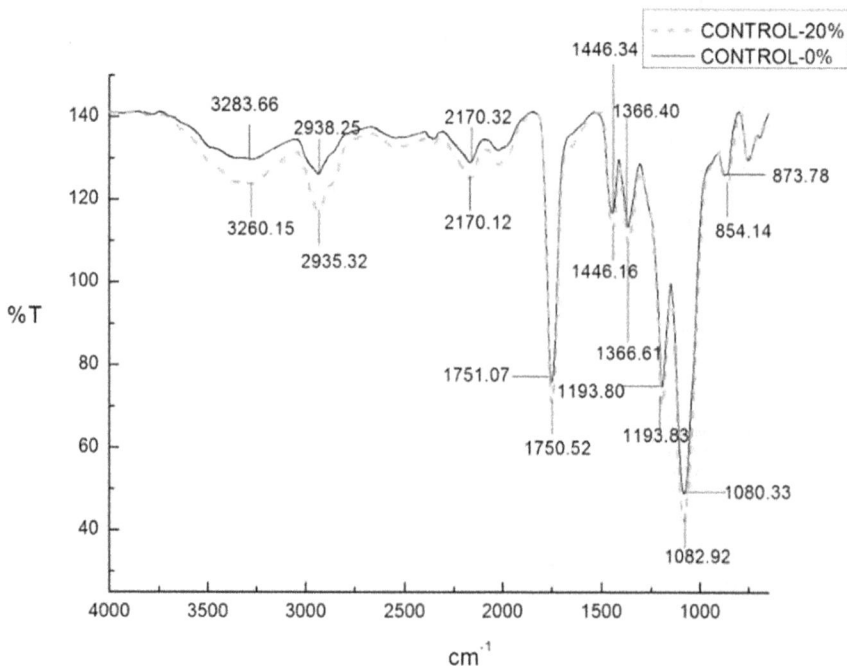

FIGURE 6.6   Infrared spectra of unvulcanized PLA/NR/TPS-0% (CONTROL-0%) and PLA/NR/TPS-20% (CONTROL-20%).

CONTROL-20% had the almost similar characteristics for the rest of the peaks. The absorption peaks at 1,750 and 1,751 cm^{-1} would be the carbonyl group C=O stretch from the PLA. The absorption peak at 1,446 cm^{-1} would be the methyl group –CH$_3$ bending vibration, whereas the absorption peaks at 1,366 and 1,193 cm^{-1} were the C–O–C stretching vibration and C–O–C asymmetrical and symmetrical valence vibrations, respectively [40]. The absorption peaks at 873 and 854 cm^{-1} were attributed to the C–C single bond in the fingerprint region, and it could also be the α-1,4-linked glucose units from TPS [41].

Another analysis was done to study the effect of vulcanization systems by using different vulcanizing agents of HVA-2 and DCP by comparing CONTROL-5%, HVA-5%, and DCP-5% with 5 wt% of TPS, as shown in Figure 6.7. Both the HVA-2-vulcanized and DCP-vulcanized samples showed variations in spectra compared to unvulcanized blends. One of the examples that could be observed was that after the cross-linking had taken place, the functional group for the free –OH stretching vibration had been reduced, proving that the hydroxyl group experienced a chemical reaction during the blending process in both vulcanized blends as the cross-linking agents might also react with the –OH groups present in starch [30]. Other than that, it was found that the HVA-2 had a peak at 2,930 cm^{-1} with highest percentage of transmittance representing C–H stretch and symmetry stretching vibrations of CH$_2$ groups, and this might be because more TPS were involved in a reaction with the

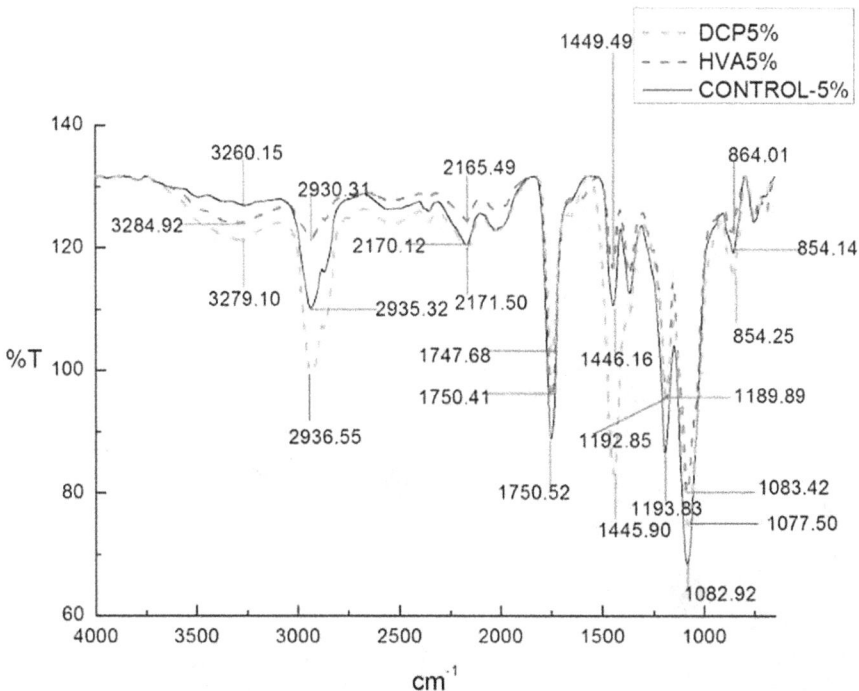

FIGURE 6.7 Infrared spectra of unvulcanized controlled, HVA-2-vulcanized, and DCP-vulcanized systems for PLA/NR/TPS-5%.

cross-linking agent so there would be lesser TPS's functional group while the reduction in percentage of transmittance in DCP was attributable to the chain scission occurring in the blend, causing the NR and TPS to be in a shorter chain so more detections at this peak were observed [33]. The absorption peak at 1,750 cm^{-1} was identified as the PLA's carbonyl group [40]. Another notable difference between the three systems was observed at the 1,445 cm^{-1} for DCP-5%. At this absorption peak, other than the functional group of –CH$_3$ bending vibration, the DCP-5% might also contain some benzene rings in the DCP structure, making it to have a lower percentage of transmittance. The CONTROL-5% had the lowest peak intensity at 1,082 cm^{-1}, which was attributable to the α-1,4-linked glucose units using TPS. It was observed that after the addition of curing agents, the %T had a higher value in both vulcanized blends, meaning that the α-1,4-linked glucose units were involved in some reactions with the cross-linking agents.

6.3.4 MORPHOLOGICAL ANALYSIS OF DCP- AND HVA-2-VULCANIZED PLA/NR/TPS BLENDS

The SEM micrographs with 300× magnification of fractured surfaces of PLA/NR are shown in Figure 6.8. As can be seen, all the PLA/NR blends showed phase-separated morphology where the rubber particles dispersedly occurred as small droplets in the PLA matrix. The phase morphology apparently showed weak interfacial adhesion evidence by the empty spherical grooves on the surface. Due to the incompatibility between the PLA and NR, this would create some voids and holes, and some large droplet sizes of rubber were also resulted. The voids formation could also be a result of rubber pull-out during the tensile testing, leaving some holes [42]. In Figure 6.8b, there was an incorporation of 5 wt% of TPS into the PLA/NR blend. It could be found that there were some TPS particles and starch granules embedding in the blend onto the surface. The incompatibility of TPS with the PLA/NR would create more and larger void formations, and this explained why there was a reduction in the mechanical properties when the TPS content was increased [27].

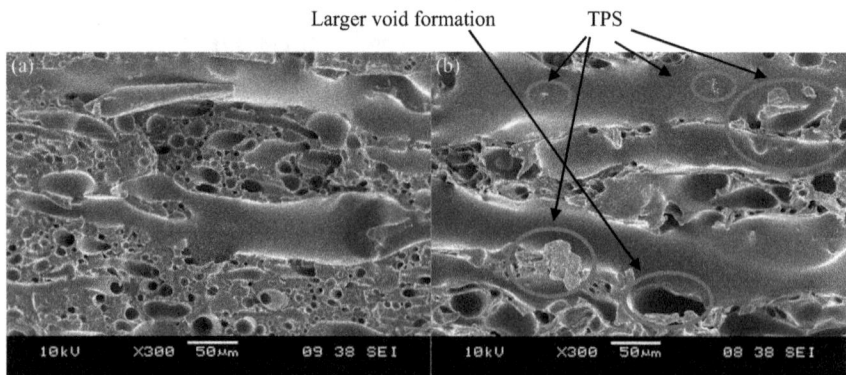

FIGURE 6.8 SEM micrographs of unvulcanized PLA/NR blends at 300× magnification for unvulcanized blend for (a) CONTROL-0% and (b) CONTROL-5%.

The same case happened in the vulcanized blends when the TPS content increased. Under the magnification of 300×, it could be observed that the HVA-5% had a smoother surface than HVA-20%, as shown in Figure 6.9. The HVA-5% had lesser void formation (smaller than 10 μm) and better interfacial interaction, giving excellent mechanical properties in tensile tests than HVA-20%. The better interfacial interaction was a result of the cross-linking formation, making the blend being more compact so the void formation would be not so prominent [24]. The HVA-20% had a rougher surface. The HVA-20% had more TPS content than HVA-5%. The rougher surfaces were a result of the TPS and starch granulations. It could be observed that the DCP-5% also had a smoother surface than DCP-20%, as shown in Figure 6.10. As the TPS content was higher, there would be more incompatibility in the phase structure due to the poor mixing, creating more void formations [43].

Another objective in this research was to study the effect of different vulcanization systems on the PLA/NR/TPS blend. To investigate that, it could be seen as shown in

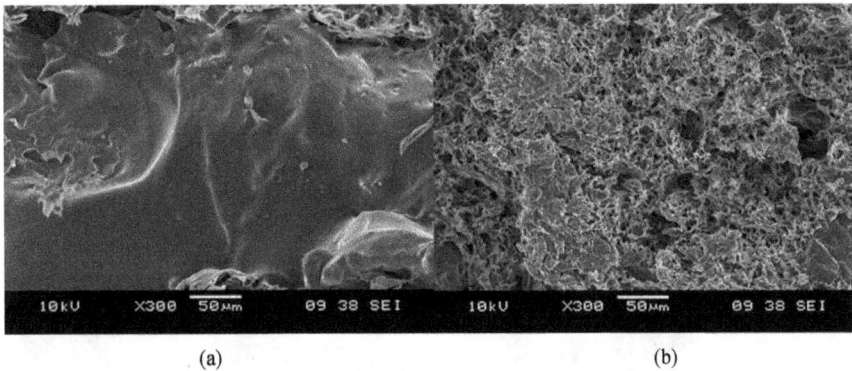

(a)   (b)

FIGURE 6.9   SEM micrographs of HVA-2-vulcanized blends at 300× magnification (a) with 5 wt% TPS content and (b) with 20 wt% TPS content.

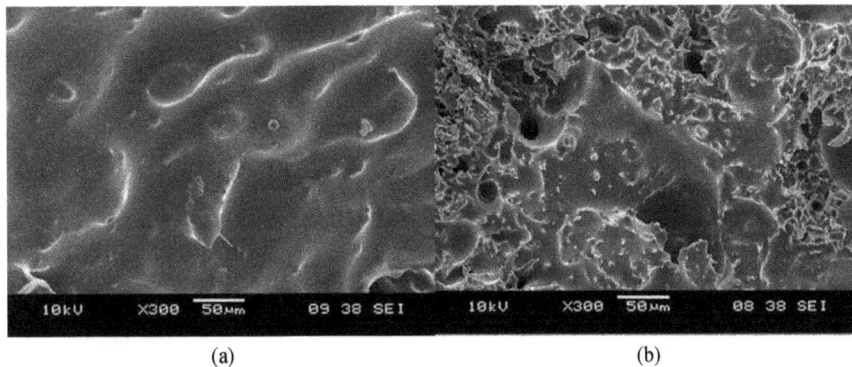

(a)   (b)

FIGURE 6.10   SEM micrographs of DCP-vulcanized blends at 300× magnification with 5 wt% TPS content and (b) with 20 wt% TPS content.

Figure 6.11. It could be observed that the CONTROL-5% had a larger void formation than HVA-5% and DCP-5%. Without the addition of curing agents, the interfacial bonding between the PLA/NR/TPS was weak. The CONTROL-5% had the
TPS particles larger than 20 μm and void formation as large as 35–40 μm. With
the incorporation of curing agent, the elastomer polymer chains underwent cross-
linking, which restricted the chain mobility and increased viscosity in the rubber
phase, resulting in high shear and elongation forces acting on the system, and the
deformation would therefore be increased so that this improved the dispersion and
distribution of TPS [44]. Based on the results of cross-linking density test, the HVA-
5% had the highest NGM with the highest crosslinking density, leading to more
compact structures so lesser hole formations were found in the morphological analysis than DCP-5% [30]. The TPS particulates in HVA-5% were smaller and harder
to identify and distribute better. So the HVA-5% had smaller TPS particles and
better distribution than the DCP-5%. The rougher surface in HVA-5% also showed
that the rubber particles were not easily being pulled out from the PLA continuous
phase. DCP-5% had a lesser void formation and the smoothest surface, but it clearly
showed that some TPS formation and dent formation were found to be higher than
those of HVA-5% so it could be deduced that the interfacial bonding in the HVA-2-
vulcanized blend was better.

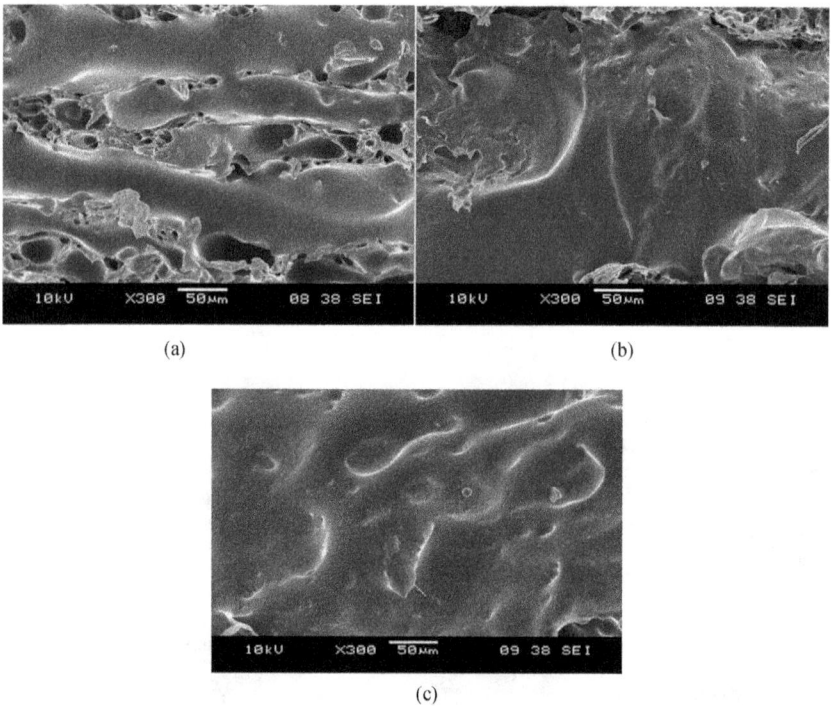

(a)   (b)

(c)

FIGURE 6.11   SEM micrographs of unvulcanized and vulcanized blends at 300× magnification with 5% TPS content for (a) unvulcanized controlled system, (b) HVA-2-vulcanized
system, and (c) DCP-vulcanized system.

6.3.5 Biodegradability Analysis

6.3.5.1 Soil Burial Test of DCP- and HVA-2-Vulcanized PLA/NR/TPS Blends

The biodegradability properties are one of the very interesting criteria of biopolymers. It is because the service lifetime of the biomaterial must be known to make sure that it will not be biodegraded and lose its performance during the application period so that it can perform its optimum usefulness for the certain period for safety. The polymer degradation is an irreversible process that can be induced by several environmental factors either in biotic factors (enzymatic action, and microorganisms) or in abiotic factors (oxidation, hydrolysis, and thermo-oxidation) [45]. One of the noticeable changes of the effect of the biodegradation would be the weight loss of the compound. The weight loss (%) of PLA/NR/TPS blends with different vulcanization systems or TPS contents against the degradation time is shown in Figures 6.12–6.14. It could be clearly observed that the unvulcanized and both the vulcanized blends were all suffering from a very great weight loss especially when the TPS content

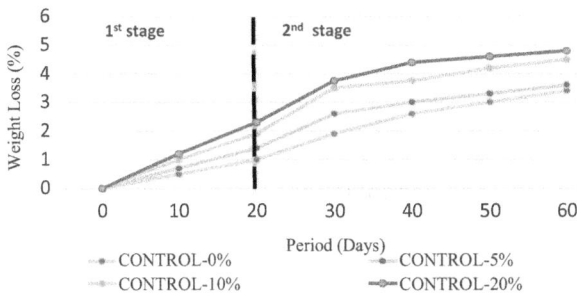

FIGURE 6.12 The effect of TPS content on the weight loss (%) of PLA/NR/TPS for controlled vulcanized system against the degradation time for every 10 days.

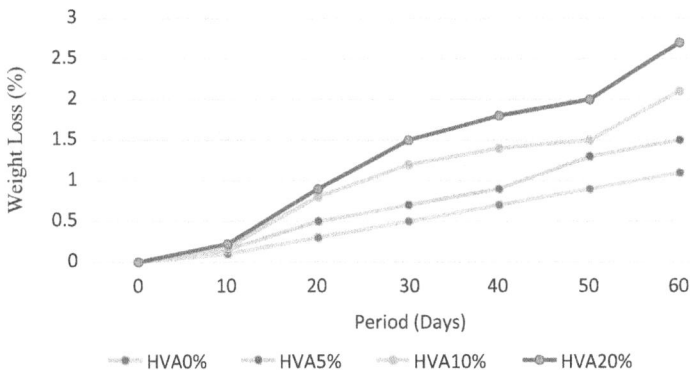

FIGURE 6.13 The effect of TPS content on the weight loss (%) of PLA/NR/TPS for HVA-2-vulcanized system against the degradation time for every 10 days.

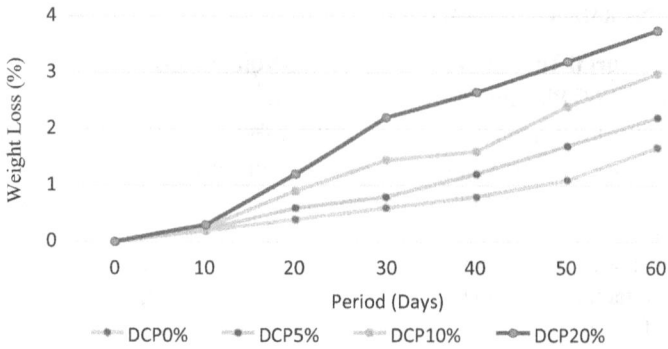

FIGURE 6.14 The effect of TPS content on the weight loss (%) of PLA/NR/TPS for DCP-vulcanized system against the degradation time for every 10 days.

was 20 wt%. The unvulcanized blend with 20 wt% of TPS achieved as high as 4.80% of weight loss after composting into the soil for 2 months, while the HVA-20% and DCP-20% were as high as 2.70% and 3.75%, respectively. These results revealed that the higher starch content enhanced the degradation kinetics and thus increased the mass loss. This could also be due to the hydrophilic nature of starch. Being hydrophilic in nature, the starch retained moisture that contributed to the degradation of the polymer [8]. The higher the starch content in the blend, the higher the moisture content that rendered faster degradation. Biodegradation of starch-based polymers occurred between the sugar groups, leading to a reduction in chain length and the splitting off of mono-, di-, and oligosaccharide units as a result of enzymatic attack at the glycosidic linkages [46]. These enzymes were produced from microorganisms that consumed organic compounds, including protein, that resided within the polymer. For the first 10 days, the biodegradation rate was slow. According to Torres et al. [47], the degradation behavior of the starch-based polymers could be represented in three stages. At the initial stage of around 24-h soil composting, it would face a certain weight loss in the polymer. The first degradation mechanism was mainly associated with the leaching of glycerol. At the second stage of about 20 days, the weight loss increased steadily and the degradation mechanism at this stage was associated with the biological activity present in the composting vessels such that the starch α-1,4-linkages would be broken down. It involved the fragmentation of larger molecules into small molecules so it would take a long time. The fragmentation process involved an enzymatic reaction to shorten and weaken the polymeric backbone chain and turned it into a shorter chain. After the fragmentation process, last it would proceed to the third step known as the assimilation (resorbing) and mineralization of those smaller fragments into carbon dioxide (aerobic), methane (anaerobic), and water. The decrease in the percent weight loss was attributable to the invasion of microorganisms into the substrate and the absorption of the moisture by the samples. The microorganisms when fed upon the substrate increased the percent weight loss of the sample. The rate of degradation increased dramatically, and the appearance changes were considerable. The starch polymers were broken into small species as a result of the degradation process [45].

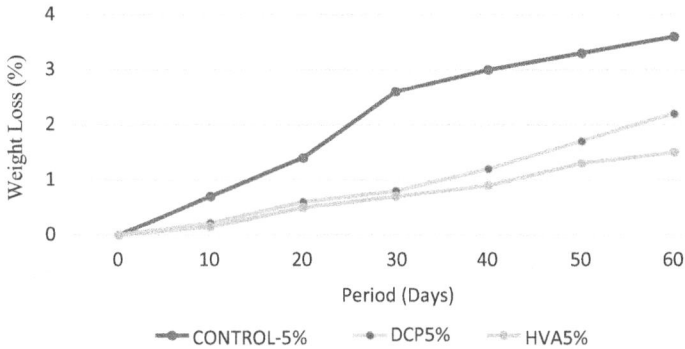

FIGURE 6.15 The effect of different vulcanization system on the weight loss (%) of PLA/NR/TPS DCP-vulcanized system against the degradation time for every 10 days.

While comparing between the three systems, as shown in Figure 6.15, it could be observed that the unvulcanized control system with 5 wt% of TPS exhibited the highest degradation rate as high as 3.6% of the weight loss after 60 days. However, it was noted that both the DCP- and the HVA-2-vulcanized systems delayed the degradation rate. The HVA-2-vulcanized system that had higher cross-linking density had a stronger chemical interfacial interaction, and this disallowed the microbial and water molecules to penetrate and attacked inside the molecular structure so it had the least weight loss of merely 1.5%, which was lower than that of DCP-vulcanized system that had lower cross-linking density in its molecular formation with 2.2%, while the unvulcanized system suffered the greatest weight loss as high as 3.6%.

6.3.5.2 Optical Microscope of DCP- and HVA-2-Vulcanized PLA/NR/TPS Blends

After 60 days of burying into the soil, the samples were washed using distilled water, dried, and observed under the optical microscope at the magnification of 150×. To study the effect of TPS on the PLA/NR blend, Figure 6.16 shows the optical microscopic image of PLA/NR at 150× magnification for 0% TPS and 20% TPS for the three respective systems after 60 days after burying into the soil. As the amount of TPS content increased up to 20%, it was apparently more obvious to observe the surface erosion by the microorganisms, as shown in Figure 6.16, as compared with those without the addition of TPS. On the surface of the sample, bacteria and fungi were obviously found as the white spots, signifying that those adhering microorganisms colonized onto the surfaces. The surface of the blend was considerably eroded with merely a few areas staying unharmed, signifying that these areas were non degraded materials. The degradation was mainly a consequence of the TPS, which enhanced the degradation kinetics. The amylopectin, which was the amorphous region of TPS, constituted about 70%–80% of the composition. The lower crystallinity and the inherent hydrophobicity of the TPS favored the absorption of moisture from the environment to allow the oxidation and hydrolysis process to take place and the microorganism could be easily accessed into the compound [8]. Starch-hydrolyzing enzymes were produced by microorganisms. These enzymes can hydrolyze the glycosidic

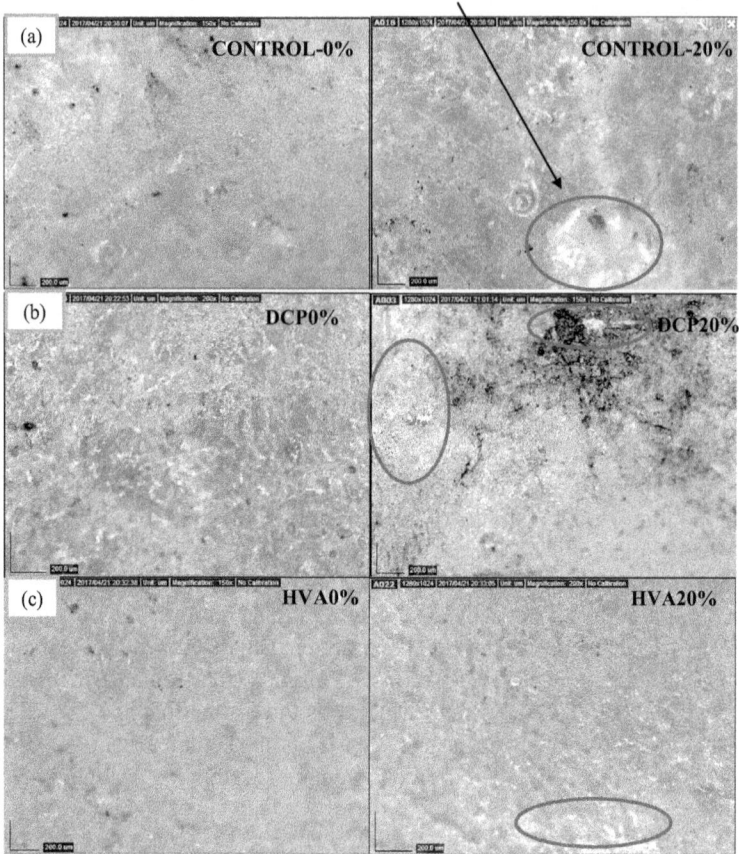

FIGURE 6.16 Optical microscopic image of PLA/NR/TPS at the magnification of 150× for 0% TPS and 20% TPS. (a) Unvulcanized blend, (b) DCP-vulcanized blends, and (c) HVA-2-vulcanized blends after 60 days after burying into the soil.

linkages and can lead to further chain rupture of starch molecules. Furthermore, the hydrolysis of amylose and amylopectin glycosidic links leads to the formation of a low molecular weight product of amylose and amylopectin structure. This involved the enzymatic reaction that shortened and weakened the polymeric backbone chain and turned into an elemental cycle such as hydrogen and nitrogen. These enzymes were produced from microorganisms that consumed organic compounds, including protein, that resided within the polymer. That is why, some white spots, cracks, and void formation were seen after 60 days of soil composting.

It could be observed from Figure 6.17 that the unvulcanized blend had the whitest spot areas being colonized by the microorganisms when compared with the other two vulcanized systems. Although there were minor infected areas in the HVA-2-vulcanized sample, its condition was better than that of the DCP-vulcanized blend. The reason was the higher cross-linking density in HVA-2 as the efficient cross-linker imparting the PLA/NR/TPS blend a compact structure. The DV process with the

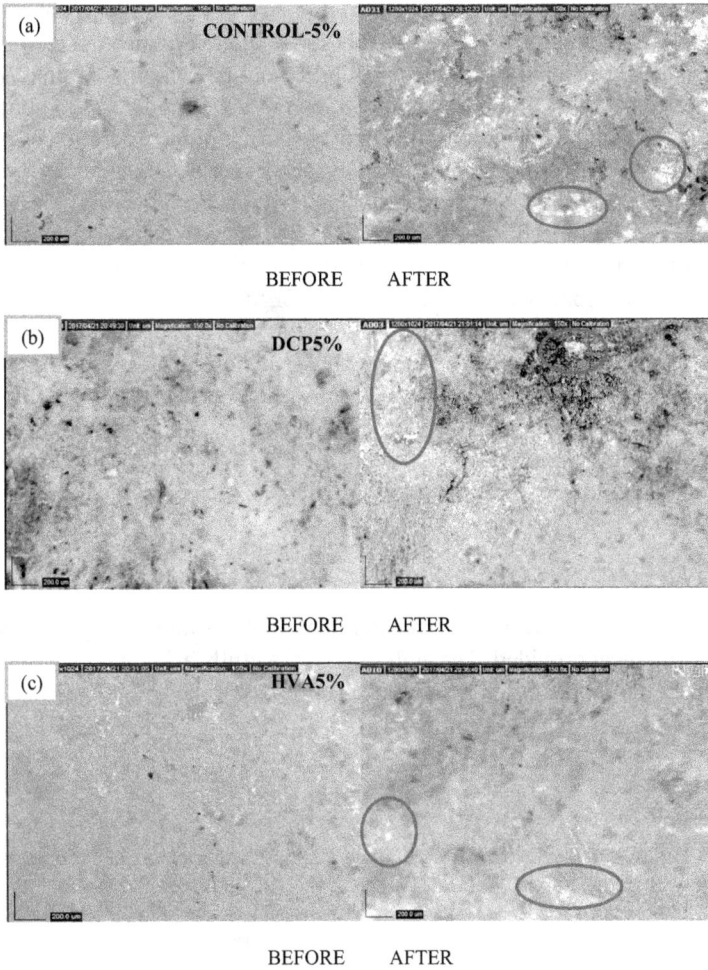

FIGURE 6.17 Optical microscopic image of PLA/NR/TPS-5% at the magnification of 150× for (a) controlled blend, (b) DCP-vulcanized blends, and (c) HVA-2-vulcanized blends before and after 60 days after burying into the soil.

use of curing agents enhanced the blend properties and prevented them from being attacked by the bacteria. The formation of cross-linking structures would have a compact structure, which had interfacial interactions to hold the polymer chains tighter with smaller surface area exposure to prevent the invasion from the bacteria [30].

6.4 CONCLUSIONS

The incorporation of TPS has been shown to be capable of improving the biodegradation rate while the materials being composted under the soil in the soil burial test. However, the most prominent unpleasant effect of increasing the TPS loading content was the deterioration in mechanical properties of the PLA/NR blends. From SEM morphology, it could be observed that the reduction in mechanical properties was primarily attributable to the incompatibility and poor interfacial interaction of TPS with the PLA/NR. The prime role of applying DV systems was to compensate for the weakness and the deterioration effect as a result of incorporation of biomaterial TPS. The DV was done to cross-link the rubber phase and the tensile properties were improved greatly as compared to the unvulcanized counterpart. The improvement in the tensile properties was mainly due to the increase in cross-linking density in the blends. There were lesser void formations with better interfacial interactions producing a more compact blend, as found in SEM micrographs. With respect to the different vulcanization agents being used, it was found out that the HVA-2-vulcanized blends were greater than the DCP-vulcanized blends in the aspect of both mechanical and morphological properties. This was because it was known that during shearing at high temperature, DCP tended to decompose. The radicals generated from the decomposition may react with the PLA and TPS molecules in the form of chain scission rather than taking place in vulcanization of NR phase.

ACKNOWLEDGMENTS

The authors would like to thank Universiti Malaysia Perlis for the supply of raw materials and equipment provided. The authors also gratefully acknowledge the financial support from the Kementerian Pendidikan Tinggi under the Grant UniMAP/RMIC/FRGS/9003-00601.

REFERENCES

1. 1. A.L. Andrady, and M.A. Neal, "Applications and societal benefits of plastics," *Phil. Trans. Roy. Soc. B Biol. Sci.*, vol. 364, pp. 1977–1984, 2009.
2. R.C. Thompson, C.J. Moore, F.S. vom Saal, and S.H. Swan, "Plastics, the environment, and human health: current consensus and future trends," *Phil. Trans. Roy. Soc. B Biol. Sci.*, vol. 364, pp. 2153–2166, 2009.
3. M.R. Gregory, "Environmental implications of plastic debris in marine settings – entanglement, ingestion, smothering, hangers-on, hitch-hiking and alien invasions," *Phil. Transact. Roy. Soc. B Biol. Sci.,* vol. 364, no. 1526, pp. 2013–2025, 2009.
4. C.M. Rochman, M.A. Browne, B.S. Halpern, B.T. Hentschel, E. Hoh, H.K. Karapanagioti, L.M. Rios-Mendoza, H. Takada, S. Teh, and R.C. Thompson. "Classify plastic waste as hazardous," *Nature,* vol. 494, pp. 169–171, 2013.
5. D.J. Tonjes, and K.L. Greene, *Degradable Plastics and Solid Waste Management Systems.* Stony Brook University, Stony Brook, 2013.
6. S.Y. Lee, and M.A. Hanna, "Tapioca starch-poly(lactic acid)-Cloisite 30B nanocomposite foams," *Polym. Compost.*, vol. 30, pp. 665–672, 2009.
7. H. Liu, F. Xie, L. Yu, L. Chen, and L. Li, "Thermal processing of starch-based polymers," *Progr. Polym. Sci.*, vol. 34, pp. 1348–1368, 2009.
8. R.F. Tester, J. Karkalas, and X. Qi, "Starch—composition, fine structure and architecture," *J. Cer. Sci.,* vol. 39, pp. 151–165, 2004.
9. M.D. da Matta, S.B.S. Jr. Sarmento, C.I.G.L. Sarantopoulos, and S.S. Zocchi, "Barrier properties of films of pea starch associated with xanthan gum and glycerol," *Polimeros-Cien. Tecnol.,* vol. 21, no. 1, pp. 67–72, 2011.

10. A.W.M. Kahar, H. Ismail, and N. Othman, "Characterization of citric acid-modified tapioca starch and its influence on thermal behavior and water absorption of high density polyethylene/natural rubber/thermoplastic starch blends," *Polym. Plast. Technol. Eng.,* vol. 50, no. 7, pp. 748–753, 2011.

11. S. Mortazavi, I. Ghasemi, and A. Oromiehie, "Effect of phase inversion on the physical and mechanical properties of low density polyethylene/thermoplastic starch," *Polym. Test.,* vol. 32, no.3, pp. 428–491, 2013.

12. A.W.M. Kahar, H. Ismail., and N. Othman, "Properties of HVA-2 vulcanized high density polyethylene/natural rubber/thermoplastic tapioca starch blends," *J. Appl. Polym. Sci.,* vol. 128, no. 4, pp. 2479–2488, 2013.

13. D.S. Rosa, M.A.G. Bardi, L.D.B. Machado, D.B. Dias, L.G.A. Silva, and Y. Kodama, "Influence of thermoplastic starch plasticized with biodiesel glycerol on thermal properties of PP blends," *J. Ther. Anal. Calorim.,* vol. 97, pp. 565–570, 2009.

14. A.J.F. Carvalho, M.D. Zambon, A.A. da Silva Curvelo, and A. Gandini, "Thermoplastic starch modification during melt processing: hydrolysis catalyzed by carboxylic acids," *Carbohyd. Polym.,* vol. 62, pp. 387–390, 2005.

15. A.W.M. Kahar, and H. Ismail, "High-density polyethylene/natural rubber blends filled with thermoplastic tapioca starch: physical and isothermal crystallization kinetics study," *J. Viny. Add. Technol.,* vol. 22, no. 3, pp. 191–199, 2016.

16. H. Li, and M.A. Huneault, "Comparison of sorbitol and glycerol as plasticizers for thermoplastic starch in TPS/PLA blends," *J. Appl. Polym. Sci.,* vol. 119, no. 4, pp. 2439–2448, 2011.

17. C. Xu, L. Fu, and Y. Chen, "Physical blend of PLA/NR with co-continuous phase structure: preparation, rheology property, mechanical properties and morphology," *Polym. Test.,* vol. 37, pp. 94–101, 2014.

18. A. Hassan, M.U. Wahit, H. Balakrishnan, A.A. Yussufa, and S.B.A. Razak, "Novel toughened polylactic acid nanocomposite: mechanical, thermal and morphological properties," *Mater. Desig.,* vol. 31, no. 7, pp. 3289–3298, 2010.

19. N. Ibrahim, and A.W.M. Kahar, "Mechanical and physical properties of polylactic acid (PLA)/thermoplastic starch (TPS) blend," *J. Polym Mater.,* vol. 33, no. 1, pp. 201–212, 2016.

20. S. Ishida, R. Nagasaki, K. Chino, T. Dong, and Y. Inoue, "Toughening of poly(L-lactide) by melt blending with rubbers," *J. Appl. Polym. Sci.,* vol. 113, no. 1, pp. 558–566, 2009.

21. A.W.M. Kahar, H. Ismail, and A. Abdul Hamid, "The correlation between crosslink density and thermal properties of high density polyethylene/natural rubber/thermoplastic tapioca starch blends prepared via dynamic vulcanisation approach," *J. Ther. Anal. Calorim.,* vol. 123, no. 1, pp. 301–308, 2016.

22. L. Elias, F. Fenouillot, J. Majeste, and P. Cassagnau, "Morphology and rheology of immiscible polymer blends filled with silica nanoparticles," *Polymer,* vol. 48, no. 20, pp. 6029–6040, 2007.

23. D. Yuan, K. Chen, C. Xu, Z. Chen, and Y. Chen, "Crosslinked bicontinuous biobased PLA/NR blends via dynamic vulcanization using different curing systems," *Carbohyd. Polym.,* vol. 113, pp. 438–445, 2014.

24. N. Bitinis, R. Verdejo, P. Cassagnau, and M.A. Lopez-Manchado, "Structure and properties of polylactide/natural rubber blends," *Mater. Chem. Phys.,* vol. 129, no. 3, pp. 823–831, 2011.

25. H. Salmah, H. Ismail, and A.A. Bakar, "The effects of dynamic vulcanization and compatibilizer on properties of paper sludge-filled polypropylene/ethylene propylene diene terpolymer composites," *J. Appl. Polym. Sci.,* vol. 107, no. 4, 2266–2273, 2007.

26. Y. Huang, C. Zhang, Y. Pan, W. Wang, L. Jiang, and Y. Dan, "Study on the effect of dicumyl peroxide on structure and properties of poly(lactic acid)/natural rubber blend," *J. Polym. Environ.,* vol. 21, pp. 375–387, 2013.

27. Y. Chen, D. Yuan, and C. Xu, "Dynamically vulcanized biobased polylactide/natural rubber blend material with continuous cross-linked rubber phase," *ACS Appl. Mater. Interf.,* vol. 6, pp. 3811–3816, 2014.

28. A.W.M. Kahar, N. Othman, and H. Ismail, "Thermoplastic elastomers from high-density polyethylene/natural rubber/thermoplastic tapioca starch: effects of different dynamic vulcanization," *Nat. Rubber Mater.,* vol. 1, pp. 242–264, 2013.

29. A. Carvalho, A.E. Job, A.A. Curvelo, and A. Gandini, "Thermoplastic starch/natural rubber blends." *Carbohyd. Polym.,* vol. 53, no. 1, pp. 95–99, 2003.

30. K. Das, D. Ray, N.R. Bandyopadhyay, A. Gupta, S. Sengupta, S. Sahoo, and M. Misra, "Preparation and characterization of cross-linked starch/poly (vinyl alcohol) green films with low moisture absorption," *Ind. Eng. Chem. Res.,* vol. 49, no. 5, pp. 2176–2185, 2010.

31. C. Thongpin, S. Klatsuwan, P. Borkchaiyapoom, and S. Thongkamwong, "Crystallization behavior of PLA in PLA/NR compared with dynamic vulcanized PLA/NR," *J. Met. Mater. Miner.,* vol. 23, no. 1, pp. 53–59, 2013.

32. J. Delville, C. Joly, P. Dole, and C. Bliard, "Solid state photocrosslinked starch based film a new family of homogeneous modified starches," *Carbohyd. Polym.,* vol. 49, no. 1, pp. 49–71, 2002.

33. P.R. Dluzneski, "Peroxide vulcanization of elastomers," *Rub. Chem. Technol.,* vol. 74, no. 3, pp. 451–492, 2001.

34. M. Awang, and H. Ismail, "Preparation and characterization of polypropylene/waste tyre dust blends with addition of DCP and HVA-2 (PP/WTD P- HVA2),", *Polym. Test.,* vol. 27, no. 3, pp. 321–329, 2008.

35. P. Juntuek, C. Ruksakulpiwat, P. Chumsamrong, Y. Raksakulpiwat, "Effect of glycidyl methacrylate-grafted natural rubber on physical properties of polylactic acid and natural rubber blends," *J. Appl. Polym. Sci.,* vol. 125, pp. 745–754, 2012.

36. K. Majdzadeh-Ardakani, A.H. Navarchian, and F. Sadeghi, "Optimization of mechanical properties of thermoplastic starch/clay nanocomposites," *Carbohyd. Polym.,* vol. 79, no. 3, pp. 547–554, 2010.

37. A.H.M. Zain, A.W.M. Kahar, and H. Ismail, "Biodegradation behaviour of thermoplastic starch: the roles of carboxylic acids on cassava starch," *J. Polym. Environ.,* vol. 26, no. 2, pp. 691–700, 2018.

38. D. Schlemmer, M.J.A. Sales, and I.S. Resck, "Degradation of different polystyrene/thermoplastic starch blends buried in soil," *Carbohyd. Polym.,* vol. 75, pp. 58–62, 2009.

39. J. Ren, H. Fu, T. Ren, and W. Yuan, "Preparation, characterization and properties of binary and ternary blends with thermoplastic starch, poly (lactic acid) and poly (butylene adipate-co-terephthalate)," *Carbohyd. Polym.,* vol. 77, no. 3, pp. 576–582, 2009.

40. A.J.R. Lasprilla, G.A.R. Martinez, and B. Hoss, "Synthesis and characterization of poly (lactic acid) for use in biomedical field," *Chem. Eng.,* vol. 24, pp. 985–990, 2011.

41. N.A. Nikonenko, D.K. Buslov, N.I. Sushko, and R.G. Zhbankov, "Analysis of the structure of carbohydrates with use of the regularized deconvolution method of vibrational spectra," *Polym. Rev.,* vol. 48, no. 1, pp. 85–108, 2016.

42. K. Pongtanayut, C. Thongpin, and O. Santawitee, "The effect of rubber on morphology, thermal properties and mechanical properties of PLA/NR and PLA/ENR blends," *Ener. Proced.,* vol. 34, pp. 888–897, 2013.

43. E. Corradini, A.J.F.D. Carvalho, A.A.D.S. Curvelo, J.A.M. Agnelli, and L.H.C. Mattoso, "Preparation and characterization of thermoplastic starch/zein blends," *Mater. Res.,* vol. 10, no. 3, pp. 227–231, 2007.

44. P. Sarazin, and B.D. Favis, "Influence of temperature-induced coalescence effects on co-continuous morphology in poly (ε-caprolactone)/polystyrene blends," *Polymers,* vol. 46, no. 16, pp. 5966–5978, 2005.

45. P.K. Annamalai, and D.J. Martin, "Can clay nanoparticles accelerate environmental biodegradation of polyolefins?," *Mater. Sci. Technol.,* vol. 30, no. 5, pp. 593–602, 2014.

46. M.Y. Ustinov, S.E. Artemenko, G.P. Ovchinnikova, G.A. Vikhoreva, A.N. Guzenko, "Composition and properties of biodegradable materials," *Fibr. Chem.,* vol. 36, no. 3, pp. 189–192, 2004.

47. F.G. Torres, O.P. Troncoso, C. Torres, D.A. Díaz, and E. Amaya, "Biodegradability and mechanical properties of starch films from Andean crops," *Inter. J. Boil. Macromol.,* vol. 48, no. 4, pp. 603–606, 2011.

48. Y.S. Yahya, A. R. Azura, and Z. Ahmad, "Effect of Curing Systems on Thermal Degradation Behaviour of Natural Rubber (SMR CV 60)," *J. Phys. Sci.,* vol. 22, no. 2, pp. 1–14, 2011.

7 Chemical Modification of *Nypa fruticans* Fiber as Filler on the Mechanical and Thermal Properties of PLA/rLDPE Biocomposites

M.S. Rasidi and P.L. Teh
Universiti Malaysia Perlis

CONTENTS

7.1 INTRODUCTION

Nowadays, plastic materials have become a part of daily usages in human life. Plastics have been chosen because they are inexpensive and have a wide range of properties, including lightweight, corrosion resistance, chemical resistance, strong, durable, electrical insulation, and ease of processing. The diversity of plastics and the versatility of their properties bring numerous societal benefits to use in making a vast array of products [1]. This projected growth is mainly due to increased demand for plastics from the public [2]. Environmental impacts of plastics cannot be overlooked but they are one of the most useful and important materials in this society. Most conventional plastics cannot be biodegraded, and their accumulation in the world was a threat to our planet [3]. Thus, due to environment and sustainability, evolving and manufacturing biodegradable plastics have become a trend to reduce plastic waste.

Polylactic acid (PLA) is one of the best bio-based polymers, which attracted the interest of many researchers, which are made from carbohydrate sources such as corn, sugarcane, or tapioca and are readily biodegradable [4]. PLA is a high-strength thermoplastic, high-modulus polymer that can be produced from renewable and sustainable resources to be used in various applications [5]. Good mechanical properties, thermal plasticity, and easy manufacture of PLA make it an alternative to petrochemical polymers in most of the applications, starting with packaging and up to horticultural materials [6]. However, PLA has drawbacks as well – low toughness and elongation at break have limited its usage. PLA has a low range of glass transition temperature, and it is brittle at room temperature [7].

The development of methods for PLA is a master topic since the advent of PLA. Blending PLA with other polymers is one approach of modifying the desired properties. Compared with copolymerization method, blending may be a much more practical and economic measure to obtain toughened PLA products. Up to now, PLA has been melt-processed with many flexible polymers to improve its toughness and flexibility [8–13]. Some interesting and noteworthy results have been reported. In general, when the softer component forms a second phase within the more brittle continuous phase, it may act as a stress concentrator which enables ductile yield and prevents brittle failure. These PLA blends displayed improved elongation at break and impact strength but reduced strength and modulus due to the addition of ductile phase. However, most of the added polymers have no biodegradability, which clearly limits the applications of the prepared blends. Thus, blends of recycled-low density polyethylene (rLDPE) with PLA will have three advantages: (1) Virgin matrix is replaced by postconsumer materials, (2) the end product will be biodegradable and cheap, and (3) brittleness of the biocomposites will be reduced [14].

The blending with rLDPE forms heterogeneous incompatible systems with PLA, resulting in limited performance improvement [15]. This often increases toughness and reduces valuable strength and modulus [16]. The introduction of natural filler in a PLA matrix has attracted increasing attention [17–20]. Reinforcing fillers are now incorporated to improve certain properties while maintaining their inherently good properties such as transparency and biodegradability. PLA reinforced with natural and man-made cellulose fibers has been synthesized and showed an enhancement in biodegradability, thermal stability, and mechanical properties [19]. Cellulose is one of the most abundant biopolymers occurring in various plant-based material, which serves as the major reinforcing phase in plant structures due to its crystalline state [21], so that it is responsible for the reinforcement effect observed due to the composites as suggested by many researchers.

Nypa fructicans is a monoecious palm with special characteristics. In Malaysia, *Nypa fructicans* palm can be found throughout the year, and it can be considered an abundant source. *Nypa fructicans* palm consists of frond, shell, husk, and leaf. Each part has a different chemical composition that is made up of cellulose (28.9–45.6 wt%), hemicellulose (21.8–26.4 wt%), and lignin (19.4–33.8 wt%) [22].

Although natural fibers can offer the resulting composites many advantages, polar fibers usually have inherently low compatibility with nonpolar polymer matrices [23,24]. As a result, many chemical and physical strategies were developed to improve the surface compatibility between PLA and fillers [25–28]. The treatment of natural fibers is beneficial to enhance the wettability of natural fibers' surface by polymers (mainly nonpolar polymers) and promote the interfacial adhesion. Chemical modification provides the means of permanently altering the nature of fiber cell walls [29] by grafting polymers onto the fibers [30,31], cross-linking the fiber cell wall [32], or by using coupling agents [33]. A coupling agent is a chemical that functions at the interface to create a chemical bridge between the filler and the matrix. It improves the interfacial adhesion when one end of the molecule is adhered to the reinforcement surface and the functional groups at the other end reacts with the polymer phase [34]. Silane is recognized as an efficient coupling agent that is extensively used in composites, adhesive formulations, or as substrate primers, giving stronger adhesion [35–39].

7.2 PREPARATION OF PLA/rLDPE/NF BIOCOMPOSITES

7.2.1 MATERIALS

PLA was supplied by TT Biotechnologies Sdn. Bhd. (Penang, Malaysia), and rLDPE was obtained from STL Plastic Sdn. Bhd. (Penang, Malaysia). Ethanol and ethylenediaminetetraacetic acid disodium salt-2-hydrate (EDTA were supplied by Fluka (Buchs, Switzerland) and Aldrich Chemical Company Inc., respectively. The *Nypa fructicans* fiber (NF) used as a filler was obtained at a plantation at Simpang Empat, Perlis, Malaysia. NF was first cleaned, ground, and dried at 80°C for 24 h. The average particle size of NF was 31 μm, as determined by a Malvern Particle Size Analyzer. The chemical composition of NF is presented in Table 7.1.

TABLE 7.1
Chemical Composition of *Nypa fructicans* Fiber

Parts (Husk)	Weight (%)
Cellulose	36.5
Hemicellulose	21.8
Lignin	28.8
Starch	0.1
Protein	1.9
Extractive	0.8
Ash	8.1
Total	98.0

7.2.2   FILLER TREATMENT

The treatment of *Nypa fruticans* was achieved by immersing the filler in solutions consisting of EDTA. Initially, the solution of EDTA (5 g/L) at a pH of 11.0 with sodium hydroxide (NaOH) at the temperature of 60°C was prepared. The NF was added into the solution and stirred at a speed of 50 rpm for 3 h. The filler-and-solution ratio was 11:1 (wt%/v) as proposed by Stuart et al. [41]. Then, the filler solution was filtered and dried at 80°C for 24 h.

7.2.3   FABRICATION OF PLA/rLDPE/NF BIOCOMPOSITES

The biocomposites were prepared using a Brabender EC Plus at a temperature of 180°C and at a rotor speed of 50 rpm. First, PLA and rLDPE were added to the mixing chamber for 2 min until they were completely melted. After 2 min of mixing, NF was added and mixed continuously for another 6 min. The total mixing time was 8 min. Next, the biocomposites were compressed using a compression molding machine (model GT 7014A). The dumbbell shape of specimen of ASTM 638 bar type IV was used. The compression of biocomposites was done at a temperature of 180°C with 1-min preheating, 4-min compressing, and, subsequently, 4-min cooling. Then, a similar procedure was done for biocomposites treated with EDTA. The formulation for PLA/rLDPE/NF biocomposites with and without EDTA treatment is shown in Table 7.2.

7.2.4   TESTING AND CHARACTERIZATION

The fabricated PLA/rLDPE/NF biocomposites underwent tensile tests according to ASTM D638 using an Instron 5569. The crosshead speed for biocomposite testing was 20 mm/min. An average of five samples for each formulation were tested. Tensile strength, elongation at break, and Young's modulus were recorded and calculated by the instrument software.

The morphological study of the tensile fracture surface of the biocomposites was carried out using a scanning electron microscope (SEM), model JSM 6260 LE JOEL. The fractured ends of specimens were mounted on aluminum stubs and sputter-coated with a thin layer of palladium to avoid electrostatic charging during examination.

TABLE 7.2

Formulation of Untreated and Treated Biocomposites

Material	Without EDTA	With EDTA
PLA (php)	70	70
rLDPE (php)	30	30
NF (php)	0, 10, 20, 30, 40	10, 20, 30, 40
EDTA (g/L)[a]	–	5

php, part per hundred of total polymer.

[a] % based on weight of NF [40].

FTIR (Fourier transform infrared) spectroscopy was used to characterize the presence of functional groups in PLA/rLDPE/NF biocomposites. The FTIR analysis was carried out using an FTIR spectroscope (Perkin Elmer, Model L1280044). The attenuated total reflectance (ATR) method was used. Four scans were recorded per sample in the frequency range of 4,000 to 650 cm^{-1} with a resolution of 4 cm^{-1}.

Thermal analysis was carried out using a Pyris Diamond TGA from Perkin Elmer. The sample (weight 7 ± 2 mg) underwent thermal scanning from 30°C to 600°C with a nitrogen airflow of 50 mL/min and a heating rate of 10°C/min.

Enzymatic testing was carried out according to method mentioned in Yoon et al. [42]. The samples were placed in a solution containing α-amylase. A buffer solution was prepared by adding 4.8 mL of 0.2 M acetic acid to 45.2 mL of 0.2 M sodium acetate solution to produce 50 mL solution. The samples were then taken out every two days, washed thoroughly, and dried in an oven at 50°C for 24 h. Then, each sample was weighed.

7.3 MECHANICAL PROPERTIES, THERMAL PROPERTIES, AND CHARACTERIZATION OF PLA/rLDPE/NF BIOCOMPOSITES

7.3.1 TENSILE PROPERTIES

Figure 7.1 presents the effect of NF contents on tensile strength of PLA/rLDPE/NF biocomposites with and without EDTA treatment. The increased NF content in PLA/rLDPE matrix had decreased the tensile strength of PLA/rLDPE/NF biocomposites without EDTA due to poor wettability, dispersion, and filler–matrix interaction between the filler and the matrix [43]. Moreover, the weak bonding between the hydrophilic filler and the hydrophobic matrix obstructed the stress propagation and

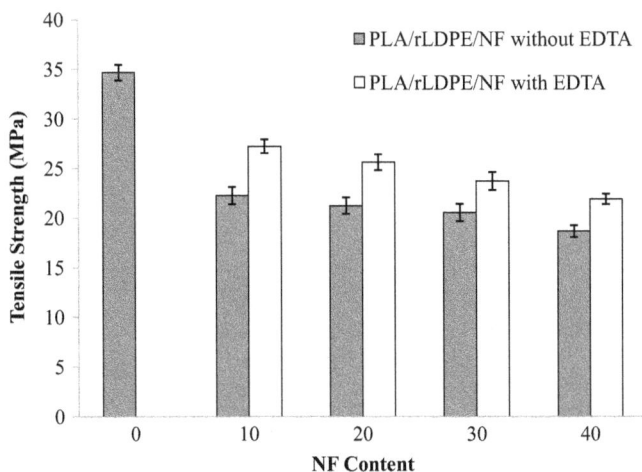

FIGURE 7.1 Effect of NF content on tensile strength of PLA/rLDPE/NF biocomposites with and without EDTA treatment.

thus caused the tensile strength to decrease when NF content increased. However, the tensile strength of PLA/rLDPE/NF biocomposites with EDTA treatment significantly improved to about 21.3% compared to the tensile strength of PLA/rLDPE/NF biocomposites without EDTA treatment. The EDTA treatment had been shown to remove noncellulosic compounds such as wax and oils, leading to filler separation from their bundles. The separation of the NF would increase the effective surface area of NF, allowing it to intermingle with the matrix.

Figure 7.2 shows the elongation at break of PLA/rLDPE/NF biocomposites with and without EDTA treatment. It can be seen that the elongation at break of PLA/rLDPE/NF biocomposites with and without EDTA treatment decreased with the increasing NF content. The decrease in elongation at break is attributed to the stiffening effect of NF due to the rigid surface of NF, thereby hindering the mobility of the matrix molecular chain (2009). The increase in NF content leads to more weak interfacial regions occurred between the filler and the matrix. Thus, cracks are easier to merge through the weak interfacial regions, and the biocomposites therefore fracture at lower degree of elongation at break as NF content increases. At the similar NF content, biocomposites with EDTA have lower elongation at break compared to PLA/rLDPE/NF biocomposites without EDTA treatment. This result indicates the increase in rigidity is a consequence of the enhanced interfacial bonding between the filler and the matrix. As the rigidity of the biocomposites increases, the deformability of a rigid interface decreases between the NF and PLA/rLDPE matrix.

Figure 7.3 shows the effect of EDTA on Young's modulus of PLA/rLDPE/NF biocomposites with and without EDTA. Both biocomposites showed the similar trend, and the Young's modulus increased as the NF content increased. The increment of the Young's modulus of biocomposites was attributable to the restriction of chain mobility in the matrix as the NF content increased, thus providing high stiffness and brittleness of the biocomposites. The biocomposites with EDTA treatment showed the increment

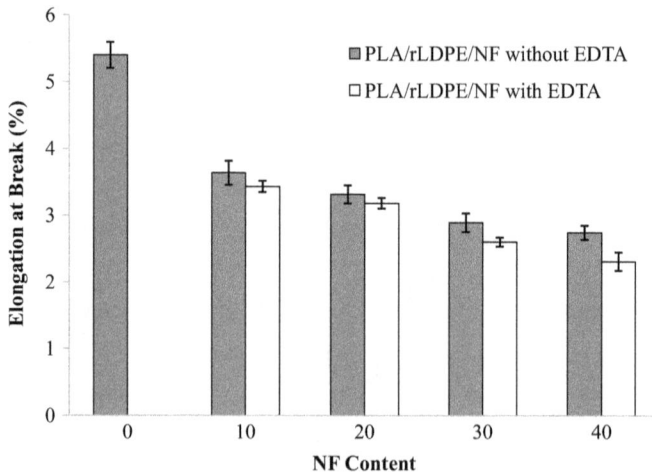

FIGURE 7.2 Effect of NF content on elongation at break of PLA/rLDPE/NF biocomposites with and without EDTA treatment.

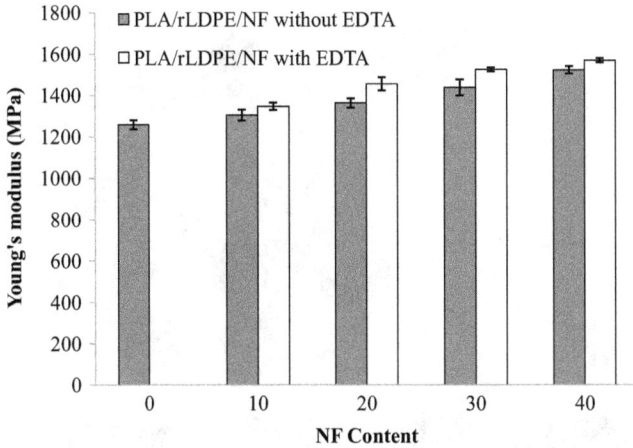

FIGURE 7.3 Effect of NF content on Young's modulus of PLA/rLDPE/NF biocomposites with and without EDTA treatment.

in Young's modulus about 7% compared to PLA/rLDPE/NF biocomposites without EDTA treatment. The stiffness of the biocomposites with EDTA treatment increased due to the improved interfacial bonding, giving good transfer of filler stiffness into the biocomposites and in turn leading to higher Young's modulus. The increment of the Young's modulus indicated that the PLA/rLDPE matrix with treated NF became stiffer and could withstand higher stress at the same strain portion.

7.3.2 MORPHOLOGICAL STUDY

Figures 7.4 and 7.5 show the SEM micrographs of tensile fracture surface of PLA/rLDPE/NF biocomposites with and without EDTA at 20 php NF content. The biocomposites filled with NF show rough fracture surface and the detachment of NF from the matrix due to the poor bonding between the NF and PLA/rLDPE matrix (Figure 7.4). This is attributable to less adhesion and interaction between them. Moreover, filler pull-out created holes in the fracture surface, evidencing the weak bonding between NF and PLA/rLDPE matrix. The formation of hole due to the filler pull-out was initially attributable to the crack propagation under tensile strength, resulting in the biocomposites having lower tensile strength at high NF content. Meanwhile, the tensile fracture surface of biocomposites with EDTA showed NF that is well dispersed in PLA/rLDPE matrix, which showed some NFs are embedded in the matrix (Figure 7.5). This is attributable to the higher interfacial adhesion between the NF and PLA/rLDPE matrix, which is achieved using the EDTA treatment.

7.3.3 FOURIER TRANSFORM INFRARED SPECTROSCOPIC ANALYSIS

Figure 7.6 illustrates FTIR spectra of PLA/rLDPE/NF biocomposites with and without EDTA. It can been seen that in the IR spectrum of biocomposites without EDTA treatment, a peak at 3,330 cm^{-1} was assigned to OH group belonging to

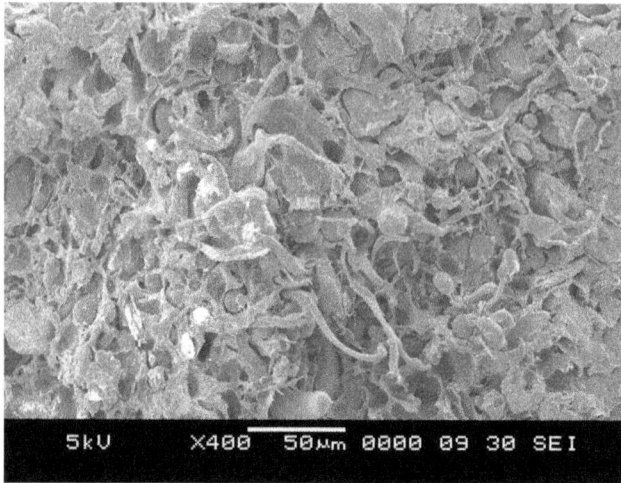

FIGURE 7.4 SEM micrograph of tensile fracture surface of PLA/rLDPE/NF biocomposite without EDTA treatment.

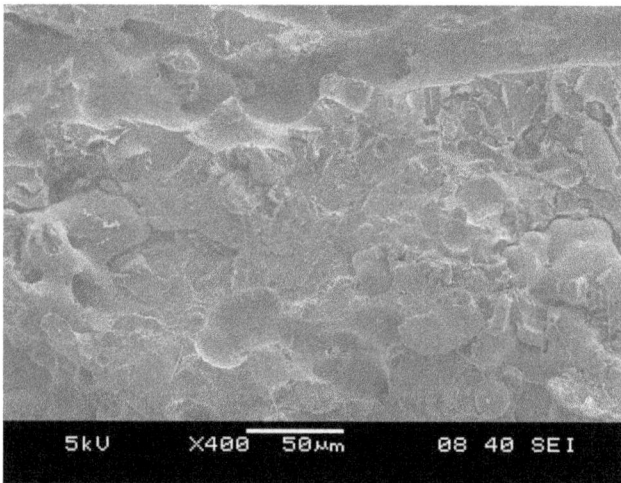

FIGURE 7.5 SEM micrograph of tensile fracture surface of PLA/rLDPE/NF biocomposite with EDTA treatment.

NF. The peak at 2,919 cm^{-1} was assigned to the C–H stretching. Moreover, peaks at 1,727 and 1,630 cm^{-1} were attributed to carbonyl (C=O) and C=C stretching from hemicellulose, respectively. Meanwhile, peaks at 1,515 and 1,440 cm^{-1} were attributable to C=C and C–H deformation from lignin, respectively. The peaks at 1,368 and 1,243 cm^{-1} were attributable to the C–H and C–O groups from cellulose and lignin, respectively. Furthermore, the peaks in the range 1,000–1,150 cm^{-1} were C–O–C and C–O groups from the main carbohydrates of cellulose and lignin. The group peaks at

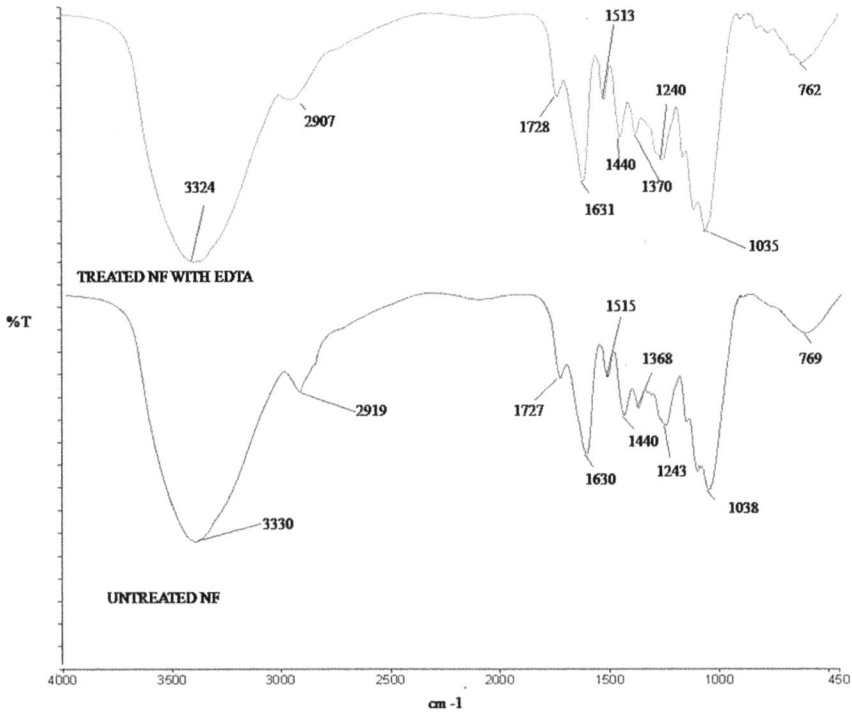

FIGURE 7.6 FTIR spectra of NF treated with and without EDTA.

700–900 cm^{-1} were C–H vibrations in lignin. Meanwhile, it can be seen that the wave absorption band for biocomposites without EDTA at 3,330 cm^{-1} was assigned to OH stretch vibration. After the EDTA treatment, the absorption band for biocomposites with EDTA decreased to 3,324 cm^{-1}. Moreover, the intensity of the peak characteristic of waxes and oils present at 2,919 cm^{-1} decreased to 2,907 cm^{-1} for the biocomposites with EDTA treatment, indicating the removal of the noncellulosic content.

7.3.4 THERMOGRAVIMETRIC ANALYSIS

Figure 7.7 illustrates the TGA curve of neat PLA/rLDPE and PLA/rLDPE/NF biocomposites treated with and without EDTA, respectively. Table 7.3 represents the percentage of weight loss of the neat PLA/rLDPE and PLA/rLDPE/NF biocomposites with and without EDTA. Based on Table 7.3, the total weight loss of PLA/rLDPE/NF biocomposites with and without EDTA is lower than that of the neat PLA/rLDPE. This result reveals that the addition of NF to PLA/rLDPE matrix effectively raises the char yields of the biocomposites, as the char acts as a protective barrier that suppresses the thermal decomposition of the PLA/rLDPE matrix, which enhances the thermal stability of the biocomposites. Furthermore, biocomposites treated with EDTA have lower total weight loss compared to biocomposites without EDTA treatment. This shows that the biocomposites with EDTA treatment have

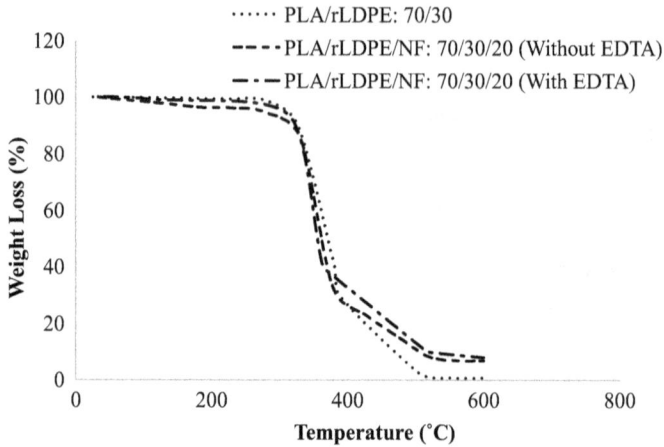

FIGURE 7.7 TGA curve of the neat PLA/rLDPE and PLA/rLDPE/NF biocomposites treated with and without EDTA.

TABLE 7.3

TGA Data of Neat PLA/rLDPE and PLA/rLDPE/NF Biocomposites Treated With and Without EDTA

		Weight Loss of PLA/rLDPE/NF			
	PLA/rLDPE	Without EDTA		With EDTA	
Temperature (°C)	70/30	70/30/20	70/30/40	70/30/20	70/30/40
0–100	0.17	1.57	2.14	0.01	0.03
100–200	0.36	2.04	3.18	0.63	1.16
200–300	0.71	2.70	4.36	0.16	0.55
300–400	72.53	62.99	53.43	34.41	38.95
400–500	21.84	21.19	20.67	36.52	27.82
500–600	4.39	8.35	11.12	22.04	22.41
Total	100.00	98.84	94.90	93.77	90.92

higher thermal stability against thermal decomposition compared to biocomposites without EDTA treatment. The higher thermal stability of biocomposites with EDTA can be attributed to the stronger interaction between the PLA/rLDPE matrix and treated NF with EDTA. Moreover, biocomposites with EDTA can promote the formation of a protective char, and the char layer can protect the biocomposites from further thermal decomposition.

7.3.5 ENZYMATIC BIODEGRADATION

Figure 7.8 shows the enzymatic biodegradation for the neat PLA/rLDPE and PLA/rLDPE/NF biocomposites treated with and without EDTA. The weight loss of the neat PLA/rLDPE was lower than that of PLA/rLDPE/NF biocomposites treated

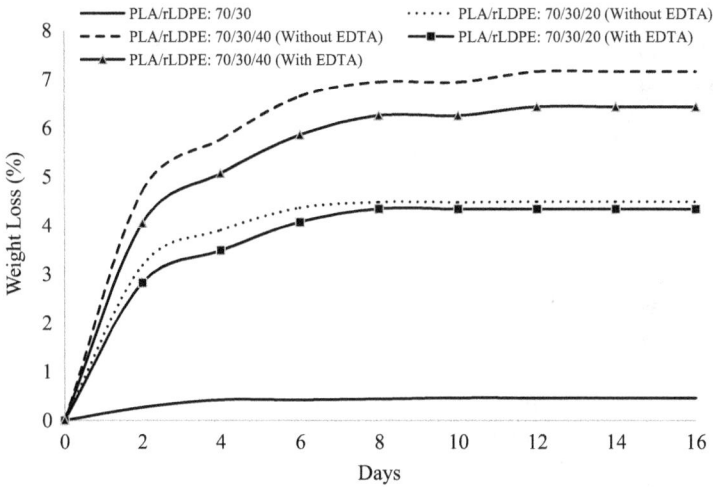

FIGURE 7.8 Weight loss of the neat PLA/rLDPE and PLA/rLDPE/NF biocomposites treated with and without EDTA on enzymatic biodegradation.

with and without EDTA. This indicated that the neat PLA/rLDPE had slower enzymatic biodegradation rate. The slower enzymatic biodegradation rate of the neat PLA/rLDPE was attributable to its hydrophobic nature. Meanwhile, both of the PLA/rLDPE/NF biocomposites treated with and without EDTA were observed to increase the weight loss with increasing NF content. The addition of NF increased the enzymatic biodegradation rate as NF is hydrophilic in nature. The presence of the hydrophilic hydroxyl group of NF was responsible for absorbing more of the amylase solution. The higher absorbing capacity of α-amylase solution facilitated the penetration of α-amylase into the PLA/rLDPE/NF biocomposites and subsequently enhanced the amylase attack on the NF. However, the treated biocomposites with EDTA exhibited a lower weight loss in enzymatic biodegradation compared to biocomposites without EDTA treatment. Treated NF with EDTA delayed the enzymatic biodegradation of the biocomposites. The biodegradability of the biocomposites was correlated with their ability to absorb the buffer solution. Biocomposites treated with EDTA improved the interfacial adhesion between the NF and matrix, thus restricting their ability to absorb α-amylase.

7.4 CONCLUSION

Biopolymers like PLA have the potential of substituting polymer matrix which is made from crude oil. Despite that, certain drawbacks of PLA are first considered in order to achieve high mechanical parameters. By blending and using natural fibers only, a full "biocomposite" can be produced. The tensile strength and elongation at break of the PLA/rLDPE/NF biocomposites were found to be decreased with increasing NF content, while Young's modulus increased. However, treated NF with EDTA enhanced the filler matrix interaction and dispersion, thus leading to an increase in tensile strength and Young's modulus of treated PLA/rLDPE/NF biocomposites compared to PLA/rLDPE/NF biocomposites without EDTA treatment. The improved interaction between the matrix and the filler was observed by the smooth surface from the tensile fracture surface of biocomposites treated with EDTA, where less fiber pull-out and detachment were observed in the SEM micrograph. Furthermore, the incorporation of NF into PLA/rLDPE blend increased the thermal stability of the biocomposites, which was shown by the lower total weight loss compared to the neat PLA/rLDPE. Chemical modification of filler using EDTA improved the adhesion between NF and matrix, which provided higher thermal stability compared to biocomposites without EDTA treatment. Meanwhile, the biodegradation of the biocomposites was evaluated based on the total weight loss when immersed in the solution containing α-amylase. Biocomposites containing NF showed a higher weight loss compared to the neat PLA/rLDPE due to the presence of the hydrophilic hydroxyl group that absorbed more of the amylase solution and accelerated the degradation. Still, the biodegradation of biocomposites treated with EDTA was less than that of biocomposites without EDTA treatment due to the improved interfacial adhesion between the fiber and the matrix, which restricted the adsorbing capacity of the buffer solution and hence delayed the biodegradation.

REFERENCES

1. R. C. Thompson, C. J. Moore, F. S. Vom Saal, and S. H. Swan, "Plastics, the environment and human health: current consensus and future trends," *Philosophical Transactions of the Royal Society B: Biological Sciences*, vol. 364, no. 1526, pp. 2153–2166, 2009.
2. A. L. Andrady and M. A. Neal, "Applications and societal benefits of plastics," *Philosophical Transactions of the Royal Society of London B: Biological Sciences*, vol. 364, no. 1526, pp. 1977–1984, 2009.
3. M. M. Syahmie, T. P. Leng, and Z. N. Najwa, "Effect of filler content and chemical modification on mechanical properties of polylactic acid/polymethyl methacrylate/ *Nypa fruticans* husk biocomposites," *IOP Conference Series: Materials Science and Engineering, IOP Publishing*, vol. 318, no. 1, p. 012011, 2018.
4. K. M. Nampoothiri, N. R. Nair, and R. P. John, "An overview of the recent developments in polylactide (PLA) research," *Bioresource Technology*, vol. 101, no. 22, pp. 8493–8501, 2010.
5. S. Farah, D. G. Anderson, and R. Langer, "Physical and mechanical properties of PLA, and their functions in widespread applications—a comprehensive review," *Advanced Drug Delivery Reviews*, vol. 107, pp. 367–392, 2016.
6. X. Wang, Y. Zhuang, and L. Dong, "Study of biodegradable polylactide/poly (butylene carbonate) blend," *Journal of Applied Polymer Science*, vol. 127, no. 1, pp. 471–477, 2013.
7. H. Tsuji, "Poly (lactic acid)," *Bio-Based Plastics: Materials and Applications.* West Sussex: John Wiley & Sons, pp. 171–239, 2013.
8. S. Aslan, L. Calandrelli, P. Laurienzo, M. Malinconico, and C. Migliaresi, "Poly (D, L-lactic acid)/poly (ε-caprolactone) blend membranes: preparation and morphological characterisation," *Journal of materials science*, vol. 35, no. 7, pp. 1615–1622, 2000.
9. F. Carrasco, O. Santana, J. Cailloux, M. Sánchez-Soto, and M. L. Maspoch, "Poly (lactic acid) and acrylonitrile–butadiene–styrene blends: influence of adding ABS–g–MAH compatibilizer on the kinetics of the thermal degradation," *Polymer Testing*, vol. 67, pp. 468–476, 2018.

10. A. R. Kakroodi, Y. Kazemi, D. Rodrigue, and C. B. Park, "Facile production of biodegradable PCL/PLA in situ nanofibrillar composites with unprecedented compatibility between the blend components," *Chemical Engineering Journal*, vol. 351, pp. 976–984, 2018.

11. A. K. Mohanty, Y. Yuryev, and M. Misra, "Durable high performance heat resistant polycarbonate (PC) and polylactide (PLA) blends and compositions and methods of making those," Google Patents, 2018.

12. Z. Qu, X. Hu, X. Pan, and J. Bu, "Effect of compatibilizer and nucleation agent on the properties of poly (lactic acid)/polycarbonate (PLA/PC) blends," *Polymer Science, Series A*, vol. 60, no. 4, pp. 499–506, 2018.

13. H. A. Topkanlo, Z. Ahmadi, and F. A. Taromi, "An in-depth study on crystallization kinetics of PET/PLA blends," *Iranian Polymer Journal*, vol. 27, no. 1, pp. 13–22, 2018.

14. A. G. Pedroso and D. d. S. Rosa, "Mechanical, thermal and morphological characterization of recycled LDPE/corn starch blends," *Carbohydrate Polymers*, vol. 59, no. 1, pp. 1–9, 2005.

15. S. Qian, Y. Tao, Y. Ruan, C. A. F. Lopez, and L. Xu, "Ultrafine bamboo-char as a new reinforcement in poly (lactic acid)/bamboo particle biocomposites: the effects on mechanical, thermal, and morphological properties," *Journal of Materials Research*, vol. 33, no. 22, pp. 3870–3879, 2018.

16. S. Qian and K. Sheng, "PLA toughened by bamboo cellulose nanowhiskers: role of silane compatibilization on the PLA bionanocomposite properties," *Composites Science and Technology*, vol. 148, pp. 59–69, 2017.

17. R. Gunti, A. Ratna Prasad, and A. Gupta, "Mechanical and degradation properties of natural fiber-reinforced PLA composites: jute, sisal, and elephant grass," *Polymer Composites*, vol. 39, no. 4, pp. 1125–1136, 2018.

18. N. Jiang, T. Yu, and Y. Li, "Effect of hydrothermal aging on injection molded short jute fiber reinforced poly (lactic acid)(PLA) composites," *Journal of Polymers and the Environment*, vol. 26, pp. 3176–3186, 2018.

19. L. Suryanegara, Y. D. Kurniawan, F. A. Syamani, and Y. Nurhamiyah, "Mechanical properties of composites based on poly (lactic acid) and soda-treated sugarcane bagasse pulp," *Sustainable Future for Human Security: Springer,* vol. 2018, pp. 277–285, 2018.

20. J. Tengsuthiwat, S. Siengchin, R. Berényi, and J. Karger-Kocsis, "Ultraviolet nanosecond laser ablation behavior of silver nanoparticle and melamine–formaldehyde resin-coated short sisal fiber-modified PLA composites," *Journal of Thermal Analysis and Calorimetry*, vol. 132, no. 2, pp. 955–965, 2018.

21. N. Maddahy, O. Ramezani, and H. Kermanian, "Production of nanocrystalline cellulose from sugarcane bagasse," *Proceedings of the 4th International Conference on Nanostructures (ICNS4)*, pp. 12–14, 2012.

22. P. Tamunaidu and S. Saka, "Chemical characterization of various parts of nipa palm (*Nypa fruticans*)," *Industrial Crops and Products*, vol. 34, no. 3, pp. 1423–1428, 2011.

23. H. Ishida, "A review of recent progress in the studies of molecular and microstructure of coupling agents and their functions in composites, coatings and adhesive joints," *Polymer Composites*, vol. 5, no. 2, pp. 101–123, 1984.

24. H. Jiang and D. P. Kamdem, "Development of poly(vinyl chloride)/wood composites. A literature review," *Journal of Vinyl and Additive Technology*, vol. 10, no. 2, pp. 59–69, 2004.

25. S. Alippilakkotte and L. Sreejith, "Benign route for the modification and characterization of poly (lactic acid)(PLA) scaffolds for medicinal application," *Journal of Applied Polymer Science*, vol. 135, no. 13, p. 46056, 2018.

26. S. Deng, J. Ma, Y. Guo, F. Chen, and Q. Fu, "One-step modification and nanofibrillation of microfibrillated cellulose for simultaneously reinforcing and toughening of poly (ε-caprolactone)," *Composites Science and Technology*, vol. 157, pp. 168–177, 2018.

27. F.-L. Jin, H. Zhang, S.-S. Yao, and S.-J. Park, "Effect of surface modification on impact strength and flexural strength of poly (lactic acid)/silicon carbide nanocomposites," *Macromolecular Research*, vol. 26, no. 3, pp. 211–214, 2018.

28. M. Jing, J. Che, S. Xu, Z. Liu, and Q. Fu, "The effect of surface modification of glass fiber on the performance of poly (lactic acid) composites: graphene oxide vs. silane coupling agents," *Applied Surface Science*, vol. 435, pp. 1046–1056, 2018.

29. C. A. S. Hill, "Wood-plastic composites: strategies for compatibilising the phases," *Journal of the Institute of Wood Science*, vol. 15, no. 3, pp. 140–146, 1999.

30. C. Daneault, B. Kokta, and D. Maldas, "Grafting of vinyl monomers onto wood fibers initiated by peroxidation," *Polymer Bulletin*, vol. 20, no. 2, pp. 137–141, 1988.

31. C. A. S. Hill, *Wood Modification: Chemical, Thermal and Other Processes*. Chichester, England: John Wiley & Sons, 2006.

32. C. K. Hong et al., "Mechanical properties of maleic anhydride treated jute fibre/polypropylene composites," *Plastics, Rubber and Composites*, vol. 37, no. 7, pp. 325–330, 2008.

33. J. Z. Lu, Q. Wu, and H. S. McNabb Jr, "Chemical coupling in wood fiber and polymer composites: a review of coupling agents and treatments," *Wood and Fiber Science*, vol. 32, no. 1, pp. 88–104, 2000.

34. Y. Xie, C. A. S. Hill, Z. Xiao, H. Militz, and C. Mai, "Silane coupling agents used for natural fiber/polymer composites: a review," *Composites Part A: Applied Science and Manufacturing*, vol. 41, no. 7, pp. 806–819, 2010.

35. K. S. Chun, H. Salmah, and O. Hakimah, "Mechanical and thermal properties of coconut shell powder filled polylactic acid biocomposites: effects of the filler content and silane coupling agent," *Journal of Polymer Research*, vol. 19, no. 5, pp. 9859–9866, 2012.

36. H. Salmah, N. A. Azieyanti, and H. Ismail, "Effect of maleic anhydride-grafted-polyethylene (MAPE) and silane on properties of recycled polyethylene/chitosan biocomposites," *Polymer - Plastics Technology and Engineering*, vol. 52, no. 2, pp. 168–174, 2013.

37. K. L. Pickering, A. Abdalla, C. Ji, A. G. McDonald, and R. A. Franich, "The effect of silane coupling agents on radiata pine fibre for use in thermoplastic matrix composites," *Composites Part A: Applied Science and Manufacturing*, vol. 34, no. 10, pp. 915–926, 2003.

38. H. Salmah, H. Ismail, and A. Abu Bakar, "Properties of paper sludge filled polypropylene (PP)/ethylene propylene diene terpolymer (EPDM) composites: the effect of silane-based coupling agent," *International Journal of Polymeric Materials and Polymeric Materials*, vol. 55, no. 9, pp. 643–662, 2006.

39. H. Salmah, A. Faisal, and H. Kamarudin, "Chemical modification of chitosan-filled polypropylene (PP) composites: the effect of 3-aminopropyltriethoxysilane on mechanical and thermal properties," *International Journal of Polymeric Materials*, vol. 60, no. 7, pp. 429–440, 2011.

40. M. Le Troedec et al., "Influence of various chemical treatments on the composition and structure of hemp fibres," *Composites Part A: Applied Science and Manufacturing*, vol. 39, no. 3, pp. 514–522, 2008.

41. T. Stuart et al., "Structural biocomposites from flax—Part I: Effect of bio-technical fibre modification on composite properties." *Composites Part A: Applied Science and Manufacturing*, vol. 37, no. 3, pp. 393–404, 2006.

42. B. S. Yoon et al., "Studies on the degradable polyethylenes: Use of coated photodegradants with biopolymers." *Journal of Applied Polymer Science*, vol. 60, no. 10, pp. 1677–1685, 1996.

43. Y. Dong et al., "Polylactic acid (PLA) biocomposites reinforced with coir fibres: Evaluation of mechanical performance and multifunctional properties." *Composites Part A: Applied Science and Manufacturing*, vol. 63, pp. 76–84, 2014.

8 Mechanical, Thermal, and Degradation Properties of Linear Low-Density Polyethylene/ Polyvinyl Alcohol/Kenaf Bast Powder Composites

A.L. Pang, H. Ismail, and A.B. Azhar
Universiti Sains Malaysia

CONTENTS

8.1 INTRODUCTION

For the past few decades, the increasing global awareness in renewable resources and environmentally compatible materials has resulted in an extensive research done on natural fibers (particularly, plant fibers/lignocellulosic) [1–3]. The combination of specific properties and environmental-friendly characteristics of natural fibers has a positive impact on their applications in polymer-based composites [4]. Furthermore, the need to produce economically feasible products has intensified the development of natural fiber-based polymer composites in various research fields and industries [5,6]. Natural fibers are generally classified according to their origins (animal, plant/lignocellulosic and mineral), and currently, many types of plant/lignocellulosic fibers are available, such as kenaf bast fiber (KNF), jute, sisal, hemp, and wood [7,8]. The attractive attributes of these natural fibers are their low price, low density, biodegradability, renewability, nonabrasive, and high specific mechanical properties [9,10]. Nevertheless, natural fibers possess several demerits, including poor fiber–matrix adhesion, high moisture absorption, low processing temperature (limiting matrix selection), and low resistance to microorganisms attack [11,12].

Among various types of plant/lignocellulosic fibers, KNF has been exploited in various sectors (particularly in academic research) over the past few years, because of its fast growing speed over a wide range of climatic conditions [13]. Moreover, KNF has been recognized as an important cellulose source for composites and other industrial applications [14,15]. The applications of KNF-based composites have been found in industries such as sports, automotive, furniture, and construction (structural and nonstructural elements) [16,17]. Similar to other natural fibers, one of the challenges encountered during the incorporation of KNF into polymer matrix is the lack of good interfacial adhesion between KNF and polymer matrix [2]. This is a result of poor compatibility between hydrophilic KNF and hydrophobic polymer matrix, thereby forming a weak filler–matrix interface [2,9]. A good interfacial adhesion between KNF and polymer matrix is essential to obtain the optimum mechanical properties, since the stress is transferred between the matrix and the filler across the interface [12].

In natural fiber-based polymer composites, the polymer matrix functions as a binder material that holds the fibers in position, protects the fiber surface from mechanical abrasion, and transfers load to fibers [12,18]. Polymeric matrices are frequently used in natural fiber-based composites due to their lightweight and low processing temperature. Both thermoplastics (polyethylene (PE), polypropylene (PP), polyvinyl chloride (PVC), and polystyrene (PS)) and thermosets (unsaturated polyester, epoxy resin, and phenol–formaldehyde) are commonly used matrix materials for natural fiber-based composites [12,19]. Thermoplastics possess many benefits over thermosets such as low processing cost with simple molding methods (extrusion, injection molding), flexibility in design, repeated heating and cooling, and better potential to be recycled [12]. However, thermoplastics such as PE, PP, PVC, and PS are nondegradable materials. In conjunction with the environmental concerns and stringent regulations and standards, blending or replacement of petroleum-based product with biodegradable matrices or renewable biofibers has been explored [11,20]. Polyvinyl alcohol (PVOH) is a synthetic biodegradable polymer that is soluble in water, and it is extensively used in agricultural mulch film or packaging applications due to its good strength and biodegradability [21,22]. PVOH has been used for the fabrication of blends and composites with synthetic polymers like PE and with natural polymers such as lignocellulosic fillers, starch, or chitosan [22].

The present work used a constant composition ratio of linear low-density polyethylene (LLDPE) and PVOH blend of 60:40 (wt%) as polymer matrices because this composition ratio gives the best overall properties as confirmed by Ismail et al. [23]. Meanwhile, KNF was selected among other natural fibers for LLDPE/PVOH blend because of several reasons. First of all, KNF is a fiber crop that is grown commercially in Malaysia (easily available) and has been proven to be environmental-friendly cellulose source by Kyoto Protocol [13,19]. Moreover, the short growing cycle of KNF (4–5 months) enables a stable supply of raw materials [19]. Hence, the highlights of this study are to explore the potential of KNF in LLDPE/PVOH matrices and its effect on the mechanical, morphological, thermal, and degradation properties of composites.

8.2 PREPARATION AND CHARACTERIZATION OF KENAF BAST POWDER FILLED WITH LINEAR LOW-DENSITY POLYETHYLENE/POLYVINYL ALCOHOL COMPOSITES

8.2.1 MATERIALS

LLDPE of grade name LL0209SC with a density of 0.92 g/cm^3 was supplied by PT. Lotte Chemical Titan Nusantara, Indonesia. KNF was obtained from National Kenaf and Tobacco Board (LKTN), Kelantan, Malaysia. KNF (as-received) was subjected to grinding and sieving processes using mini grinder from Rong Tsong Precision Technology Co., Taiwan (Model RT-34), and Endecotts sieve from Syarikat Saintifik Jaya, Malaysia (Model: PRADA), respectively. The average particle size of KNF measured by Malvern Particle Size Analyzer Instrument (Model: Mastersizer 3000) was 17 µm.

8.2.2 PREPARATION OF COMPOSITES

Prior to the preparation of composites, KNF was dried in vacuum oven at 80°C for 24 h to remove moisture. The composition of LLDPE/PVOH was fixed at 60:40 (wt%) throughout the research work, due to the optimization in previous work by Ismail et al. [23]. Meanwhile, the loadings of KNF varied from 10 to 40 parts per hundred parts of resin (phr). All the materials were melt-mixed using an internal mixer (Thermo Haake Rheomix, Model: R600/610), at the temperature and rotor speed of 150°C and 50 rpm, respectively. The total melt mixing time was 10 min, and the torque readings were recorded during the processing. The mixtures were then subjected to hot pressing in an electrically heated hydraulic press (GoTech Testing Machine, Model: KT-7014 A) at the temperature of 150°C into a 1-mm-thick sheet. The compression molding cycles involved 4-min preheating, 4-min hot pressing, and 3-min cool pressing. The pressure used for hot pressing and cool pressing was 10 MPa. The molded sheet of composites formed was used for further testing.

8.2.3 Characterizations

Tensile testing was determined in agreement with ASTM D638 Type IV. The test was completed using Universal Testing Machine (Instron, Model: 3366) with a constant speed of 5 mm/min. A total of five dumbbell-shaped samples were tested, and the test was performed at the temperature of 25°C ± 3°C. The average values of tensile strength, tensile modulus, and elongation at break of all samples were calculated. Similarly, the samples retrieved after degradation test were subjected to tensile test in order to obtain the average values of tensile strength, tensile modulus, and elongation at break. The retention ratio of tensile properties was determined as follows:

$$\text{Retention }(\%) = \frac{\text{Value after degradation}}{\text{Value before degradation}} \times 100 \tag{8.1}$$

SEM (scanning electron microscopy) micrographs of tensile fracture surfaces of composites were obtained by Zeiss Supra-35VP field emission scanning electron microscope (FESEM) operating at 5 kV. Prior to examination, a thin layer of gold was sputter-coated on the sample surfaces by an electrodeposition technique to impart electrical conduction and to avoid poor image resolution.

Thermogravimetric (TG) and derivative thermogravimetric (DTG) analyses of composites were performed via Perkin Elmer Pyris 6 TGA analyzer. Particularly, the sample of weights between 5 and 10 mg were heated from 30°C to 600°C under nitrogen atmosphere at a heating rate of 20°C/min. The TG-DTG thermograms were analyzed to obtain the temperature at 5% weight loss ($T_{5\%}$), temperature at maximum weight loss (T_{max}), and char residue (%).

8.3 TENSILE, MORPHOLOGICAL AND THERMAL PROPERTIES OF LINEAR LOW-DENSITY POLYETHYLENE/POLYVINYL ALCOHOL/KENAF BAST POWDER COMPOSITES

8.3.1 Tensile Properties

Figures 8.1 and 8.2 display the effect of KNF loading on the tensile strength and tensile modulus of LLDPE/PVOH/KNF composites. Based on Figure 8.1, the tensile strength of the composites was found to be decreased by approximately 6.8%, 10.2%, 17.1%, and 23.9%, with the addition of 10, 20, 30, and 40 phr KNF loadings into LLDPE/PVOH matrices, respectively. Basically, the introduction of KNF in LLDPE/PVOH matrices has resulted in a reduction of tensile strength of the composites. This is because the hydrophilic KNF was not compatible with the hydrophobic LLDPE from LLDPE/PVOH matrices, thereby forming a weak interface between them. Moreover, composites with higher KNF loading (i.e., 40 phr KNF) show greater reduction in tensile strength most probably due to an excessive amount of KNF loaded into the system, thereby causing poor KNF surface wetting by LLDPE/PVOH matrices. Consequently, the applied stress was not efficiently transferred from polymer matrices to fillers, subsequently leading to poor tensile strength.

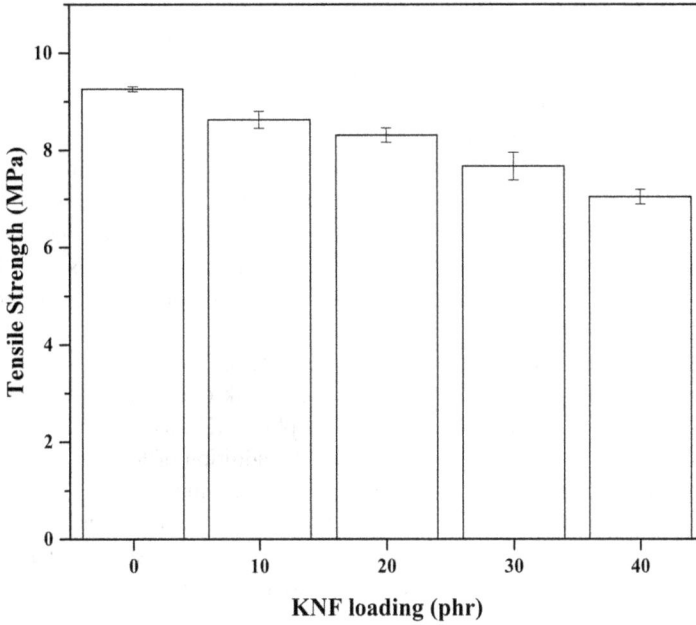

FIGURE 8.1 Tensile strength of LLDPE/PVOH/KNF composites with different KNF loadings.

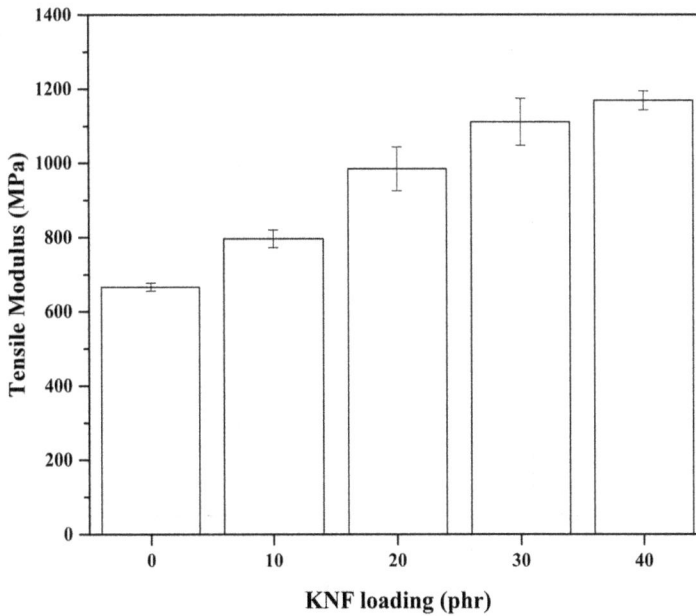

FIGURE 8.2 Tensile modulus of LLDPE/PVOH/KNF composites with different KNF loadings.

However, LLDPE/PVOH/KNF composites demonstrated an increasing trend of tensile modulus when the KNF loading was increased from 10 to 40 phr, as shown in Figure 8.2. Tensile modulus indicates the stiffness of the composites; i.e., the higher the value of tensile modulus, the stiffer the composites are [24]. From Figure 8.2, the results show that the addition of KNF was found to impart higher stiffness to LLDPE/PVOH matrices. It was suggested that the presence of KNF has restricted the mobility of LLDPE/PVOH matrices, thereby improving the stiffness of composites. Additionally, tensile modulus was observed to be increased from 796.5 to 1,169.0 MPa when 10–40 phr of KNF was added in the composite system. This was attributed to an increase in the stiffening effect from KNF in the composites [25].

Figure 8.3 shows the elongation at break of LLDPE/PVOH/KNF composites with different KNF loadings. The elongation at break of composites was found to be decreased from 6.5% to 3.0% when 10–40 phr KNF was incorporated into the system. This observation is expected because the addition of KNF has reduced the deformability of composites through the restriction on mobility of LLDPE/PVOH matrices. The elongation at break was found to be significantly decreased for composites with 40 phr KNF loading, as the capacity of composites to deform upon stress application was greatly reduced at high filler loading.

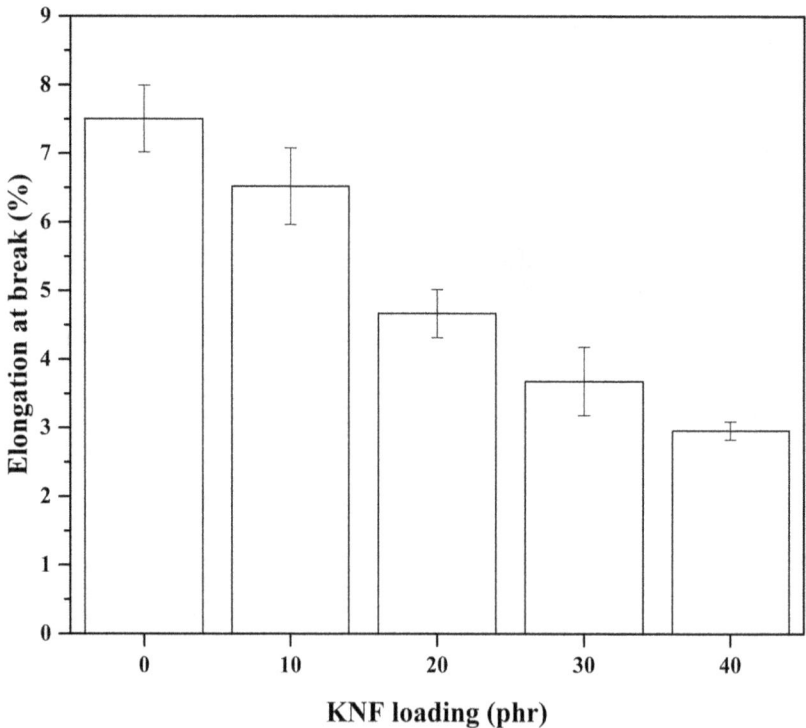

FIGURE 8.3   Elongation at break of LLDPE/PVOH/KNF composites with different KNF loadings.

8.3.2 MORPHOLOGICAL PROPERTIES

Figure 8.4a–d illustrates the SEM micrographs of fractured surfaces of LLDPE/PVOH/KNF composites with different KNF loadings. Based on Figure 8.4a, long fibrils were observed on the fractured surface of LLDPE/PVOH matrices and such fibrillar structure was assigned to LLDPE due to its high flexibility. In the case of LLDPE/PVOH/KNF composites (Figure 8.4b–d), it was noted that the fibrillar structures become shorter with increasing KNF loading up to 40 phr. This finding justifies the lower elongation at break of LLDPE/PVOH/KNF composites at higher loading of KNF as explained in Section 8.3.1. Furthermore, as illustrated in Figure 8.4b–d, the SEM micrographs revealed poor KNF surface wetting by LLDPE/PVOH matrices as detachments of KNF can be seen on the fractured surfaces of composites. This finding indicates poor filler–matrix interfacial adhesion between KNF and LLDPE/PVOH matrices, thereby decreasing the composite's tensile strength.

8.3.3 THERMOGRAVIMETRIC ANALYSIS (TGA)

TGA was carried out to evaluate the effect of KNF loading on the thermal stability and degradation temperature of LLDPE/PVOH/KNF composites. Figures 8.5 and 8.6 demonstrate the TG-DTG thermograms of LLDPE/PVOH/KNF composites

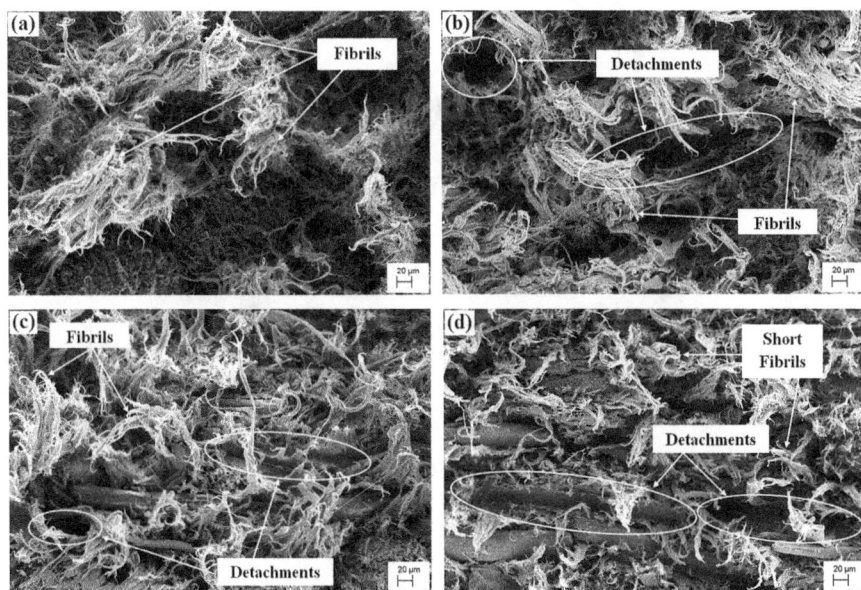

FIGURE 8.4 SEM micrographs of LLDPE/PVOH/KNF composites with (a) 0 phr KNF, (b) 10 phr KNF, (c) 20 phr KNF, and (d) 40 phr KNF loadings (200× magnification).

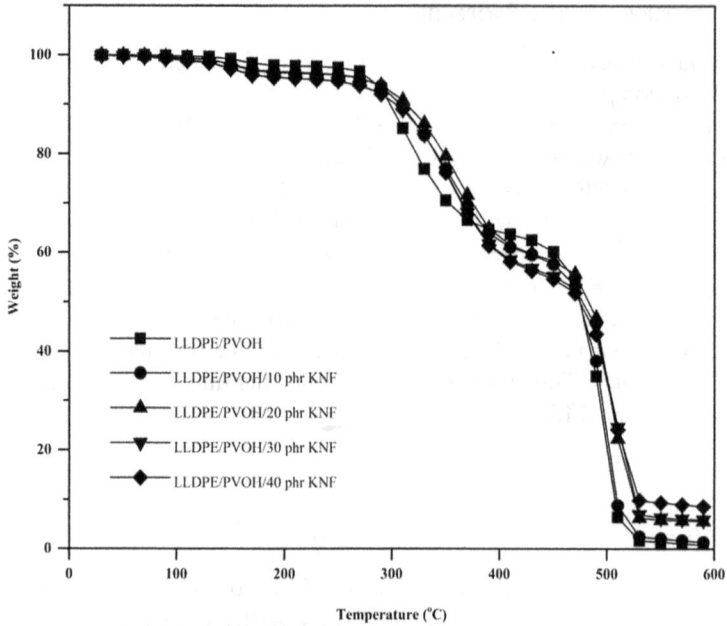

FIGURE 8.5 TG thermograms of LLDPE/PVOH/KNF composites with different KNF loadings.

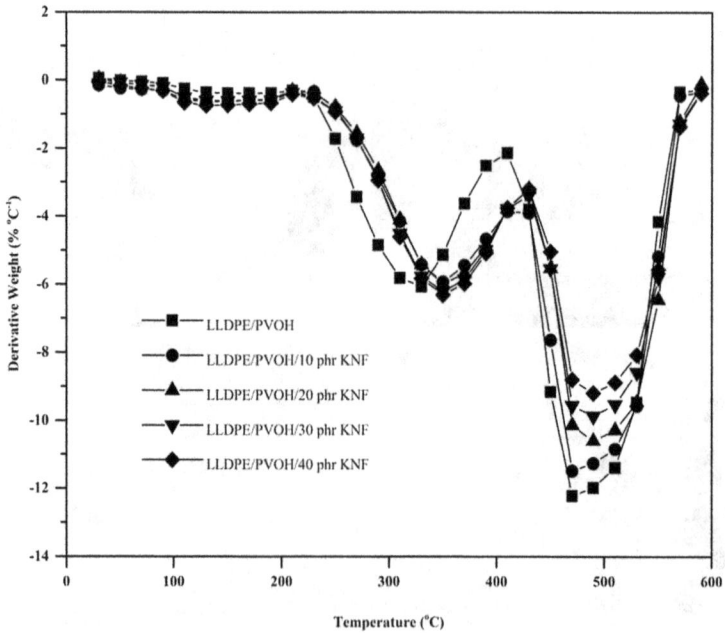

FIGURE 8.6 DTG thermograms of LLDPE/PVOH/KNF composites with different KNF loadings.

TABLE 8.1

Thermal Parameters of LLDPE/PVOH/KNF Composites with Different KNF Loadings

KNF Loading (phr)	T_{maxI} (°C)	T_{maxII} (°C)	Char Residue at 590°C (%)
0	326	470	0.76
10	344	470	1.31
20	349	479	5.57
30	349	479	5.87
40	349	480	8.56

with different KNF loadings. A detailed evaluation of TG-DTG thermograms on the maximum degradation temperatures (T_{maxI} and T_{maxII}) and the percentage of char residue is summarized in Table 8.1. The T_{maxI} and T_{maxII} were obtained from the first and second peaks in Figure 8.6. Basically, all of the composite samples experienced two main degradation stages, as seen from the presence of two peaks in Figure 8.6. Moreover, an initial weight loss in the temperature range of 100°C–200°C was also observed, which was related to the gradual evaporation of moisture absorbed in the composites.

Based on Figure 8.6 and Table 8.1, T_{maxI} and T_{maxII} at 326°C and 470°C of LLDPE/PVOH matrices were attributed to the decomposition of PVOH and LLDPE, respectively [21]. In the case of LLDPE/PVOH/KNF composites, T_{maxI} corresponded to the decomposition of PVOH and cellulosic components of KNF. According to previous studies by Fisher et al. [26] and Peng et al. [27], the degradation of hemicelluloses occurred between 200°C and 260°C, celluloses between 240°C and 350°C, and lignin between 280°C and 500°C. Meanwhile, T_{maxII} of LLDPE/PVOH/KNF composites was related to the thermal decomposition of LLDPE and lignin from KNF.

T_{maxI} and T_{maxII} of LLDPE/PVOH/KNF composites (at all loadings of KNF) were higher than those of LLDPE/PVOH matrices, as shown in Table 8.1. Moreover, T_{maxI} and T_{maxII} increased from 344°C to 349°C and 470°C to 480°C with increasing KNF loadings from 10 to 40 phr, respectively. This shows that the addition of KNF improved the thermal stability of LLDPE/PVOH/KNF composites, particularly at higher KNF loading. This finding is in agreement with the study by Sdrobis et al. [28] and El-Shekeil et al. [29].

During the degradation process, volatile materials (such as CO and CH_4) and char residue are formed [30,31]. The char formation may act as a barrier between the heat source and the polymeric materials, which inhibits the emission of volatile decomposition products from the composites [30]. From Table 8.1, the char residue at 590°C of LLDPE/PVOH matrices was 0.76%, while it was increased to 1.31% and 8.56% when 10 and 40 phr KNF were added into LLDPE/PVOH composites, respectively. This shows an apparent increase in the thermal stability of LLDPE/PVOH/KNF composites.

8.4 INFLUENCE OF NATURAL WEATHERING ON TENSILE AND MORPHOLOGICAL PROPERTIES OF LINEAR LOW-DENSITY POLYETHYLENE/POLYVINYL ALCOHOL/KENAF BAST POWDER COMPOSITES

8.4.1 TENSILE PROPERTIES

Figure 8.7 presents the tensile strength of LLDPE/PVOH/KNF composites at different KNF loadings after natural weathering exposure of 3 and 6 months. As expected, all the composites experienced a progressive decline in tensile strength with prolonged weathering exposure up to 6 months. The decrement in tensile strength after weathering is attributed to poor stress transfer caused by the degradation of LLDPE/PVOH matrices and KNF, resulting from the combined effects of ultraviolet (UV) radiation, oxygen, heat, and water. During weathering exposure, the absorption of UV radiation and oxygen by polymer matrices can probably induce polymer chain scission and oxidation process that lead to the formation of oxidized groups such as carbonyl and carboxyl groups [4]. Additionally, degradation of KNF is attributed to the effect of water and UV absorption during weathering exposure. The water and UV absorption by KNF caused the breakdown of its components (cellulose, hemicelluloses, and lignin), thereby reducing the interfacial adhesion between KNF and LLDPE/PVOH matrices.

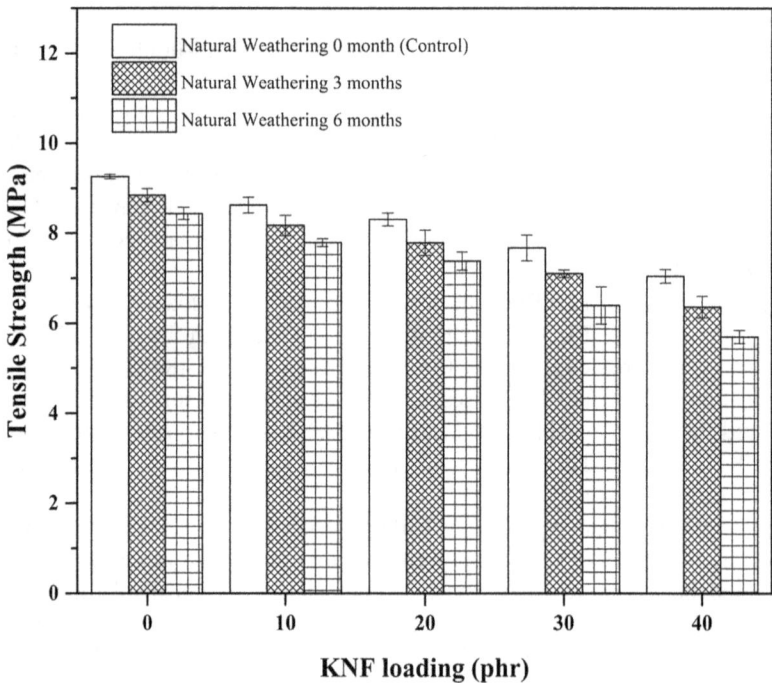

FIGURE 8.7 Tensile strength of LLDPE/PVOH/KNF composites at different KNF loadings before and after natural weathering exposure of 3 and 6 months.

In fact, greater reduction in tensile strength was found for composites after 6 months of weathering exposure in comparison with composites after 3 months of weathering exposure. This is supported by the lower retention values of tensile strength, as shown in Table 8.2. When composites were exposed to longer weathering time, severe surface cracks were developed as a consequence of cyclic expansion and contraction of the composite sample due to unstable temperature [32]. These surface cracks are vulnerable to moisture absorption and may further impair the tensile strength of composites. These findings are in agreement with the earlier report by Thirmizir et al. [33], Abdullah et al. [28], and Silva et al. [34].

Referring to Figure 8.8 and Table 8.2, the tensile modulus of composites and its retention values decrease with increasing weathering exposure time from 3 to 6 months. For instance, at 10 phr KNF loading, the tensile modulus of composites was 796.5, 711.8, and 662.1 MPa for unweathered, and after 3 and 6 months of weathering exposures, respectively. The decrement in tensile modulus after weathering is because of the fact that when composites are exposed to weathering, the hydrophilic KNF and PVOH show a tendency to interact with the water present in the environment. Subsequently, the swollen KNF may have detached and washed away (by rainwater) from composites, thereby decreasing the stiffness of composites. Indeed, greater detachment of KNF might occur when the composites were exposed for a longer time in environment, as revealed by the SEM morphology in Figure 8.10. Furthermore, it is worth to note that composites with higher KNF loading (20–40 phr) exhibited greater reduction in tensile modulus compared to composites with lower KNF loading (10 phr KNF) in the first 3 months of weathering. This observation suggested that a significant amount of KNF and PVOH may have diminished caused by weathering. Over the next 6 months, a gradual decrease in tensile modulus was probably due to the loss of KNF.

Elongation at break is one of the most important tensile properties when studying polymer degradation, as it provides an early indication of mechanical failure [35]. Referring to Figure 8.9, elongation at break of weathered composites and its retention values show a decline trend, which is similar to that of tensile strength.

TABLE 8.2
Retention of Tensile Properties of LLDPE/PVOH/KNF Composites at Different KNF Loadings Before and After Natural Weathering Exposure of 3 and 6 Months

| | Retention (%) | | | | | |
| | Tensile Strength | | Tensile Modulus | | Elongation at Break | |
KNF Loading (phr)	3 months	6 months	3 months	6 months	3 months	6 months
0	95.53	91.14	96.16	98.35	78.24	71.69
10	98.70	90.30	89.37	83.12	70.17	53.30
20	93.71	88.83	85.52	73.93	69.54	50.63
30	92.56	83.38	83.46	72.15	67.20	45.07
40	90.33	80.94	82.16	71.29	61.89	43.58

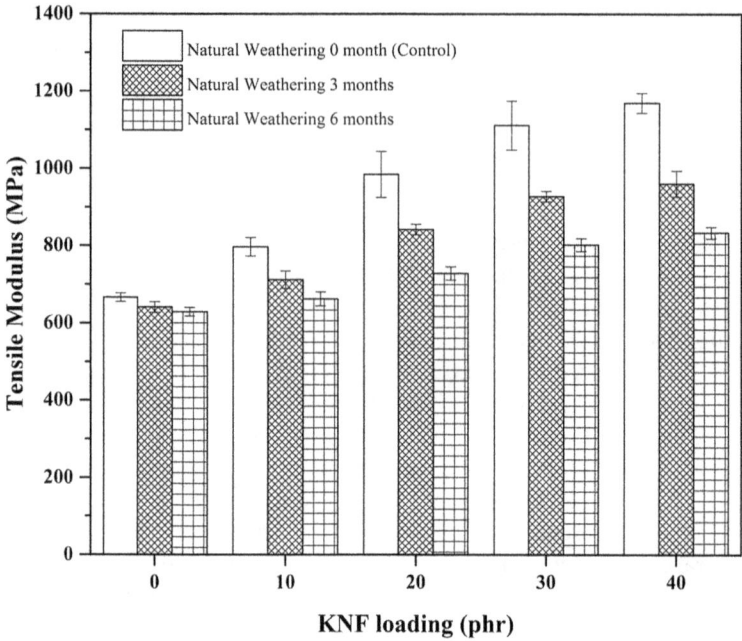

FIGURE 8.8 Tensile modulus of LLDPE/PVOH/KNF composites at different KNF loadings before and after natural weathering exposure of 3 and 6 months.

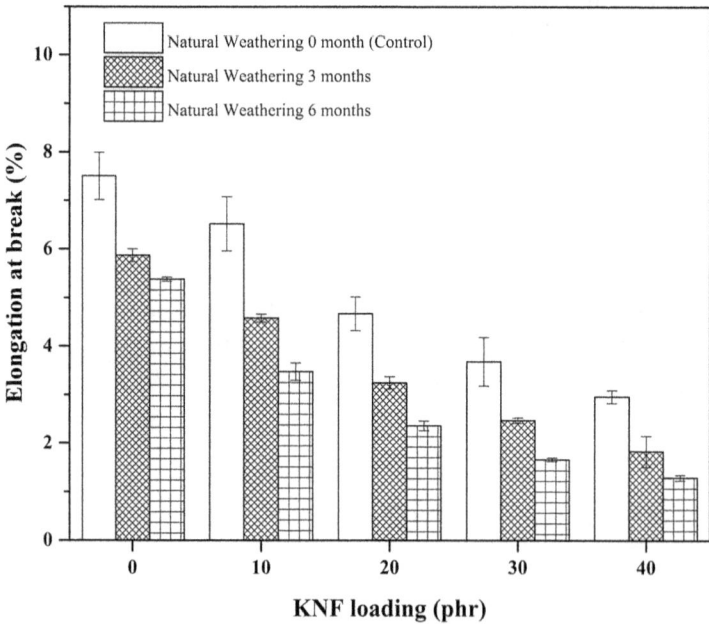

FIGURE 8.9 Elongation at break of LLDPE/PVOH/KNF composites at different KNF loadings before and after natural weathering exposure of 3 and 6 months.

The reduction in elongation at break is due to the development of surface cracks on composites during weathering. In fact, these surface cracks become deeper at longer weathering time (i.e., 6 months) and the inner composite's layer is exposed to degradation factors. This will promote the loss of KNF and PVOH due to the erosion of composites by the combined effect of UV radiation and water (rain). This is similar to the earlier reports by Sarifuddin et al. [36] and Zaaba et al. [37].

8.4.2 Morphological Properties

Figure 8.10a–f visualizes the changes in surface morphology of LLDPE/PVOH/KNF composites at different KNF loadings after exposed to 3 and 6 months of natural weathering. It can be seen that the surface of composites becomes rougher and severely damaged with increasing KNF loading and prolonged weathering exposure.

FIGURE 8.10 SEM micrographs of LLDPE/PVOH/KNF composites with (a) 0 phr KNF, (b) 10 phr KNF, (c) 40 phr KNF after 3 months of weathering exposure, and (d) 0 phr KNF, (e) 10 phr KNF, and (f) 40 phr KNF after 6 months of weathering exposure (1000× magnification).

As seen in Figure 8.10a, there are many spots of microcracks on the LLDPE/PVOH matrix surface after 3 months of weathering exposure. Moreover, these microcracks were observed to have propagated with increasing weathering exposure time up to 6 months (Figure 8.10b). The formation of microcracks after weathering is due to the cyclic expansion and contraction of polymer matrices resulting from the fluctuation in temperature [36,38]. LLDPE/PVOH/KNF composites with 40 phr KNF (Figure 8.10c and f) exhibited a highly damage surface with deeper cracks as compared to composites with 10 phr KNF (Figure 8.10b and e) after 3 and 6 months of weathering exposure, respectively. This revealed that the composites are more vulnerable to environmental factors at higher KNF loading. Sarifuddin et al. [36] and Zaaba et al. [37] also found that composites with higher filler loading are more likely to increase the degradation rate of composites.

Compared to Figure 8.10a, weathered LLDPE/PVOH/KNF composites (after 3 months) displayed larger and propagated surface cracks with KNF protruded from the surface (Figure 8.10b and c). The larger surface cracks after weathering were probably initiated by the swelling of KNF due to the absorption of water from environment [33]. Furthermore, a closer examination of weathered composites after 6 months showed that the composite surface was remarkably changed, as shown in Figure 8.10e and f. The surface of weathered composites looks rougher with remarkable matrix loss, particularly at 40 phr KNF loading (Figure 8.10f). Additionally, there are many cavities formed, which are attributed to the detachment of KNF from the surface of composites. These results indicated that the degradation of composites during natural weathering was a result of combined degradation factors such as UV radiation, oxygen, heat, and water [33,39]. Thus, the weak interfacial adhesion between LLDPE/PVOH matrices and KNF may lead to poor tensile properties.

8.5 CONCLUDING REMARKS

The volume of thermoplastics utilized in the packaging, automotive, building and construction, as well as other low-cost and high-volume applications is expanding. In order to reduce the use of petroleum-based plastic and minimize the harm to the environment, there is an increasing awareness in maximizing the use of renewable materials in composite materials. This chapter focuses on the use of natural fiber (KNF) in thermoplastic composite materials. The attractive characteristics about KNF are its positive environmental impact and cost benefits. However, KNF and other natural fibers, in general, portrayed several limitations that render them incompatible with most polymeric systems. The poor interface adhesion between KNF and polymer matrix leads to poor stress-transfer efficiencies, thereby limiting their application range at the industrial scale. Nevertheless, the potential of using KNF in thermoplastic composite materials is promising but requires appropriate research on the interface quality of these materials, mechanical properties, thermal stability, and eventual degradation properties of these composite materials.

In summary, new composites based on LLDPE, PVOH, and KNF were obtained by melt mixing followed by compression molding methods. The tensile test results showed that tensile modulus increased, whereas tensile strength and elongation at break of LLDPE/PVOH/KNF composites decreased as KNF loading increased.

SEM results revealed the poor interfacial adhesion between KNF and LLDPE/PVOH. LLDPE/PVOH/KNF composites with higher KNF loading showed better thermal stability than composites with lower KNF loading.

ACKNOWLEDGMENT

The authors wish to appreciate Universiti Sains Malaysia (USM) for the Research University Grant for Cluster (RUC) (No: 1001/PKT/8640014) and also Malaysian Higher Education for financial support (MyBrain15) that have made this research work possible.

REFERENCES

1. A. Sdrobis, R.N. Darie, M. Totolin, G. Cazacu, and C. Vasile, "Low density polyethylene composites containing cellulose pulp fibers," *Compos. Part. B-Eng.,* vol. 43, pp. 1873–1880, 2012.
2. L.H. Zaini, M. Jonoobi, P.M. Tahir, and S. Karimi, "Isolation and characterization of cellulose whiskers from kenaf (*Hisbiscus cannabinus* L.) bast fibers," *J. Biomater. Nanobiotechnol.,* vol. 4, pp. 37–44, 2013.
3. R. Ayadi, M. Hanana, R. Mzid, L. Hamrouni, M.I. Khouja, and A. Salhi, "*Hisbiscus cannabinus L.* - kenaf: a review paper," *J. Nat. Fibers.,* vol. 14, pp. 466–484, 2017.
4. V. Fiore, T. Scalici, F. Nicoletti, G. Vitale, M. Prestipino, and A. Valenza, "A new eco-friendly chemical treatment of natural fibers: effect of sodium bicarbonate on properties of sisal fiber and its epoxy composites," *Compos. Part. B-Eng.,* vol. 85, pp. 150–160, 2016.
5. T.W. Yee, L.T. Sin, W.A.W.A. Rahman, and A.A. Samad, "Properties and interactions of poly (vinyl alcohol)-sago pith waste biocomposites," *J. Compos. Mater.,* vol. 45, pp. 2199–2209, 2011.
6. U. Nirmal, S.T.W. Lau, and J. Hashim, "Interfacial adhesion characteristics of kenaf fibers subjected to different polymer matrices and fiber treatments," *J. Compos.,* vol. 2014, 12 pages, 2018.
7. H.M. Akil, M.F. Omar, A.A.M. Marzuki, S. Safiee, Z.A.M. Ishak, and A. Abu Bakar, "Kenaf fiber reinforced composites: a review," *Mater. Des.,* vol. 32, pp. 4107–4121, 2011.
8. M. Farsi, "Thermoplastic matrix reinforced with natural fibers: a study on interfacial behavior," in *Some Critical Issues for Injection Molding*, J. Wang, Ed. InTech, Rijeka, Croatia, 2012, pp. 225–250.
9. A.L. Pang, and H. Ismail, "Tensile properties, water uptake, and thermal properties of polypropylene/waste pulverized tire/kenaf (PP/WPT/KNF) composites," *Bioresources,* vol. 8, pp. 806–817, 2013.
10. M. George, P.G. Mussone, and D.C. Bressler, "Modification of the cellulosic component of hemp fibers using sulfonic acid derivatives: surface and thermal characterization," *Carbohyd. Polym.,* vol. 134, pp. 230–239, 2015.
11. L.H. Carvalho, E.L. Canedo, S.R. Farias Neto, A.G. Barbosa de Lima, and C.J. Silva, "Moisture transport process in vegetable fiber composites: theory and analysis for technological applications," in *Industrial and Technological Applications of Transport in Porous Materials, Advanced Structured Materials*, J.M.P.Q. Delgado, Ed. Springer, Berlin, Heidelberg, 2013, pp. 37–62.
12. K.L. Pickering, M.G.A. Efendy, and T.M. Le, "A review of recent developments in natural fiber composites and their mechanical performance," *Compos. Part. A-Appl S.,* vol. 83, pp. 98–112, 2016.

13. M. Ramesh, "Kenaf (*Hibiscus cannabinus* L.) fiber based bio-materials: a review on processing and properties," *Prog. Mater. Sci.,* vol. 78–79, pp. 1–92, 2016.
14. A.L. Pang, and H. Ismail, "Studies on the properties of polypropylene/(waste tire dust)/ kenaf (PP/WTD/KNF) composites with addition of phthalic anhydride (PA) as a function of KNF loading," *J. Vinyl. Addit. Techn.,* vol. 20, pp. 193–200, 2014.
15. J. Datta, and P. Kopczynska, "Effect of kenaf fiber modification on morphology and mechanical properties of thermoplastic polyurethane materials," *Ind. Crop. Prod.,* vol. 74, pp. 566–576, 2015.
16. R. Mahjoub, J.M. Yatim, A.R.M. Sam, and S.H. Hashemi, "Tensile properties of kenaf fiber due to various conditions of chemical fiber surface modifications," *Constr. Build. Mater.,* vol. 55, pp. 103–113, 2014.
17. N. Saba, M.T. Paridah, and M. Jawaid, "Mechanical properties of kenaf fiber reinforced polymer composite: a review," *Constr. Build. Mater.,* vol. 76, pp. 87–96, 2015.
18. M.M. Kabir, H.Wang, K.T. Lau, and F. Cardona, "Chemical treatments on plant-based natural fiber reinforced polymer composites: an overview," *Compos. Part. A-Appl S.,* vol. 43, pp. 2883–2892, 2012.
19. I. Kamal, M.Z. Thirmizir, G. Beyer, M.J. Saad, N.A. Abdul Rashid, and Y. Abdul Kadir, "Kenaf for biocomposite: an overview," *J. Sci. Technol.,* vol. 6, pp. 41–66, 2014.
20. B. Tajeddin, R.A. Rahman, and L.C. Abdullah, "The effect of polyethylene glycol on the characteristics of kenaf cellulose/low-density polyethylene biocomposites," *Int. J. Biol. Macromol.,* vol. 47, pp. 292–297, 2010.
21. R. Nordin, H. Ismail, Z. Ahmad, and Abu A. Bakar, "Performance improvement of (linear low-density polyethylene)/poly(vinyl alcohol) blends by in situ silane crosslinking," *J. Vinyl. Addit. Techn.,* vol. 18, pp. 120–128, 2012.
22. B.K. Tan, Y.C. Ching, S.C. Poh, L.C. Abdullah, and S.N. Gan, "A review of natural fiber reinforced poly(vinyl alcohol) based composites: application and opportunity," *Polymers,* vol. 7, pp. 2205–2222, 2015.
23. H. Ismail, Z. Ahmad, R. Nordin, and A. Rashid, "Processibility and miscibility studies of linear low-density polyethylene/poly(vinyl alcohol) blends," *Polym.-Plast. Technol.,* vol. 48, pp. 1191–1197, 2009.
24. M.J. John, and S. Thomas, "Biofibres and biocomposites," *Carbohyd. Polym.,* vol. 71, pp. 343–364, 2008.
25. R.A. Majid, H. Ismail, and R.M. Taib, "Effect of polyethylene-g-maleic anhydride on properties of low density polyethylene/thermoplastic sago starch reinforced kenaf fiber composites," *Iran. Polym. J.,* vol. 19, pp. 501–510, 2010.
26. T. Fisher, M. Hajaligol, B. Waymack, and D. Kellogg, "Pyrolysis behavior and kinetics of biomass derived materials," *J. Anal. Appl. Pyrol.,* vol. 62, pp. 331–349, 2002.
27. Y. Peng, R. Liu, and J. Cao, "Effects of antioxidants on photodegradation of wood flour/polypropylene composites during artificial weathering," *Bioresources,* vol. 9, pp. 5817–5830, 2014.
28. M.Z. Abdullah, Y. Dan-mallam, and P.S.M.M. Yusoff, "Effect of environmental degradation on mechanical properties of kenaf/polyethylene terephthalate fiber reinforced polyoxymethylene hybrid composite," *Adv. Mater. Sci. Eng.,* vol. 2013, 8 pages, 2013.
29. Y.A. El-Shekeil, S.M. Sapuan, M. Jawaid, and O.M. Al-Shuja's, "Influence of fiber content on mechanical, morphological and thermal properties of kenaf fibers reinforced poly (vinyl chloride)/thermoplastic polyurethane poly-blend composites," *Mater. Des.,* vol. 58, pp. 130–135, 2014.
30. C.L. Beyler, and M.M. Hirschler, "Thermal decomposition of polymers," in *SFPE Handbook of Fire Protection Engineering,* P.J. DiNenno, Ed. NFPA, Quincy, MA, 2001, pp. 110–131.

31. D.G. Dikobe, and A.S. Luyt, "Comparative study of the morphology and properties of PP/LLDPE/wood powder and MAPP/LLDPE/wood powder polymer blend composites," *Exp. Polym. Lett.*, vol. 4, pp. 729–741, 2010.

32. Z.N. Azwa, B.F. Yousif, A.C. Manalo, and W. Karunasena, "A review on the degradability of polymeric composites based on natural fibers," *Mater. Des.*, vol. 47, pp. 424–442, 2013.

33. M.Z.A. Thirmizir, Z.A.M. Ishak, R.M. Taib, S. Rahim, and S.M. Jani, "Natural weathering of kenaf bast fiber-filled poly(butylenes succinate) composites: effect of fiber loading and compatibilizer addition," *J. Polym. Environ.*, vol. 19, pp. 263–273, 2011.

34. C.B. Silva, A.B. Martins, A.L. Catto, and R.M.C. Santana, "Effect of natural ageing on the properties of recycled polypropylene/ethylene vinyl acetate/wood flour composites," *Revista Materia*, vol. 22, pp. 11835–11851, 2017.

35. M.A.S. Spinace, and M.A. De Paoli, "Biocomposite of a multilayer film scrap and curaua fibers: preparation and environmental degradation," *J. Thermoplast. Compos.*, vol. 2015, pp. 1–16, 2015.

36. N. Sarifuddin, H. Ismail, and Z. Ahmad, "Incorporation of kenaf core fiber into low density polyethylene/thermoplastic sago starch blends exposed to natural weathering," *Mol. Cryst. Liq. Cryst.*, vol. 603, pp. 180–193, 2014.

37. N.F. Zaaba, H. Ismail, and M. Jaafar, A study of the degradation of compatibilized and uncompatibilized peanut shell powder/recycled polypropylene composites due to natural weathering. *J. Vinyl. Addit. Techn.*, vol. 23, pp. 290–297, 2015.

38. C. Homkhiew, T. Ratanawilai, and W. Thongruang, "Effects of natural weathering on the properties of recycled polypropylene composites reinforced with rubberwood flour," *Ind. Crop. Prod.*, vol. 56, pp. 52–59, 2014.

39. Y. Chen, N.M. Stark, M.A. Tshabalala, J.M. Gao, and Y.M. Fan, "Weathering characteristics of wood plastic composites reinforced with extracted or delignified wood flour," *Materials*, vol. 9, pp. 610–621, 2016.

9 Mechanical Properties and Degradation Behavior of Polyvinyl Alcohol/Starch Blend

N.F. Zaaba and H. Ismail
Universiti Sains Malaysia

CONTENTS

9.1   INTRODUCTION

Usages of plastic materials keep increasing years ahead. It was begun in 1930 once the main thermoplastics such as polyolefins, polystyrene, poly(vinyl chloride), and poly(methyl methacrylate) were developed [1]. Since then, plastic has been used in various applications including packaging, medical appliances, transportation, and communications. A global consumption of these plastics around the region is of about 22–23 million tonnes yearly. Unfortunately, the petrochemical process to produce plastics is a tricky target for naturally occurring decomposition (i.e., bacteria and fungi). This is primarily due to the macromolecular structure of plastic. Plastic materials usually generate difficulties and pollution to the environment because of their non-degradable substances and long-lasting life [2–5]. As a result, the accumulated waste in ecosystems causing environmental risks such as shortage of landfill, air pollution due to open burning of the plastic materials, and also plastic waste requires higher cost for recycling [6]. Because of that, alternative approaches such as blending the other biodegradable synthetic polymeric materials or natural polymer with starch have been explored extensively in various applications [7,8].

9.2   POLYVINYL ALCOHOL/STARCH BLEND

In polymer science, blending is a common process to reduce cost as well as to enhance the insufficient physical properties of the present plastic materials. It is probable to use properties by varying the composition and processing of blends. In fact, starch-based materials are the leading compostable blends. Besides, the aim of blending is to combine the higher cost of polymers with low cost of starch with better physical properties [9,10]. Blending of biodegradable polymers is one of the strategies implemented in producing compostable polymer materials [11–14]. Among all, polyvinyl alcohol (PVA)/starch blends are the most significant biodegradable plastic materials [15]. It is an ecologically biodegradable polymer and broadly implemented in packaging and agricultural mulch films [16]. However, PVA and starch is a physical blend which is highly desirable to further improve their tensile strength, elongation, and water resistance properties for other applications [8,17–19]. Numerous studies and patented formulation have been improved especially in the formulation of PVA/starch blend. This leads to enhancement of strength and modulus, as well as the compatibility of the film [20,21]. Adding a compatibilizer or chemical modification improves the compatibility between starch and synthetic polymers [22–25]. In another way, oxidation, etherification esterification, and cross-linking can be implemented for starch modification. Among all, methylation was generally used as one of the esterification methods in interpreting the substitution pattern in polymer chains and the structure of polysaccharides. It has been demonstrated as an effective method to enhance the properties of the blend [26,27]. Apart from that, cross-linking is another way of improving the physical and mechanical properties of the starch/PVA blends [28]. Cross-linking agents, such as borax [29], glutaraldehyde [30,31], and tetraethylene glycol diacrylate [32], were used to create intermolecular linkages by reacting with the hydroxyl groups in starch. Thus, the macromolecular networks were formed.

Once the PVA is blended with starch, improved properties are achieved. However, in terms of biodegradability, the PVA/starch blends have lower biodegradability than other biodegradable plastics. Bastioli et al. [33] stated that only 75% weight loss was detected in 300 days of degradation test on amylose–PVA composites (PVA–starch blend) with activated sludge. In another finding, Iwanami and Uemura [34] reported about 50% weight loss of PVA–starch blend in a waste-composting process. Besides that, Tudorachi et al. [35] stated significant weight losses as well as a deterioration of the physical–mechanical characteristics of polymeric materials based on PVA, starch, glycerin, and urea in the presence of microorganisms. The nature of the microorganisms used and the composition of the blend effected the weight loss of the material, whereas the higher content of starch contributes to the highest values of weight loss of the samples [8,36]. The appearance of spherical holes throughout biodegradation showed the colonization of the degrading microorganisms. The amorphous part of

PVA was consumed by the microorganisms once after the starch had been consumed [35]. After a complex mechanism of many decomposition stages, the thermal degradation of the systems took place. Several stages were overlapped particularly in the temperature range between 160°C and 370°C where the second and the third stages occurred [37].

9.3 PREPARATION AND CHARACTERIZATION OF BIODEGRADABLE POLYMER FILM BLENDS BASED ON POLYVINYL ALCOHOL AND STARCH

In polymer science and engineering, the multicomponent and multiphase polymers have become a very important research field. By blending different polymers by physical or chemical blending techniques, new and comprehensive materials are produced. Additionally, in polymer applications, at least 80% of polymer materials occurred in the blend form. Polymer blends are helpful in designing tailor-made materials with good processability, properties, and price/performance ratio [38].

9.3.1 MATERIALS

Granulated PVA with an average molecular weight of 145,000, partially hydrolyzed grade with a pH of 4.5–7 and a density of 1.3 g/cm³, was supplied by Merck (M) Sdn Bhd. This PVA is soluble in hot water. The melting temperature is above 200°C, and thermal decomposition temperature is above 300°C. Corn starch and tapioca starch of round and polygonal shapes with a particle size of approximately 5–26 μm were supplied by Thye Huat Chan (M) Sdn. Bhd. Glutaraldehyde with a density of 1.331 g/cm³ was used as a cross-linking agent. It is a water-soluble material which was supplied by Polyorganic. Glycerol with a molecular weight of 92.11 g/cm³ and polyethylene glycol (PEG) with a molecular weight ranging from 300 to 10,000,000 g/mol were used as plasticizers generally for plasticizing thermoplastic starch. The glycerol used was produced by BDH Company and supplied by VWR International. The PEG was produced by BDH Company and supplied by Advanced Sterilization Products.

9.3.2 SAMPLE FABRICATION

About 2.5 g of PVA was taken in a round bottom flask with 50 mL de-ionized water. It was then stirred for 10 min at a temperature of 95°C and a speed of >1,000 rpm in a water bath. Next, 2.5 g of corn starch was dissolved in 50 mL de-ionized water and stirred using a glass rod for few seconds before adding it to the round bottle flask. The blend was stirred for another 10 min. Then, the additives were added according to the Table 9.1. Stirring was continued for another 10 min. The mixtures were cast onto a glass plate which was placed on a leveled flat surface. Care must be taken to eliminate the bubbles which are the by-product of preparation. The films were dried at room temperature for 12 h before cured in an oven at 90°C for 30 min. The films were then peeled off and reserved. The whole procedure was repeated again but with tapioca starch instead of corn starch.

TABLE 9.1

The Formulation for Film Preparation

Blends	Glycerol (g)	Glutaraldehyde (g)	PEG (g)
1	2	–	–
2	2	0.2	–
3	2	1	–
4	2	2	–
5	4	0.4	–
6	2	–	0.2
7	2	–	0.6
8	4	–	0.4
9	4	–	0.6
10	6	–	0.6
11	2	0.4	0.4
12	2	0.6	0.6

9.3.3 CHARACTERIZATIONS

The films were subjected for tensile test according to ASTM D882. The results of tensile strength (TS), elongation at break (Eb), and modulus for each film were obtained using the Instron 3366 testing machine. Six dumbbell-shaped specimens were cut with a Wallace die cutter having a width of 6.4 mm and an average thickness of 0.0910 mm. The gauge length and the grip distance were set as 50.0 mm. A crosshead speed of 50 mm/min was used, and the test was performed at a temperature of 25°C ± 3°C with a relative humidity of 60% ± 5%.

The Fourier transform infrared radiation (FTIR) analysis was recorded by Perkin-Elmer FTIR spectrometer with a wavelength range from 4,000 to 450 cm^{-1}. The films were cut into small about 10 mm × 10 mm pieces from each sample and placed on MIRacle ATR Accessory (MIRacle base optics assembly). The spectra were recorded in reflection. The information on the chemical structure of the sample material is provided by FTIR technique. The bonds that make up a chemical compound absorbed the infrared light and produced frequencies that can be measured by the infrared spectroscopy. Each bond provides an energy of vibrations or natural frequency. Higher frequency is usually associated by a stiffer bond. The IR spectrum is compared with some known compound.

The water absorption (W_a) of films was performed as described by Yun et al. [19]. Dried starch/PVA blend films were immersed in distilled water at room temperature (25°C). Then, moisture on the surface of the films was removed after the equilibrium (24 h), and the weight of the films was measured. The water absorption, from which the swelling behavior can be studied, in starch/PVA blend films, was calculated as follows:

$$\text{Water absorption } (\%) = \frac{W_e - W_o}{W_o} \times 100 \qquad (9.1)$$

where W_e is the weight of starch/PVA blend film at the adsorbing equilibrium, and W_o is the first dry weight of starch/PVA blend film.

Field emission scanning electron microscope (SEM, model ZEISS Supra 35VP) was used to study the morphology of fractured surface of the sample after subjected for tensile test. Morphology study was carried out in order to observe the starch and PVA distribution and starch aggregation. The test specimens were sputtered coated on a Polaron SC 515 sputter coater with a layer of carbon to eliminate the electrostatic charging effects. Images of SEM micrograph for sample were taken at certain magnification.

Soil burial degradation was performed as described in the study by Thakore et al. [39] with a minor modification. The polybag with an approximate capacity of 10 L was filled with soil taken from a local area and exposed to the environment naturally. The plastic samples were cut into dumbbell and buried in the soil at the depth of 10 cm. The degradation of the specimen was determined after 14 days by taking the specimen carefully from the soil and washing it gently with distilled water to remove the soil. The specimen was analyzed via SEM to observe the morphology of the sample.

9.3.4 TENSILE PROPERTIES OF POLYVINYL ALCOHOL/STARCH BLEND

Tensile strength is known as the maximum of a stress–strain curve and, in general, indicates when necking will occur. It also defined as the maximum load applied in breaking a tensile test piece divided by the original cross-sectional area of the test piece. The definition of elongation at break is the elongation recorded at the moment of rupture of the specimen, often expressed as a percentage of the original length. It corresponds to the breaking or maximum load.

Figure 9.1 shows the tensile strength of PVA/corn starch and PVA/tapioca starch blend films. From the Figure 9.1, the tensile strength of PVA/corn starch showed a higher value than PVA/tapioca starch blend films. This was because of the particle size of the corn starch which was smaller and made the blending more compatible [40]. Starch particles with smaller size had larger surface area exposed [41]. Thus, the interaction of PVA with starch particle was increased, and the tensile strength

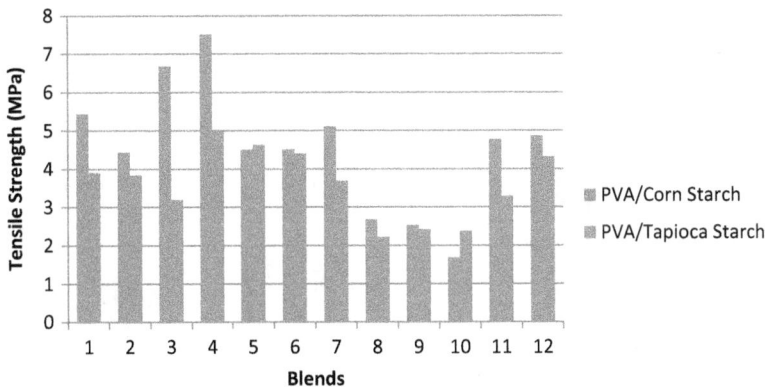

FIGURE 9.1   Tensile strength of PVA/corn starch and PVA/tapioca starch blend films.

also increased. Comparable results were stated by Azahari et al. [8] where the intro-duction of corn starch in PVA matrix might enhance the tensile strength of the blend. Blend 10 showed the lowest tensile strength. However, it has high elongation at break and modulus (Figures 9.2 and 9.3). This was due to the formulation of the blend with the presence of plasticizer only. The increment of plasticizer levels in the blend led to the elongation at break of the film increased. Thus, the addition of plasticizer was important to improve the film brittleness, which is effected by high intermolecular forces [42]. Plasticizers increased the mobility of the polymer chains by reducing these forces, thereby increasing the flexibility and extensibility of the film [43,44].

Blend 4 has the highest tensile strength, which was due to the presence of glu-taraldehyde in the blend. Glutaraldehyde is a cross-linking agent. It was used to introduce the cross-linking process and increased the tensile strength of the film [45]. However, as the film has high tensile strength, it showed low elongation at break and modulus. So, the presence of cross-linking agent reduced the extensibility and flexibility of the films.

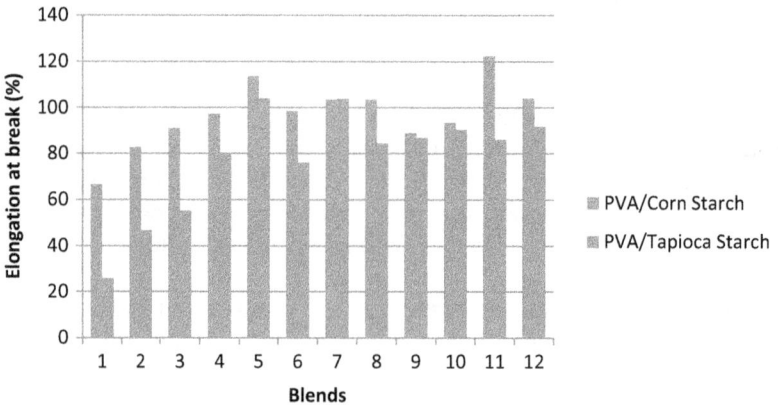

FIGURE 9.2   Elongation at break of PVA/corn starch and PVA/tapioca starch blend films.

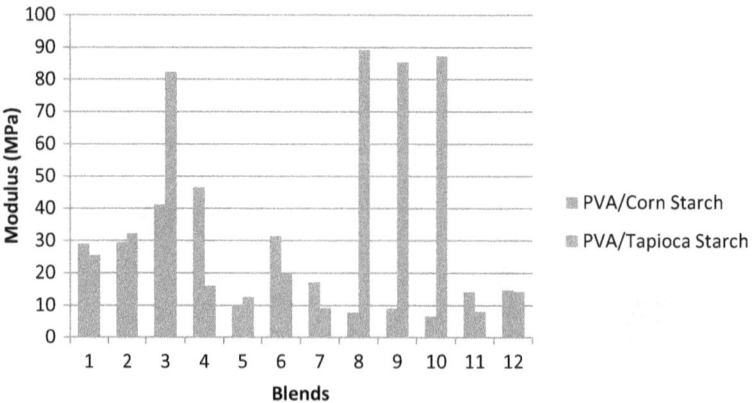

FIGURE 9.3   Modulus of PVA/corn starch and PVA/tapioca starch blend films.

Figure 9.4 shows the optimum value for glycerol, PEG, and glutaraldehyde. As we can see, the tensile strength of PVA/corn starch blend films decreased as the amount of glycerol increased. This was due to the effect of glycerol as a plasticizer which reduced the intermolecular forces between the PVA and starch particles. Tensile strength

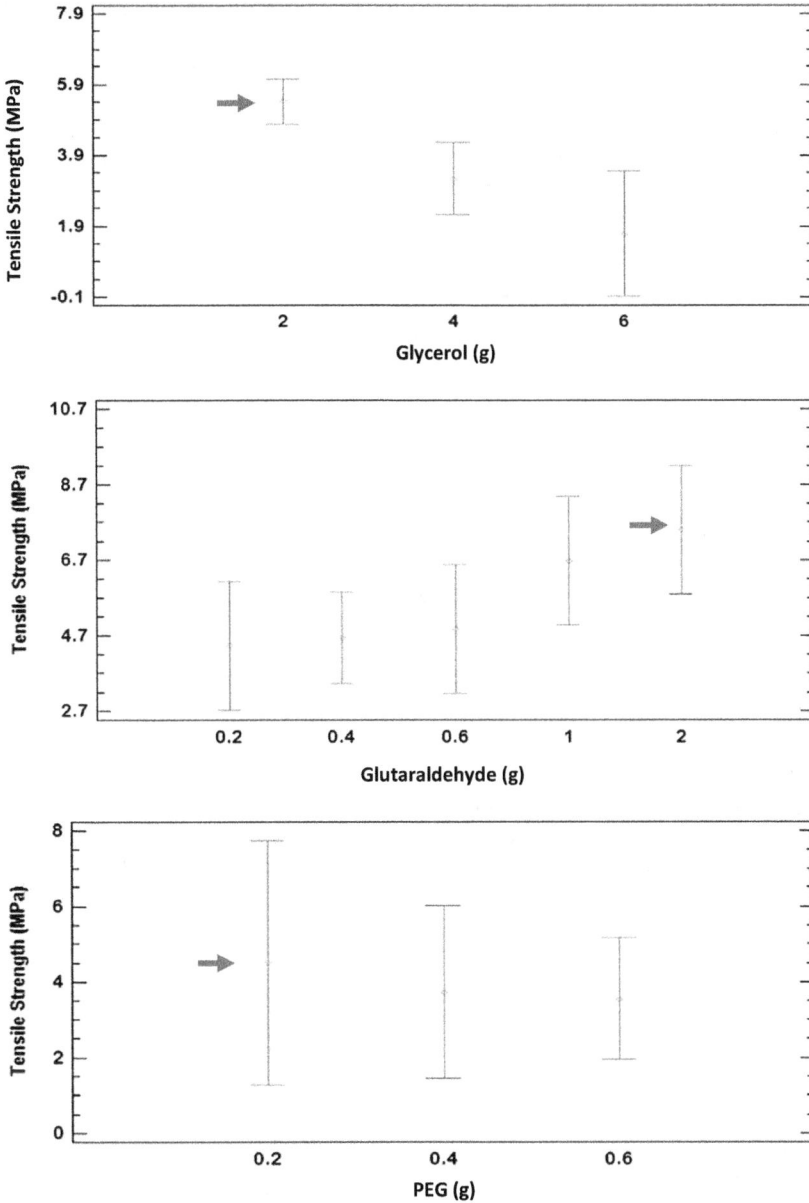

FIGURE 9.4 Tensile strength of PVA/corn starch blend films as a function of glycerol, glutaraldehyde and PEG.

of films has the highest value at the level of 2 g of glycerol. However, PEG (plasticizer) showed a different result as the tensile strength decreased because the amount of PEG increased. About 0.2 g of PEG showed the highest value of tensile strength. Arvanitoyannis et al. [46] reported that the increment of soluble starch blends and water/polyol concentrations in gelatin led to decrease the tensile strength of the blends.

Besides that, the tensile strength of films increased as the glutaraldehyde increased. Glutaraldehyde is a cross-linking agent that increases the intermolecular forces and cross-linking process between the particles; 2 g of glutaraldehyde showed the highest tensile strength of films.

Figure 9.5 shows the tensile strength of PVA/tapioca starch blend films as a function of glycerol, glutaraldehyde and PEG. For PVA/tapioca starch blend films, the optimum value of glycerol and PEG was similar to that of PVA/corn starch blend films. PVA/tapioca blend with 2 g of glycerol and 0.2 g of PEG showed the highest value of tensile strength. However, the optimum value of glutaraldehyde was different. As shown in Figure 9.5, the tensile strength of films was similar even the amount of glutaraldehyde added is increased. Thus, the optimum amount of glutaraldehyde for PVA/tapioca starch is 2.0 g.

9.3.5 CHEMICAL STRUCTURE ANALYSIS OF POLYVINYL ALCOHOL/STARCH BLEND

Numerous characteristic peaks were identified from the FTIR spectra of the PVA with corn and tapioca starch (blends 4, 7, and 10) blend films. From these FTIR spectra, every peak is corresponding to the functional groups of monomers in the polymeric chains. According to IR spectra of Figures 9.6 and 9.7, the strong and broad absorption bands of around 3,270.08 and 32,720.01 cm^{-1} of the B10 and B4 samples, respectively, indicated that there was a characteristic absorption peak of the stretching vibration of –OH among amylose chains and between amylose and amylopectin [47]. Further, C–H stretching of CH_2 was observed at 3,000 to 2,800 cm^{-1} [48]. Another peak was observed around 1,610 and 1,031 cm^{-1} in the films. Those peaks were attributed to C=O stretching of PVA residue and C–O–H groups, respectively. Around 1,008–1,136 cm^{-1} was attributed to the C–O stretching in starch and PVA chains. When comparing the blend films, it can be noticed that the –OH peaks of B7 and B10 have a similar IR peak. This was due to the similar formulation of the films which both films contained glycerol/PEG as plasticizer. While B4 used glycerol (plasticizer) and glutaraldehyde (cross-linking agent) in the films, higher absorbance was resulted from the higher concentration of the bond in the polymer. Similar peak of PVA/tapioca starch films can be observed from Figure 9.7.

9.3.6 WATER ABSORPTION BEHAVIOR OF STARCH/PVA BLEND

Water absorption can be determined by the total water absorbed in specified circumstances. It can also be defined as the amount of water absorbed by a material when immersed in water for a stipulated period of time. There are some elements that influencing water absorption which are additives used, type of plastic, temperature, and length of exposure. The absorption of moisture by organic polymeric materials causes swelling, leaching, dissolving, plasticizing, and/or hydrolyzing, thus resulting

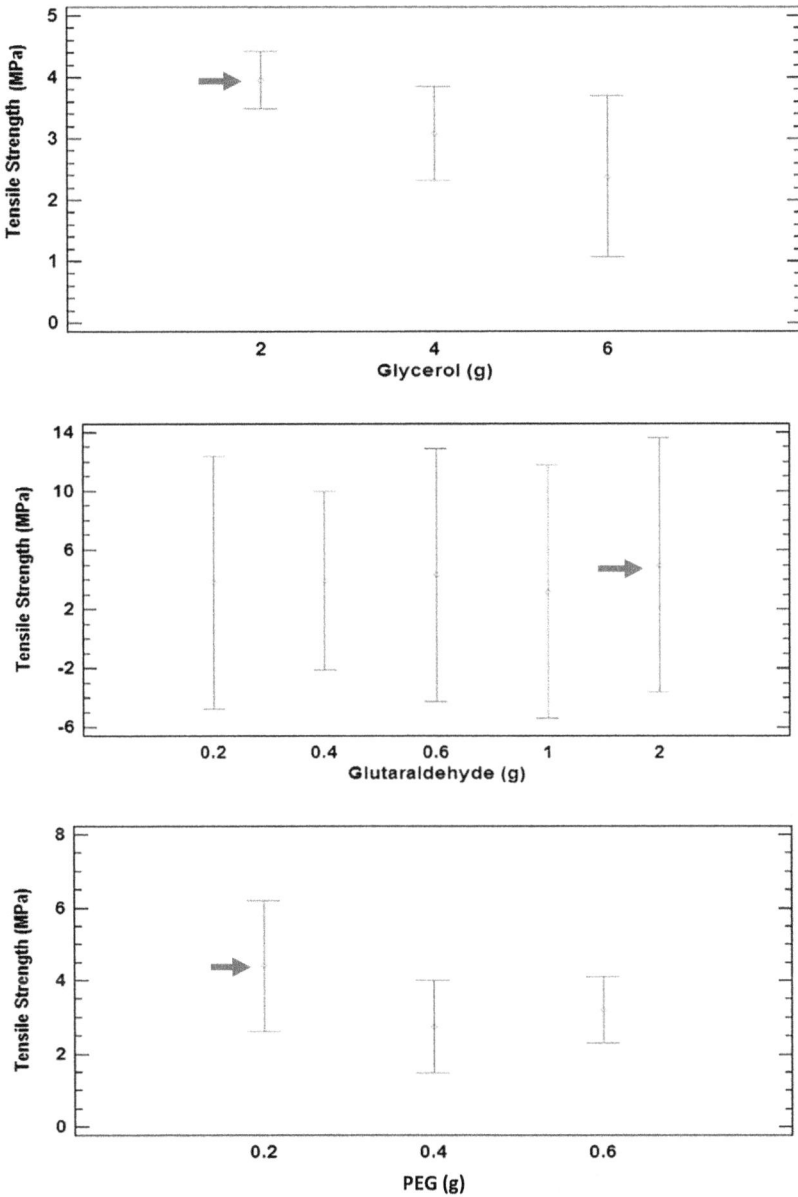

FIGURE 9.5 Tensile strength of PVA/tapioca starch blend films as a function of glycerol, glutaraldehyde and PEG.

in embrittlement, discoloration, stress cracking, and loss of mechanical and electrical properties [49,50].

Figure 9.8 shows water absorption for both types of PVA/starch blends. Blend 10 has the highest value of water absorption. This was due to the different formulation of the blend as shown in Table 9.1. Films that contain more hygroscopic plasticizers

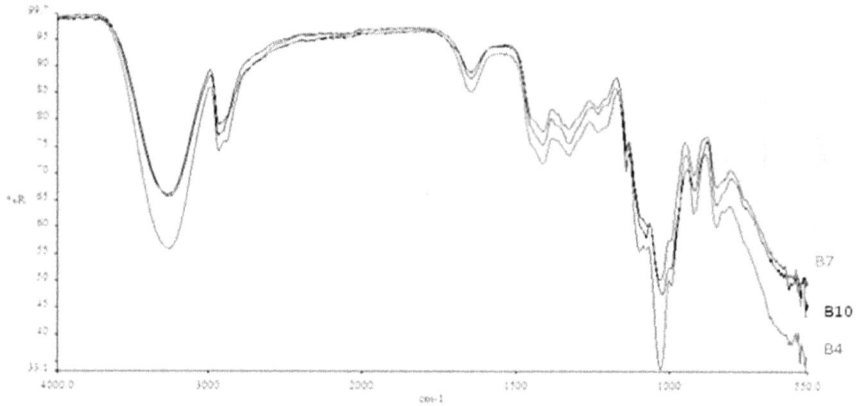

FIGURE 9.6 Intensity differences of FTIR spectrum for blend 4 (B4), blend 7 (B7), and blend 10 (B10) of PVA/corn starch blend films.

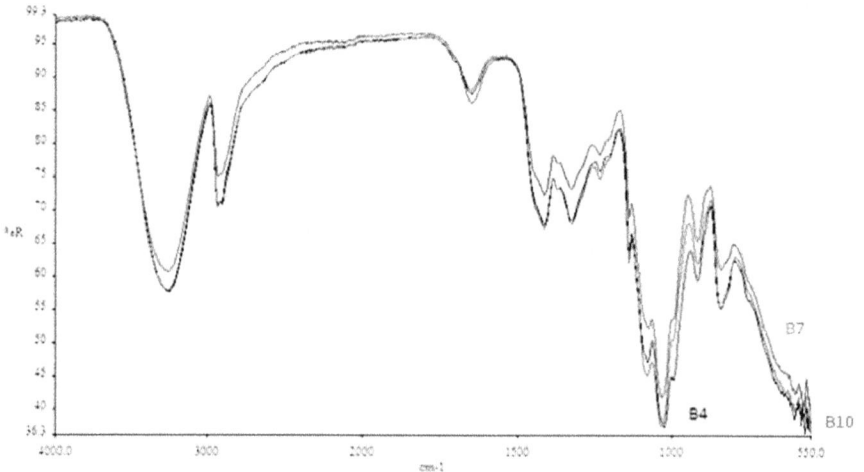

FIGURE 9.7 Intensity differences of FTIR spectrum for blend 4 (B4), blend 7 (B7), and blend 10 (B10) of PVA/tapioca starch blend films.

(glycerol) absorbed more water [44]. Therefore, blend 10 which has the highest amount of glycerol tends to absorb more water, whereas blend 4 which has less glycerol tends to absorb less water. Heterogeneous cross-linking with glutaraldehyde (blends 2, 3, 4, 11, and 12) showed more hydrophobic structures, which then interfered in the interaction with water and ions and changed their mechanic characteristics.

Besides that, PVA with tapioca starch has a higher value of water absorption compared to PVA with corn starch. This was due to the different particle size of the starch. Tapioca starch has a bigger particle size compared to corn starch [40]. Therefore, it can absorb more water than corn starch. The presence of hydroxyl groups that are available in both PVA and starch would interact with water molecules [51].

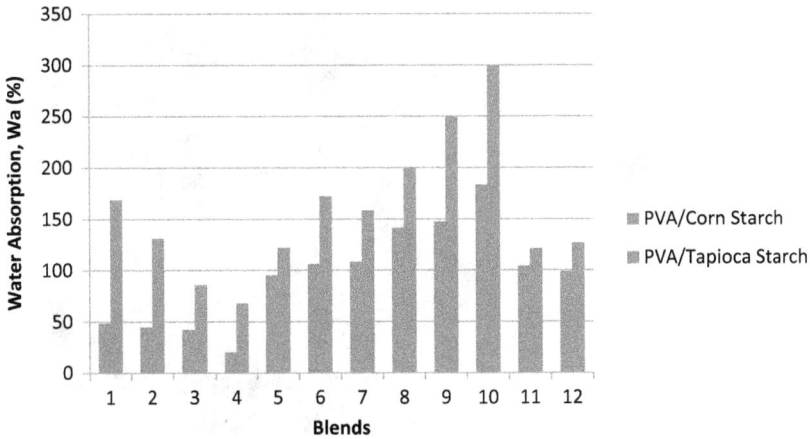

FIGURE 9.8 Water absorption test of different blends with corn and tapioca starch.

The hygroscopic nature of starch would make it prone to moisture, whereas water molecules may act as a natural plasticizer by creating the starch more flexible compared to its characteristic as a hard and rigid filler in a completely dry state [52].

9.3.7 MORPHOLOGICAL STUDIES OF POLYVINYL ALCOHOL/STARCH BLEND

The surface morphology of PVA/starch blend films was analyzed using FESEM under 300× magnification. The incompatibility and agglomeration between the PVA and the starch were evaluated from micrographs. In Figure 9.9, the PVA/starch blend showed an irregular and rough surface. The observations are consistency with the micrographs that has been analyzed by Xiong et al. [53], which indicates that PVA/starch had several unequalized pores/holes, indicating poor compatibility and miscibility of the film. However, if PVA blending is compared with both starches, PVA/corn starch (Figure 9.9a–c) has better surface morphology compared to PVA/tapioca starch (Figure 9.9d–f). It seems that PVA/tapioca blend films exhibited formation of holes along with blooming/blushing on the surface of the structure which made the surface morphology poorer compared to PVA/corn starch. This was might be due to the different particle size of the starch used in the blend. Similar findings were reported by Phetwarotai et al. [40] in their study, whereas the SEM micrograph of corn starch blends showed smaller particle size than tapioca starch blends. Therefore, PVA/corn starch blends presented better compatibility and smooth surface structure compared to PVA/tapioca starch blends. The physical properties of the thermoplastic starch are greatly influenced by the amount of plasticizer present. Figure 9.9c exhibited that PVA/corn starch film has a smooth surface structure compared to the other structures. This was due to the presence of more plasticizer content in the blend. Plasticizer reduced the intermolecular forces and increased the mobility of polymer chains [44,54]; thus, a smooth surface structure of the film can be observed. Conversely, films with the addition of glutaraldehyde showed a rough surface as shown in Figure 9.9a. The same goes for PVA/

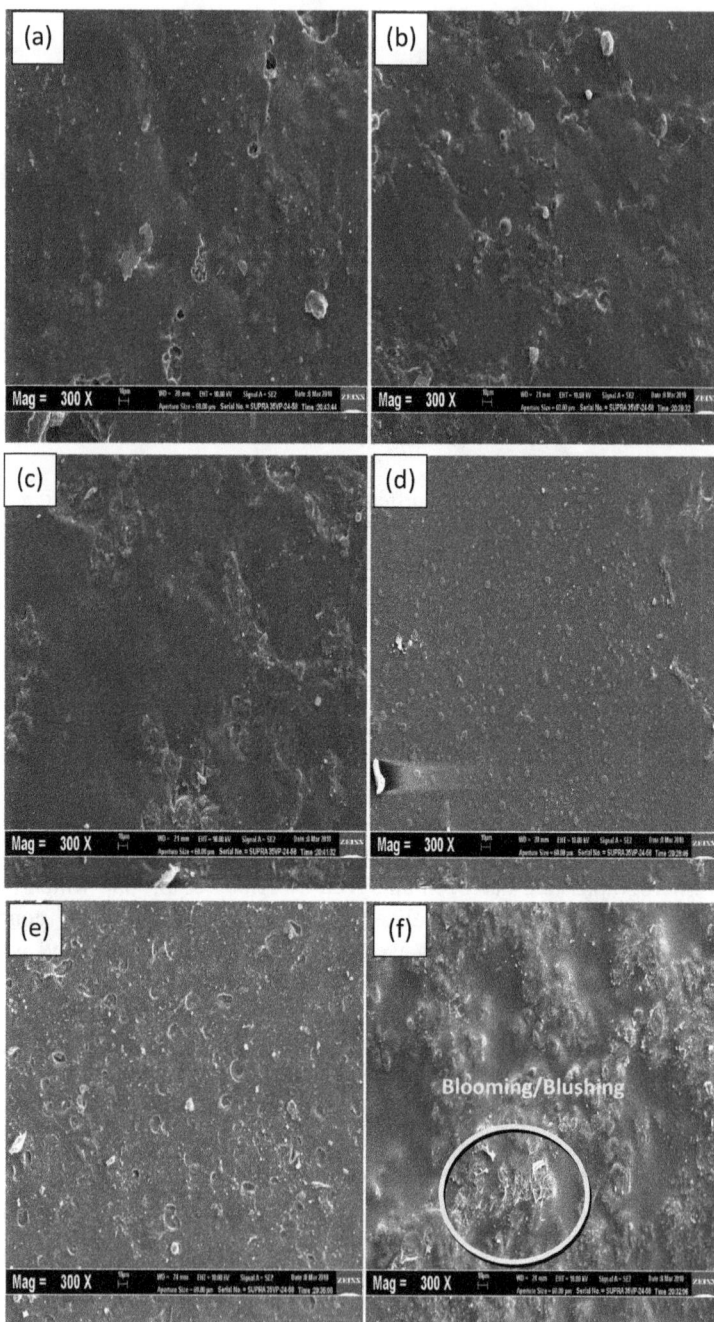

FIGURE 9.9 FESEM micrographs of PVA/corn starch (a) blend 4, (b) blend 7, (c) blend 10; and PVA/tapioca starch (d) blend 4, (e) blend 7, (f) blend 10 under 300× magnification.

tapioca films. Figure 9.9b and d shows that PVA/corn starch blend 4 has a smooth surface compared to PVA/tapioca starch blend 7.

As stated by Sakellariou et al. [55], the appearance of a white residue on PVA/ starch blend films containing plasticizers has been referred to as blooming/blushing. This occurs when the plasticizer concentration exceeds its compatibility limit in the polymer, causing phase separation and physical exclusion of the plasticizer. Figure 9.9f shows the appearance of blooming/blushing as about 6 g of glycerol was used in this blend. This amount of glycerol was exceeding its compatibility limit, which affected the surface of the film.

9.3.8 MORPHOLOGICAL STUDIES OF PVA/STARCH BLEND AFTER SOIL BURIAL TEST

Soil burial degradation test was carried out using soil taken from a local area and exposed to the environment naturally. Biodegradability and surface morphology of the films were analyzed by using SEM after 14 days of burial time. Figure 9.10 shows that there are a lot of unequalized holes observed on the surface structure. PVA/tapioca starch shown in Figure 9.10d–f presents better surface structure compared to PVA/corn starch blends shown in Figure 9.10a–c. This was due to the different type of starch blended with the PVA. It seems that PVA/corn starch films degraded faster and more badly compared to PVA/tapioca starch films. Corn starch has smaller particle size compared to tapioca starch. Therefore, the microorganisms tend to easily attack corn starch compared to tapioca starch. For PVA/corn starch, blend 10 (Figure 9.10c) showed most imperfect surface structure followed by blend 7 (Figure 9.10b) and blend 4 (Figure 9.10a). This was due to the different formulation of the blends. Blend 4 has more cross-linking agent compared to blends 7 and 10 which both films contained plasticizer instead of cross-linking agent. As mentioned earlier, cross-linking agent was used to introduce cross-linking process between the structures. Thus, blends with cross-linking agent presented better tensile properties and more resistance from the microorganism. Chiellini et al. [56] reported the effect of cross-linking agents such as glutaraldehyde on thermal properties and degradation rate of gelatin and PVA blends. The results reported by the authors demonstrated a complete and a slower degradation rate of these films.

9.4 CONCLUDING REMARKS

PVA/starch blends have shown a favorable prospective to overwhelm the limitations of biodegradable blends. The synergistic effects presented by the PVA matrices and starches were accountable for the enhancement in overall performances and properties of the blends. The addition of different types of plasticizers and cross-linking agents such as glycerol, PEG and glutaraldehyde has produced a promising result. They have provided pronounce potential in mechanical properties as well as degradation behavior. In particular, corn starch has shown an excellent result compared to tapioca starch due to efficient interaction caused by bigger surface area. However, the availability of hydrogen bonds among the starch macromolecules leads to a poor processing ability which decreases their movements during processing and limiting their use on an industrial scale. Thus, there are anxieties concerning poor stress-transfer

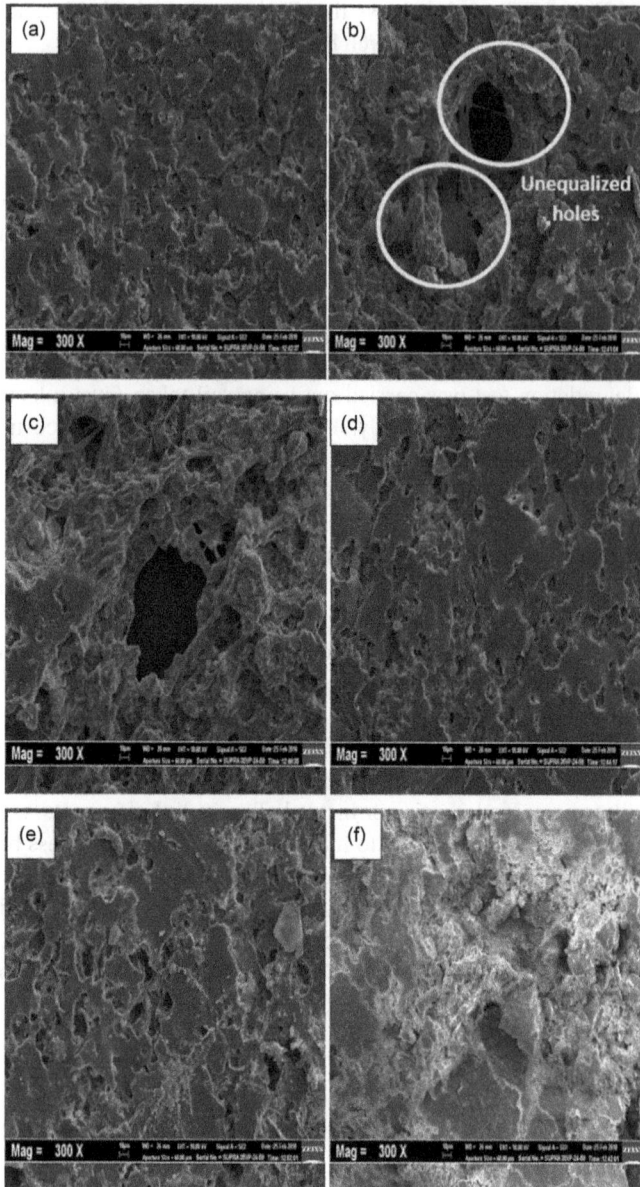

FIGURE 9.10 FESEM micrographs of PVA/corn starch (a) blend 4, (b) blend 7, (c) blend 10; and PVA/tapioca starch (d) blend 4, (e) blend 7, (f) blend 10 under 300× magnification after soil burial test conducted.

efficiencies, poor interface quality between starches, and polymer, as well as high water permeability which should be apprehensive. Until now, current studies on the properties of novel biodegradable/starch blends have someway revealed some pronounced possibility to play a positive role in the certain application.

ACKNOWLEDGMENT

The authors wish to thank Universiti Sains Malaysia (USM) Postdoctoral Fellow Scheme for their assistant and financial support.

REFERENCES

1. G. Sangeeta and K. J. Asim, "Studies on the properties and characteristics of starch-LDPE blends films using crosslinked glycerol modified starch," *European Polymer Journal,* vol. 43, pp. 3976–3987, 2007.
2. Y. J. Wang, W. Liu, and Z. Sun, "Effects of glycerol and PE-g-MA on morphology, thermal and tensile properties of LDPE and rice starch blends," *Journal of Applied Polymer Science,* vol. 92, pp. 344–350, 2004.
3. H. Ismail, A. H. Abdullah, and A. A. Bakar, "Influence of acetylation on the tensile properties, water absorption, and thermal stability of (High density polyethylene)/(soya powder)/(kenaf core) composites," *Journal of Vinyl and Additive Technology,* vol. 17, no. 2, pp. 132–137, 2011.
4. N. F. Zaaba, H. Ismail, and M. Mariatti, "Utilization of polyvinyl alcohol on properties of recycled polypropylene/peanut shell powder composites," *Procedia Chemistry,* vol. 19, pp. 763–769, 2016.
5. N. F. Zaaba, H. Ismail, and M. Mariatti, "Effect of peanut shell powder content on the properties of recycled polypropylene (RPP)/peanut shell powder (PSP) composites," *Bioresources,* vol. 8, no. 4, pp. 5826–5841, 2013.
6. N. F. Zaaba, H. Ismail, and M. Jaafar, "A study of the degradation of compatibilized and uncompatibilized peanut shell powder/recycled polypropylene composites due to natural weathering," *Journal of Vinyl and Additive Technology,* vol. 23, no. 4, pp. 290–297, 2017.
7. K. Leja and G. Lewandowicz, "Polymer biodegradation and biodegradable polymers – a review," *Polish Journal of Environmental Studies,* vol. 19, no. 2, pp. 255–266, 2010.
8. N. A. Azahari, N. Othman, and H. Ismail, "Biodegradation studies of polyvinyl alcohol/corn starch blend films in solid and solution media," *Journal of Physical Science,* vol. 22, no. 2, pp. 15–31, 2011.
9. D. R. Lu, C. M. Xiao, and S. J. Xu, "Starch-based completely biodegradable polymer materials," *Express Polymer Letters,* vol. 3, no. 6, pp. 366–375, 2009.
10. E. Schwach and L. Avérous, "Starch-based biodegradable blends: morphology and interface properties," *Polymer International,* vol. 53, no. 12, pp. 2115–2124, 2004.
11. J.-F. Zhang and X. Sun, "Mechanical properties of poly(lactic acid)/starch composites compatibilized by maleic anhydride," *Biomacromolecules,* vol. 5, pp. 1446–1451, 2004.
12. R. Narayan, "Rationale, drivers and technology examples." In K. C. Khemmani and C. Scholz, editors. *Biobased and biodegradable polymer materials.* Washington, DC: ACS Sym Ser 939, pp. 282–306, 2006.
13. B. Y. Shin and D. H. Han, "Compatibilization of PLA/starch composite with electron beam irradiation in the presence of a reactive compatibilizer," *Advanced Composite Materials,* vol. 22, no. 6, pp. 411–423, 2013.
14. K. Leja and G. Lewandowicz, "Polymer biodegradation and biodegradable polymers – a review," *Journal of Environmental Studies,* vol. 19, no. 2, pp. 255–266, 2010.
15. T. Nakashima, C. Xu, Y. Bin, and M. Matsuo, "Morphology and mechanical properties of poly(vinyl alcohol) and starch blends prepared by gelation/crystallization from solutions," *Colloid and Polymer Science,* vol. 279, no. 7, pp. 646–654, 2001.

16. T. Ishigaki, Y. Kawagoshi, M. Ike, and M. Fujita, "Biodegradation of polyvinylalcohol-starch blend plastic film," *World Journal of Microbiology and Biotechnology,* vol. 15, no. 3, pp. 321–327, 1999.

17. Z. Guohua, L. Ya, F. Cuilan, Z. Min, Z. Caiqiong, and C. Zongdao, "Water resistance, mechanical properties and biodegradability of methylated-cornstarch/poly(vinyl alcohol) blend film," *Polymer Degradation and Stability,* vol. 91, no. 4, pp. 703–711, 2006.

18. S. Tang, P. Zou, H. Xiong, and H. Tang, "Effect of nano-SiO$_2$ on the performance of starch/polyvinyl alcohol blend films," *Carbohydrate Polymers,* vol. 72, no. 3, pp. 521–526, 2007.

19. Y.-H. Yun, Y.-W. Wee, H.-S. Byun, and S.-D. Yoon, "Biodegradability of chemically modified starch (RS4)/PVA blend films: part 2," *Journal of Polymer and the Environment,* vol. 16, no. 1, pp. 12–18, 2008.

20. X. Huang and A. Netravali, "Biodegradable green composites made using bamboo micro/nano fibrils and chemically modified soy protein resin," *Composite Science and Technology,* vol. 69, no. 7, 2009.

21. G.-X. Zou, P.-Q. Jin, and L.-Z. Xin, "Extruded starch/PVA composites: water resistance, thermal properties and morphology," *journal of Elastomers and Plastics,* vol. 40, no. 4, pp. 303–316, 2008.

22. H. Wang, X. Sun, and P. Seib, "Strengthening blends of poly(lactic acid) and starch with methylenediphenyl diisocyanate," *Journal of Applied Polymer Science,* vol. 82, no. 7, pp. 1761–1767, 2001.

23. S. Kalambur and S. S. H. Rizvi, "An overview of starch-based plastic blends from reactive extrusion," *Journal of Plastic Film and Sheeting,* vol. 22, no. 1, pp. 39–58, 2006.

24. R. Mani, J. Tang, and M. Bhattacharya, "Synthesis and characterization of starch-graft-polycaprolactone as compatibilizer for starch/polycaprolactone blends," *Macromolecular Rapid Communications,* vol. 19, pp. 283–286, 1998.

25. Halimatudahliana, H. Ismail, and M. Nasir, "The effect of various compatibilizers on mechanical properties of polystyrene/polypropylene blend," *Polymer Testing,* vol. 21, no. 2, pp. 163–170, 2002.

26. P. Mischnick and G. Kuhn, "Model studies methyl amyloses: correlation between reaction conditions and primary structure," *Carbohydrate Research,* vol. 290, pp. 199–207, 1996.

27. T. Kondo, C. Sawatari, R. S. J. Manley, and D. G. Gray, "Characterization of hydrogen bonding in cellulose-synthetic polymer blends systems with regioselectively substituted methycellulose," *Macromolecules,* vol. 27, no. 1, pp. 210–215, 1994.

28. Z. Liu, Y. Dong, H. Men, M. Jiang, J. Tong, and J. Zhou, "Post-crosslinking modification of thermoplastic starch/PVA blend films by using sodium hexametaphosphate," *Carbohydrate Polymers,* vol. 89, pp. 473–477, 2012.

29. B. Sreedhar, M. Sairam, D. K. Chattopadhyay, S. Rathnam, and M. Rao, "Thermal, mechanical, and surface characterization of starch–poly(vinyl alcohol) blends and borax-crosslinked films," *Journal of Applied Polymer Science,* vol. 96, pp. 1313–1322, 2005.

30. K. Pal, A. K. Banthia, and D. K. Majumdar, "Effect of heat treatment of starch of starch on the properties of the starch hydrogels," *Materials Letters,* vol. 62, no. 2, pp. 215–218, 2008.

31. B. Ramaraj, "Crosslinked poly(vinyl alcohol) and starch composite films. II. Physicomechanical, thermal properties and swelling studies," *Journal of Applied Polymer Science,* vol. 103, no. 2, pp. 909–916, 2007.

32. P. Marques, A. Lima, G. Bianco, J. Laurindo, R. Borsali, and J. Le Meins, "Thermal properties and stability of cassava starch films crosslinked with tetraethylene glycol diacrylate," *Polymer Degradation and Stability,* vol. 91, pp. 726–732, 2006.

33. C. Bastioli, V. Bellotti, L. D. Giudice, and G. Gilli, "Mater-bi: properties and biodegradability," *Journal of Environmental Polymer Degradation,* vol. 1, pp. 181–191, 1993.

34. T. Iwanami and T. Uemura, "Properties and applications for packaging of starch/modified PVA alloys properties and positive use of biodegradable new polymer 'Mater-bi'," *Polymer Preprints Japan,* vol. 41, pp. 2210–2212, 1992.

35. C. N. Tudorachi, M. Cascaval, M. Rusu, and M. Pruteanu, "Testing of polyvinyl alcohol and starch mixtures as biodegradable polymeric materials," *Polymer Testing,* vol. 19, pp. 785–799, 2000.

36. A. Cano, M. Cháfer, A. Chiralt, and C. González-Martínez, "Physical and antimicrobial properties of starch-PVA blend films as affected by the incorporation of natural antimicrobial agents," *Foods (Basel, Switzerland),* vol. 5, no. 1, p. 3, 2015.

37. M. T. Taghizadeh and N. Sabouri, "Thermal degradation behavior of polyvinyl alcohol/starch/carboxymethyl cellulose/clay nanocomposites," *Universal Journal of Chemistry,* vol. 1, no. 2, pp. 21–29, 2013.

38. B. Imre and B. Pukánszky, "Compatibilization in bio-based and biodegradable polymer blends," *European Polymer Journal,* vol. 49, no. 6, pp. 1215–1233, 2013.

39. I. M. Thakore, S. Desai, B. D. Sarawade, and S. Devi, "Studies on biodegradability, morphology and thermomechanical properties of LDPE/modified starch blends," *European Polymer Journal,* vol. 37, no. 1, pp. 151–160, 2001.

40. W. Phetwarotai, P. Potiyaraj, and D. Aht-Ong, "Properties of compatibilized polylactide blend films with gelatinized corn and tapioca starches," *Journal of Applied Polymer Science,* vol. 116, pp. 2305–2311, 2010.

41. Md. S. Hossen, I. Sotome, M. Takenaka, S. Isobe, M. Nakajima, and H. Okadome, "Effect of particle size of different crop starches and their flours on pasting properties," *Japan Journal of Food Engineering,* vol. 12, no. 1, pp. 29–35, 2011.

42. F. O. Faria, A. E. S. Vercelheze, and S. Mali, "Physical properties of biodegradable films based on cassava starch, polyvinyl alcohol and montmorillonite," *Química Nova,* vol. 35, no. 3, pp. 487–492, 2012.

43. X. V. Cao, H. Ismail, A. A. Rashid, T. Takeichi, and T. V. Huu, "Mechanical properties and water absorption of kenaf powder filled recycled high density polyethylene/natural rubber biocomposites using mape as a compatibilizer," *Bioresources,* vol. 6, no. 3, pp. 3260–3271, 2011.

44. M. G. A. Vieira, M. A. da Silva, L. O. dos Santos, and M. M. Beppu, "Natural-based plasticizers and biopolymer films: a review," *European Polymer Journal,* vol. 47, no. 3, pp. 254–263, 2011.

45. D. F. Parra, C. C. Tadini, P. Ponce, and A. B. Lagäo, "Mechanical properties and water vapor transmission in some blends of cassava starch edible films," *Carbohydrate Polymer,* vol. 58, pp. 475–481, 2004.

46. I. Arvanitoyannis, E. Psomiadou, A. Nakayama, S. Aiba, and N. Yamamoto, "Edible film made from gelatin, soluble starch and polyols," *Food Chemistry,* vol. 60, pp. 593–604, 1997.

47. H. Judawisastra, R. D. R. Sitohang, L. Marta, and Mardiyati, "Water absorption and its effect on the tensile properties of tapioca starch/polyvinyl alcohol bioplastics," *IOP Conference Series: Materials Science and Engineering,* vol. 223, no. 012066, pp. 1–10, 2017.

48. A. M. Shehap, "Thermal and spectroscopic studies of polyvinyl alcohol/sodium carboxy methyl cellulose blends," *Egyptian Journal of Solids,* vol. 31, no. 1, pp. 75–91, 2008.

49. M. Asem, W. M. F. W. Nawawi, and D. N. Jimat, "Evaluation of water absorption of polyvinyl alcohol starch biocomposite reinforced with sugarcane bagasse nanofibre: optimization using two-level factorial design," *IOP Conference Series: Materials Science and Engineering,* vol. 368, no. 012005, pp. 1–10, 2018.

50. H. Ismail and N. F. Zaaba, "The mechanical properties, water resistance and degradation behaviour of silica-filled sago starch/PVA plastic films," *Journal of Elastomers and Plastics,* vol. 46, no. 1, pp. 96–109, 2014.

51. G. H. Yew, A. M. Mohd Yusof, Z. A. Mohd Ishak, and U. S. Ishiaku, "Water absorption and enzymatic degradation of poly(lactic acid)/rice starch composites," *Polymer Degradation and Stability,* vol. 90, no. 3, pp. 488–500, 2005.

52. D. Preechawong, M. Peesan, P. Supaphol, and R. Rujiravanit, "Characterization of starch/poly (ε-caprolactone) hybrids foams," *Polymer Testing,* vol. 23, no. 6, pp. 651–657, 2004.

53. H. Xiong, S. Tang, P. Zou, and H. Tang, "Effect of nano-SiO_2 on the performance of starch/polyvinyl alcohol blend films," *Carbohydrate Polymers,* vol. 72, no. 3, pp. 521–526, 2008.

54. U. Shah, A. Gani, B. A. Ashwar, and A. Shah, "A review of the recent advances in starch as active and nanocomposite packaging films," *Cogent Food & Agriculture,* vol. 1, no. 1, p. 1115640, 2015.

55. P. Sakellariou, R. C. Rowe, and E. F. T. White, "An evaluation of the interaction and plasticizing efficiency of the polyethylene glycols in ethylcellulose and hydrokypropyl methylcellulose films using the torsional braid pendulum," *International Journal of Pharmaceutics,* vol. 31, pp. 55–64, 1986.

56. E. Chiellini, A. Corti, and R. Solaro, "Biodegradation of poly(vinyl alcohol) based blown films under different environmental conditions," *Polymer Degradation and Stability,* vol. 64, p. 305, 1999.

10 Ubi gadong (*Dioscorea hispida*) as Potential Biocomposite Material
A Comprehensive Review

K.Z. Hazrati, S.M. Sapuan and M.Y.M. Zuhri
Universiti Putra Malaysia

R. Jumaidin
Universiti Teknikal Malaysia Melaka

CONTENTS

10.1 INTRODUCTION: NATURAL FIBER COMPOSITES

Over the years, the utilization of bio-fibers as reinforcement in composites has become renowned in engineering applications due to the rising environmental concerns and the obligation for growing sustainable materials. The US Department of Agriculture and the US Department of Energy have indicated the aim is that at least 10% of all building constructions are from bio-based sources in 2020 and should then be expanded to 50% by 2050 [1]. The exponential growth in bio-composites

TABLE 10.1

Types of Plant-Based Natural Fibers and Their Advantages

Types of Plant Fibers	Advantages of Natural Fibers
• Basf fiber – flax, hemp, kenaf	• Low density
• Leaf fiber – pineapple, sisal, abaca	• Low cost
• Seed fiber – cotton, milkweed	• Renewability
• Stalk fiber – wheat, rice, barley	• Recyclability
• Grass fiber - bamboo, bagasse	• Low energy consumptions
• Wood fiber – hard, soft wood	• Wide distribution
	• CO_2 neutral
	• No health risk when inhale

Source: Adapted from [3].

shows that their broad application will take place in future as they will be the next generation of renewability materials. Bio-fiber materials are economical, biodegradable, and generate less greenhouse gas emissions with low density [2].

Natural fibers are categorized according to their origin, from either plants, animals, or minerals [4]. Despite that, natural fibers from plants are the most popular reinforcement materials in composites. Table 10.1 shows various types of plant fibers which include basf, leaf, seed, stalk, grass, and wood. Plant fibers can be classified based on their types or parts extracted [5]. Recent studies have largely been exploratory on the usage of non-wood fibers such as ubi gadong (*D. hispida*), kenaf (*Hibiscus cannabinus*), sugar palm fiber (*Arenga pinnata*), sugarcane (*Saccharum officinarum*), sago palm (*Cycas revoluta*), pineapple leaf (*Ananas comosus*), and oil palm fiber (*Elaeis guineensis*). Currently, numerous studies on natural fibers have reported their convenient and potential fabrication for new properties in composite products [6].

Several studies on natural fibers have shown significant increases in properties which may vary depending on their classified species, geographical location, method of fabrication, etc. [4]. Sanyang et al. [7] had found that natural fibers extracted from a plant, commonly referred to as lignocellulose fibers, possess different structural parameters from one plant to another. The structure of natural fibers consists of primary and secondary cell walls created from three basic components: cellulose, hemicellulose, and lignin [8]. It has been noted that the mechanical properties will be influenced by the cellulose content, which are different aspects of fiber loading or volume fraction of fibers, fiber length, fiber orientation, and fiber adhesion between the fiber-matrix [9].

10.2   *DIOSCOREA HISPIDA*

10.2.1   History of *Dioscorea hispida*

Ubi gadong, scientifically known as *D. hispida*, is a seasonal herb that originates and grows in tropical areas such as China, India, Thailand, Myanmar, and Indonesia. With tubular and thorny stems, it may grow up to 20 m in height. It has been estimated that there are about 600 *Dioscorea* species around the world which can be consumed using various cooking methods such as boiling, frying, and baking. It is a wild tuber plant that possesses a tuber-shaped pole with the range of 5–20 cm in length. It comes in various shapes, such as round or oval, with yellowish-brown skin. The skin thickness is within 0.15–0.3 cm, as shown in Figure 10.1 [10]. Previous research has documented the genus of *D. hispida* as belonging to the Dioscoreaceae family which includes many other types of species such as *D. daemona*, *D. hirsute*, *D. lunata*, and *D. virosa*.

This plant is known as "Ubi gadong" among the Malaysians and is not commonly consumed by today's society due to its toxic nature. It consists of alkaloid toxins of low molecular weight known as *dioscorea*. *D. hispida*-based foods have a complicated preparation process since it contains dioscorine toxins which can be harmful to health [11] (Figure 10.2).

Recent work by historians had established that *D. hispida* is an herbal plant with lobed and poisonous tubers and a yellowish-white flesh. The plant has a prickly stem and trifoliate leaves with attributes such as rapid growth, rapid maturity, considerable

FIGURE 10.1   *D. hispida* plants.

FIGURE 10.2   *D. hispida's* irregular shaped tubers with stiff rootlets.

distance between tubers and soil surface, and easy harvest [12]. There are quick and simple methods to remove dioscorine before it is manufactured as a food product. The sonication method is not only quick and easy, but also the level of hygiene during the handling process and the quality of the unoxidized tubers are better compared to the traditional method. The use of *D. hispida* is not only limited to food sources but also widely used especially in traditional medicine, cosmetics, and as a basic ingredient in the starch and pharmaceutical industry [13].

In Malaysia, there are 19 species of *Dioscorea* such as *D. hispida*, *D. piscatorum*, and *D. alata*. Several investigations have revealed that each species is different in every physical aspect and is commonly known as from the *Dioscorea* species since it contains dioscorine toxic substances. However, this tuber plant is currently an abandoned species due to being poisonous, as well as causing damage to the forest or human habitations [14]. Detoxification can be accomplished by watering the bulbs from 1 to 2 weeks or by the process of heating or drying under the sun [12]. The bulb can be peeled, sliced, and soaked underwater for seven days. It can also be soaked in a salt solution for five days before being dried. The author has reported the distribution of *D. hispida* in several tropical and subtropical areas such as within Peninsular Malaysia, in Terengganu. Terengganu is an exceptional state where beaches stretch across from the north to the south regions [14] (Table 10.2).

TABLE 10.2
Removing Toxicity from *D. hispida*

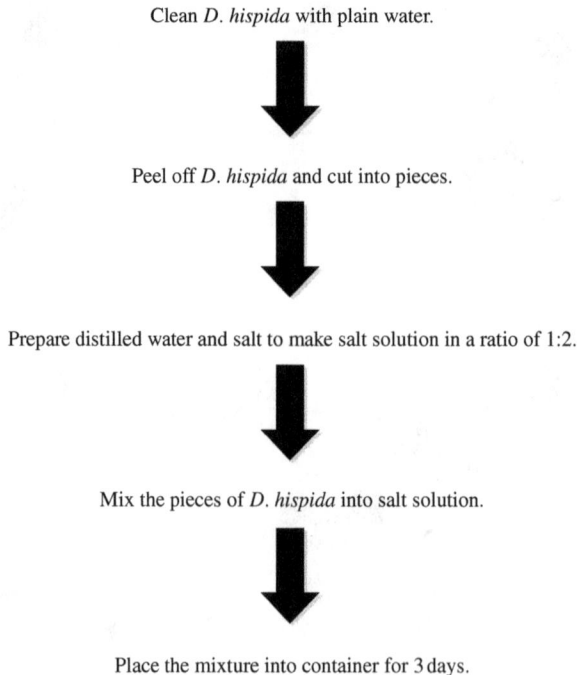

Clean *D. hispida* with plain water.

⬇

Peel off *D. hispida* and cut into pieces.

⬇

Prepare distilled water and salt to make salt solution in a ratio of 1:2.

⬇

Mix the pieces of *D. hispida* into salt solution.

⬇

Place the mixture into container for 3 days.

(Continued)

TABLE 10.2 (*Continued*)
Removing Toxicity from *D. hispida*

Clean the pieces of *D. hispida,* and leave them in flowing water for 3 days.

Place pieces of *D. hispida* into oven/dryer for 24 h at 65°C.

Source: Adapted from [15].

The traditional method for removing the poison requires at least four days of processing, whereby the *D. hispida* is cut into small pieces before being placed and soaked in flowing water (such as a river or canal) to eliminate the effects of poison. In addition, the tubers harbor an unpleasant odor due to the fermentation process that occurs during the soaking process. Dioscorine is a type of toxic alkaloid found in *D. hispida* and is usually concentrated in the skin area [16] (Table 10.3).

10.2.2 CHEMICAL COMPOSITION OF *DIOSCOREA HISPIDA*

Several studies have documented various lignocellulosic materials in the tuber of *D. hispida*, such as cellulose, hemicellulose, and complex polymer of lignin. It has been reported that cellulose molecules are the main structure of a plant, whereas hemicellulose and lignin carry out fibers of the plant [17]. The plant species are influenced by the content of lignocellulose since it has a diverse genetic composition in the plant. Currently, a number of students have shown that lignocellulosic contents of a plant are 35%–50% cellulose, 20%–35% hemicellulose, and 15%–25% lignin [18] (Figure 10.3).

Thus far, previous studies have indicated that the chemical composition of *D. hispida* may vary, not just relying on the plant components, plant age, weather growing environment, soil condition, and plant height. It has been thought that phenolic compounds found in *D. hispida* tuber peels have several acidic contents such as chlorogenic acid, *p*-hydroxybenzaldehyde, methylester of protocatechuic acid, and caffeic acid. However, the only methylester of protocatechuic acid was discovered in the tuber's flesh [19]. According to [16], lignin has a stable structure, while the hemicellulose component is the fastest lignocellulosic content to combust compared to lignin which is hard to decompose. The process of biomass pyrolysis includes four stages: the first stage is called thermolysis as moisture progresses, and the second stage is hemicellulose decomposition, followed by the third stage which is cellulose decomposition, and the last stage which is lignin decomposition. All stages

TABLE 10.3
Vernacular Names of Ubi gadong *(D. hispida)*

Country	Vernacular Names of *D. hispida*
Chinese	Bai Shu Liang
Burmese	Kywe
Malaysia	Gadog, Gadong, Gadong Lilin, Gadong Mabok, Taring Pelanduk, Tuba Ubi, Ubi Arak, Ubi Akas, Ubi Cerok, Ubi Kendudok, Ubi Kipas, Ubi Nasi (Malay), Ha, Ha-U (Pagan), Bigap, Gadongan, Gang, Gong, Kedut, Kuoi, Kuoe, Ki-E, Sulur Gadong (Sakai), Ha, Ha'u, Hubi Gak (Semang), Bekoi, Bekoya, Gakn, Ubi Bekoi (Tembe)
French	Igname Épineuse Amčre, Morsure De Cobra
India	Bolkande, Podava-Kizhangu, Podava Kelengu, Podukkilangu, Venni (Malayalam), Baichandi, Bhul Kand, Dukar Kand (Marathi) Baikanda, Banya Alu, Bainya Alu, Hasar Sanga, Khulu Sanga, Kolhua, Kolokanda, Kulia Kulia Kanda (Oriya), Hastyaluka, Marpa Shpoli (Sanskrit), Kavalakodi, Pei Perendai, Periperendai (Tamil), Chanda Gadda, Puli Dumpa, Tellaagini Geddalu, Thella Chanda Gadda, Thella Gadda (Telugu), Magasirigadda, Peccheruvu (Andhra Pradesh)
German	Bittere Yamswurzel, Bittere Yamswurzel, Giftige Yams
Laos	Houo Koi
Japanese	Mitsuba Dokoro
Indonesia	Ghadhung (Madurese), Gadung, Gadung Ketan, Gadung Kuning, Gadung Padi
Philippines	Mamo (Bikol), Gigos, Kalot, Orkot (Bisaya), Karot (Iloko), Bagai (Manobo)
Taiwan	Bai, Da
Thailand	Khli, Koi, Kloi, Kloi-Nok, Kloy Kao Niaw, Man Kloi
Vietnamese	Cû Năn, Cû Nân, Cû Nê, Cû Nâu Trăng, Dây Năn, Mài Lông

Source: Adapted from [11].

FIGURE 10.3 Fibers of *D. hispida*.

reported depend on the temperature for decomposing. Most studies on future sources of energy will benefit in understanding the chemical conversion of biomass [20].

Table 10.4 shows that the chemical composition of the cellulose in the stockpile sample is higher compared to the fresh sample [16]. It has been observed that the stockpile sample contains a much harder cell wall than the fresh sample. The solvent ethanol benzene is proficient for extracting the cellulose compared to ethanol toluene and water. The lignin content was greater for the stockpile sample compared to the fresh sample, demonstrating that the ethanol benzene solvent extracted the highest percentage of lignin composition. The difference in percentage between the stockpile sample and fresh sample is due to the lignification of the cell wall in the tuber [18]. Tattiyakul and Naksriarporn had identified that lignin can be applied in the bio-plastic industry due to its eco-friendly features [21].

Recent studies have reported that the contents of cellulose in stockpile *D. hispida* are increasing as the number of storage period increases. This is due to the wall expanding and lignifying, which indirectly affects the cellulose content [22]. According to Bhandari and Kawabata [23], *D. hispida* has a high content of cellulose with vast potential in the ethanol production industry. Indirectly, this is encouraging for sustainable energy manufacturer.

10.2.3 PHYSICOCHEMICAL PROPERTIES OF *DIOSCOREA HISPIDA* TUBERS

D. hispida tuber has been used as the main food source during World War II and can easily be found in Asian regions, especially in tropical and subtropical areas. The starch contained in this food encourages a slow digestion in the human gastrointestinal tract, and it is also gluten-free [24]. Tattiyakul and Naksriarporn found that the tuber of *D. hispida* contains a toxic alkaloid called dioscorine which can be harmful if the raw tuber is consumed [21]. To date, several studies have investigated the process of removing the toxic substances. It has been suggested that the tuber should undergo several steps such as getting soaked in flowing water for up to 7 days, or soaked in salted water for 5 days before being dried [15] (Table 10.5).

TABLE 10.4
Chemical Composition of *D. hispida* Tubers

Sample	Solvent	Cellulose (%)	Hemicellulose (%)	Lignin (%)
Fresh *D. hispida*	Ethanol:toluene	44.6	13.4	7.0
	Ethanol:benzene	48.2	14.7	9.3
	Water	39.3	12.7	6.3
Stockpile *D. hispida*	Ethanol:toluene	54.7	14.9	9.6
	Ethanol:benzene	59.9	15.9	10.3
	Water	46.9	13.4	8.6

Source: Adapted from [16].

TABLE 10.5

Physicochemical Composition of *D. hispida* Tuber and Flour (100 g Basis)

Component	*D. hispida* Tuber	*D. hispida* Flour
Carbohydrate (g)	18	92.3
Protein (g)	1.81	1.1
Fat (g)	1.6	0.9
Fiber (g)	1.9	2.4
Ash (g)	0.7	0.3
Moisture (%)	77	3

Source: Adapted from [24].

Previous studies on tuber flour have demonstrated that the raw material of starches has been used to produce various products. Starch is applied in numerous functions and has several capabilities such as stabilizing, gelling, thickening, film forming, moisture retention, and long shelf life [25]. Harju found that the development of natural starch-based products is the main factor for creating water-resistant materials by using the starch modification method [26].

According to this study, *D. hispida* solid waste production can be useful for providing a cheap substrate with respect to the manufacture of bio-products. By using different application ranges, better yields with high starch can be obtained with good quantity and thick peel. Several investigations of factors that influence the yield of starch have been identified such as the drying of raw materials in the oven compared to drying under the sun's heat [27].

10.3 APPLICATION OF *DIOSCOREA HISPIDA*

There are various applications of biodegradable polymer materials, e.g., used as sources of food and in packaging, medicine, and agriculture. The awareness of environmental responsibility through the good development of biopolymer materials is increasing in regard to industries as well as consumers.

10.3.1 *DIOSCOREA HISPIDA* AS A SOURCE OF FOOD

D. hispida produces good quality starch since it has an available source of carbohydrates content. Furthermore, the tubers contain protein, fiber, ash, dioscorine, water, and fat, as shown in Table 10.6 [28]. Recent studies had proven that tubers have a low shelf life because of the maximum water content, making them easy to spoil. It was reported that the increasing water content level will make microorganisms grow easily. An attempt to reduce the harm is by making a beneficial product, such as flour, using the tubers. *D. hispida* tuber flour is convenient for consumption because it has small amounts of water content, thereby giving it a longer shelf life. Additionally, the flour can be used to make bread, cake, noodles, and many types of meals. The tuber flour is produced using the drying technique following a specified temperature using

TABLE 10.6
Nutrients of *D. hispida* (100 g)

Nutrient	Value of Composition
Calorie	101 cal
Water	78%
Carbohydrate	18%
Protein	1.81%
Fiber	0.93%
Ash	0.69%
Dioscorine	0.41%
Fat	0.16%

Source: Adapted from [28].

an oven or naturally by sunlight. Consequently, the shape, physical and chemical properties, degradation, etc. are affected [20]. In 2013, Jambi found that during the drying process, the optimal temperature must be used since it affects the nutrient contained within as well as the shelf life of the tuber flour [28].

Bhandari and Kawabata had revealed that *D. hispida* tubers can be eaten after undergoing the extraction process to remove the toxic alkaloid dioscorine [23]. The cooking methods to prepare it are boiling, frying, pressure boiling, or baking. According to *Laporan Program Penyelidikan Ubi Gadong 2010 (Dioscorea hispida* Dennst.), many people stated that the taste of *D. hispida* after cooking is better than the taste of other species [29]. They know that the tubers must be detoxified before consumption; otherwise, food poisoning will result. Some people know that to overcome this poisoning, plenty of coconut water should be consumed. A great deal of previous research on *D. hispida* is present for making traditional local foods such as *kuih puteri mandi*, *kuih onde-onde*, and *pengat*. Many people eat glutinous rice and grated coconut together with detoxicated yam during breakfast or rainy season [29].

Currently, a number of studies have demonstrated how to identify and test for detoxification on yam by feeding it to domestic animals such as goats, chickens, and fish. Maneenoon et al. reviewed the literature and found some evidence for this detoxification from the Sakai people in Thailand who use the boiling method to remove the toxin before eating. It was reported that people in Papua New Guinea also apply the slice and boil technique to remove the toxin [30].

10.3.2 *DIOSCOREA HISPIDA* AS NATURAL MEDICINE

According to the World Health Organization (WHO), almost 80% of the world population use *D. hispida* as a traditional medicine and for producing natural products for maintaining individual health. *D. hispida* is an herbal plant in tropical *Dioscoreaceae* families and consists of alkaloids, saponins called "dioscorine," flavonoid, tannins, carbohydrates, proteins, glycosides, phenol, steroid, and phytosterols [31]. To date, the research has been concerned with diosgenin which is the main element for synthesizing steroid drugs. In other ways, it can be used as a painkiller or rheumatic drug. In the Philippines, tubers are used as expectorant and adjuvant in circumstances of tumors or swollen lymph nodes. They are also used against arthritic and rheumatic pain cases [32]. Burkill et al. reported that tubers are used for external application as medicine for irritated feet or killing worms in harmful areas [33]. The literature has emphasized the importance of *Dioscorea* species and stated that the mucus from the tubers can heal warts, treat fevers and asthma, and can also be used to poison fish. In 2018, researchers reported that there are a variety of natural medicines used in the local market to cure diabetes, high blood pressure, cough, kidney, injury, and cancer, and also to maintain health [34]. It is believed that natural medicine has numerous advantages based on the type of disease treated [35].

Nashriyah et al. have found that the leaves of the plant or the battered leaves can be used to cure diseases, applied as anti-parasite medicine, or placed on the stomach for bloating [31]. The tubers of the plant are applied to treat arthritis pain, heartburn, vomiting, or used as a laxative. Studies have investigated the plant's benefits when applied to cure hernia, rheumatism, swelling, and asthma illnesses. According to Roslan et al., *D. hispida* can also be used as fish poison, pesticide, and bait by the village people to make it easy for them to trap fish [36]. Pharmacological studies have identified the biological aspects of tubers extracted from the plant which displayed promising features such as anticancer, antimicrobial, antioxidant, analgesic, anti-inflammatory, thrombolytic activities, and cardioprotective agents. Recent studies have been accomplished to determine the different quantities of the plant to identify the bioactive composition. Current studies highlighted the biological properties of *D. hispida* based on varied fraction composition. It is often recommended by traditional medicine since it consists of several components and possesses moderate active pharmacological effects [37].

10.4   STARCH

10.4.1   Properties of Starch

In the past two decades, many researchers have sought to determine the uses of starch, the main component of carbohydrates stored in natural plants, and fiber which contains cellulose [38]. Starch can be categorized in natural biopolymer's comprehensive growth in roots, tubers, leaves, fruits, and seed. In the midst of biopolymers, starch is deemed as the most encouraging due to its features of biodegradability, availability, renewability, and low-cost consumption [39]. It is now well established that the popular development of starches is typically extracted from natural sources such as from tubers, cereals, roots, and legumes [40]. Recent studies have identified the common commercial starch extracted from another source of stems, such as sago palm and sugar palm starches. The production of sago starch has been investigated by Sahari et al. who determined that sago is fractioned from two heights: the base (1 m from the bottom) and middle (5 m from the bottom) of the sago trunk [41].

A large and growing body of literature had investigated starch constructed from a combination of two major macromolecular components, linear amylose and amylopectin, as shown in Figure 10.4. The studies have reported that amylose is the principal

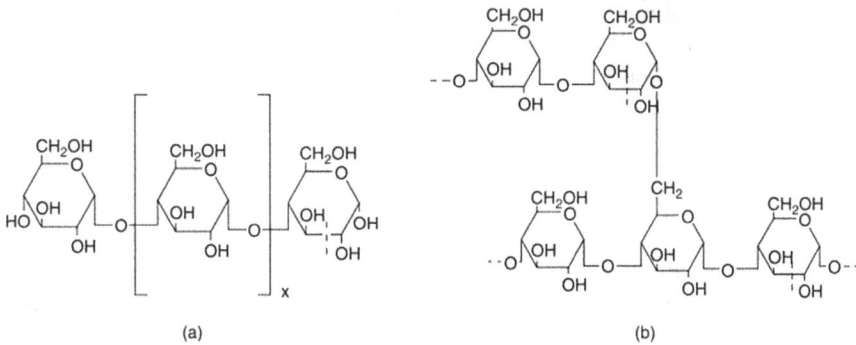

FIGURE 10.4 Unit structure of (a) amylose and (b) amylopectin. (Adapted from Pérez et al. [42].)

linear component and has a degree of polymerization (DP) reaching 600. These molecules contain around 200–20,000 glucose units which shape a helix resulting in bond angles within the glucose units [42]. Amylopectin is the main component of the granule and is well known as the largest natural polymer. It is a strongly branched polymer that consists of slight side chains of 30 glucose units fixed together with 20–30 glucose units throughout the chain. The amylopectin molecules may consist of up to 2 million glucose units [43].

A considerable amount of literature has been published on starches from various sources. It was reported that the starch granules are heterogeneous due to their size, shape, and molecule constituents [44], and the ratio of polysaccharides was identified starch properties [45]. The majority of amylose molecules are created in a linear chain, and their molecular weight, including their DP, can diverge which will influence the mixture's viscosity. Furthermore, processing method and recrystallization are significant for the quality of the outcome [46]. However, amylopectin indicates a high molecular weight with the widely branched polysaccharide element of starch and contains numerous short-chain-shaped α-D-glucopyranosyl residues through (1→4) linkages. Due to the widely branched molecular structure of amylopectin, the viscosity of starch was reported to be lower [47].

The percentage composition of amylose and amylopectin in starch will affect the physicochemical and physicomechanical properties of the substance. In an investigation into the amylopectin composition, studies showed that the values of elongation at break properties will be influenced by amylopectin content [48]. In addition, the advancement of biopolymer derived from starch will be affected by the branched structure and polymer network of amylopectin [49].

A fraction of amylose/amylopectin is the main factor which can affect the properties of starch. According to previous studies, there are different natural plant sources that contain several amounts of amylose and amylopectin in starch granules, apart from small quantities of proteins, lipids, and phosphorus [50]. The evidence presented in these studies shows that the characteristic of starch is hydrophilic, the moisture content is based on environmental humidity, and the capability of starch to pass through the transformation process is called gelatinization. This process is where amylose and amylopectin are filtered out until the granule is interrupted into a mixed polymer solution. The temperature required is between 52°C and 170°C, depending on the amylose content [51].

10.4.2 THERMOPLASTIC STARCH

In recent years, there has been an increasing interest in starch development since it has a rapid biodegradation speed and is a low-cost material. Thermoplastic starch (TPS) is processed starch under high temperature and shear to form a moldable thermoplastic [52]. TPS is an important aspect of plasticized starch which is commonly processed under heat and pressure to break the crystalline structure of starch and form an amorphous TPS. Throughout the gelatinization process, the addition of plasticizers and water composition in starch are necessary factors for creating hydrogen bonds within the starch. In other words, strong interactions occur among the hydroxyl groups in the starch molecules, hence producing TPS [53]. Together, these studies provide important insights into plasticizers that are commonly used such as fructose, glycerol, and sorbitol [54]. The studies presented thus far provide evidence that plasticizers can increase the flexibility and processing ability to produce good material properties. This also indicates that the properties of TPS depend on moisture and relative humidity of the air (Figure 10.5).

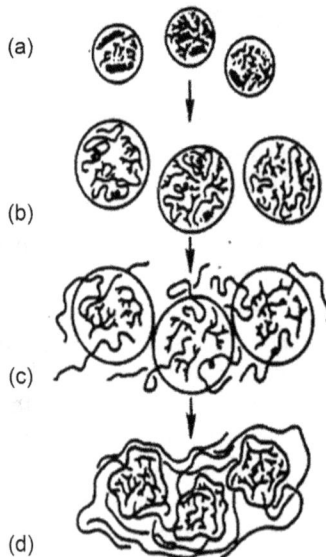

FIGURE 10.5 Starch gelatinization process; (A) starch granule formed from amylose and amylopectin; (B) addition of water separated amylose crystallinity and induced granule swell; (C) addition of heat and increasing water content resulted in more swelling, causing amylose to spread out of the granule; (D) starch granules that mainly contain amylopectin have crashed and kept in the amylose matrix, forming a gel. (Adapted from [55].)

Numerous studies have attempted to explain how the plasticizer content will influence the properties of the TPS product, such as water solubility, water absorption, and mechanical strength. It has been reported that glycerol is the most frequently applied plasticizer; the capability of glycerol in biodegradable films is effective for TPS [56]. Previous studies have shown that glycerol is more effective compared to sorbitol in plasticizing starch, with a glycerol to starch ratio of 30%/70%; this is considered the optimum value for the mixing method [57]. Several reports have shown that sorbitol-plasticized TPS/polylactic acid (PLA) blends present higher tensile strength results and modulus as well as less volatility in thermogravimetric valuation. It was further demonstrated that sorbitol-plasticized TPS can constantly spread in the PLA matrix even though various additives are present [58].

10.4.3 PLASTICIZATION OF STARCH

Recent attention has focused on composites where a group of matrix polymers contain various conventional fillers such as fiber. The properties of the composite material are dependent on the interfacial adhesion, size, shape, and volume fraction of the filler [59]. Current studies have focused on TPS composite for improving mechanical properties compared to pure starch films. The formulation of TPS composites with natural fiber has been established, and the resulting products are low in cost, renewable, biodegradable, and non-toxic [60]. Thus far, previous researches have confirmed the effectiveness of TPS. The biodegradable matrices can be reinforced with natural fiber to upgrade the composite's features, such as good mechanical properties and lightweight [61]. There are numerous types of plasticizers that are usually used to produce TPS such as glycerol, sorbitol, urea, fructose, glycols, and water [62]. The main criteria must be considered when choosing a suitable plasticizer; for example, it should be hydrophilic in nature, polar, have a high boiling point with increased temperatures during processing (130°C–170°C), and possess small molecules. It has been reported that glycerol is commonly used as a plasticizer due to its promising features such as availability in liquid form, making it smooth during processing [63]. Further studies have indicated that sorbitol plasticized film has higher tensile strength than glycerol plasticized film. However, the film is brittle compared to the glycerol plasticized film. In another major study, the author found that better mechanical properties can be attained by blending glycerol and sorbitol compared to having them separated. It has been conclusively shown that the plasticizer quantity in starch/plasticizer mixture must be within 20%–40%. A higher percentage may result in the decrease of mechanical strength and rigidity of TPS, whereas a lower percentage will affect the brittle structure due to the anti-plasticization effect [64]. Prior studies have demonstrated that the plasticizer function is to interrupt the starch granules by separating the inter-molecular and intra-molecular hydrogen bonds of starch. The de-polymerization process occurs when the plasticizer molecule removes the starch interactions and exchanges them with starch–plasticizer interactions. As a result, the melting temperature of starch may decrease [65]. The studies presented thus far provide evidence that TPS is an environmentally friendly material which requires various types of modifications and treatments to enhance future properties.

10.5   STARCH OF *DIOSCOREA HISPIDA*

D. hispida is a starchy tuber in the *Dioscoreaceae* family which is also known in Malaysia as "Ubi gadong." It has been reported that *D. hispida* is a toxic plant with a thorny stem, white to yellow flesh, and with a pale-yellow flower, creating a bulky tuber. It also contains highly toxic dioscorine which is the water-soluble alkaloid that exists in the tuber of *D. hispida*. The extraordinary feature of this tuber is that it can be transformed to a consumable food after undergoing the detoxification process by boiling, roasting, or soaking in flowing water for 7–14 days. According to the WHO, tubers should be consumed in safe quantities below 10 mg of hydrogen cyanide (HCN) per kilogram of body weight; otherwise, the person will be poisoned [66]. Data from several sources have identified the starch extraction process of *D. hispida* after removing the toxin. The tubers of *D. hispida* were washed, peeled, and sliced into small pieces before being blended using the selected blender. They were then soaked in distilled water for 30 min and filtered using selected cotton cheesecloth filter. The soak lasted around 24 h to remove the starch, and the starch was dried completely until it can be collected [67] (Figure 10.6).

Recent studies have investigated the properties of *D. hispida* starch that contains the amount of 11.46% ± 0.08% and has lower moisture content with water-binding capacity (WBC) [67]. The author revealed the result due to the ability of creating the hydrogen and covalent bonds with the starch structure [68]. The swelling power and solubility of *D. hispida* starch can be identified through several steps. The temperature

(a)

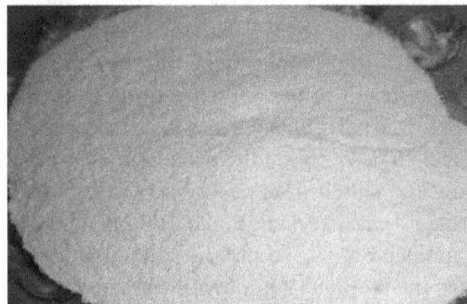

(a)

FIGURE 10.6   (a) Sliced tubers of *D. hispida*. (b) The starch of *D. hispida*.

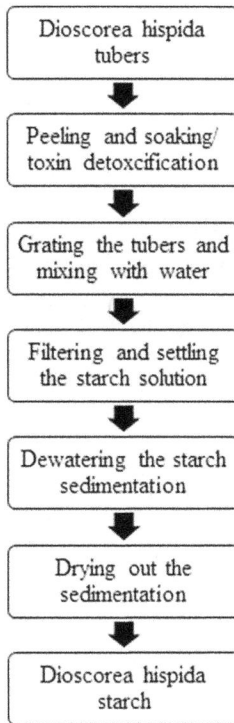

FIGURE 10.7   The process of *D. hispida* starch. (Adapted from [67,71].)

is steadily increased from 60°C to 90°C within 30 min, and the weight of the swollen starch is calculated. It has been observed that the swelling power and starch solubility have increased due to the breakup of intermolecular bonding, enabling the hydrogen bonds to attach more with water molecules [69]. In 1988, Takahashi et al. published a paper in which they described how the amylose starch can be available in a crystal structure at a temperature range of 50°C–60°C. This prevents the granules from tremendously swelling. More energy was absorbed by the increasing temperature, demonstrating the reason why the rate of swelling had increased [70] (Figure 10.7).

10.6   CONCLUSION

This study extensively reviewed the available information on *D. hispida* for its potential properties as a biopolymer when converted into TPS. The concept of waste and recycled *D. hispida* starch is sustainable and economical, and has numerous applications. The properties of the material can be enhanced through starch modifications such as the inclusion of fiber/filler to maximize the benefits of the material. The theoretical implications of these findings are unclear regarding several techniques and methods to be implemented for starch modification in biocomposite materials. Even though there are previous works that discussed the modification of TPS, no work has been accomplished on TPS *D. hispida* composite materials.

ACKNOWLEDGMENT

The authors would like to thank Universiti Putra Malaysia for the financial support provided through the Universiti Putra Malaysia Grant scheme as well as the German Malaysian Institute for providing support to the principal author in this project.

REFERENCES

1. L. Yan, N. Chouw, and K. Jayaraman, "Flax fibre and its composites - a review," *Compos. Part B Eng.*, vol. 56, pp. 296–317, 2014.
2. H. J. Kim, S. Miyamoto, Y. Takada, and K. Takemura, "Effect of surface modification on flexural properties of jute fiber green composites," *ICCM Int. Conf. Compos. Mater.*, vol. 6300, pp. 2–6, 2011.
3. K. Arun Kumar, S. M. Sudhanan, K. M. Kumar, and G. Ranjith Kumar, "A study on properties of natural fibres -a review," *Int. Res. J. Eng. Technol.*, vol. 10, pp. 2395–2356, 2017.
4. M. R. Ishak, S. M. Sapuan, Z. Leman, M. Z. A. Rahman, U. M. K. Anwar, and J. P. Siregar, "Sugar palm (*Arenga pinnata*): its fibres, polymers and composites," *Carbohydr. Polym.*, vol. 91, no. 2, pp. 699–710, 2013.
5. D. R. Mulinari, H. Jacobus, and C. Voorwald, "Natural fiber in polymer composites," 2007.
6. M. R. Mansor, S. M. Sapuan, E. S. Zainudin, A. A. Nuraini, and A. Hambali, "Hybrid natural and glass fibers reinforced polymer composites material selection using analytical hierarchy process for automotive brake lever design," *Mater. Des.*, vol. 51, pp. 484–492, 2013.
7. M. L. Sanyang, S. M. Sapuan, M. Jawaid, M. R. Ishak, and J. Sahari, "Recent developments in sugar palm (*Arenga pinnata*) based biocomposites and their potential industrial applications: a review," *Renew. Sustain. Energy Rev.*, vol. 54, pp. 533–549, 2016.
8. R. Ranjan, P. K. Bajpai, and R. K. Tyagi, "Mechanical characterization of banana/sisal fibre reinforced PLA hybrid composites for structural application," *Eng. Int.*, vol. 1, no. 1, pp. 39–48, 2013.
9. N. Saba, M. T. Paridah, and M. Jawaid, "Mechanical properties of kenaf fibre reinforced polymer composite: a review," *Constr. Build. Mater.*, vol. 76, pp. 87–96, 2015.
10. S. Tajuddin, N. Mat, A. G. Yunus, and A. R. Shamsul Bahri, "Anatomical study of stem, petiole, leaf, tuber, root and flower of *Dioscorea hispida* Dennst. (Dioscoreaceae) by using optical microscope, SEM and TEM," *J. Agrobiotechnol.*, vol. 4, no. 1, pp. 32–41, 2013.
11. T. K. Lim, "Edible medicinal and non-medicinal plants," *Edible Med. Non-Med. Plants*, vol. 10, pp. 1–659, 2016.
12. K. K. Behera, S. Sahoo, and A. Prusti, " Effect of Plant Growth Regulator on *in vitro* micropropagation of 'Bitter Yam' (*Dioscorea hispida Dennst.*)," *Int. J. Integr. Biol.*, vol. 4, no. 1, pp. 50–54, 2008.
13. N. J. Amanze, N. J. Agbo, D. N. Njoku, and N. Root, "Selection of yam seeds from open pollination for adoption in yam (*Dioscorea rotundata* Poir) production zones in Nigeria," *J. Plant Breed. Crop Sci.*, vol. 3, no. 4, pp. 68–73, 2011.
14. N. Mat *et al.*, "The distribution of *Dioscorea hispida* Dennst. germplasm in Terengganu and phylogenetic relationships of *Dioscorea* spp. using internal transcribed spacer (ITS)," *J. Agrobiotechnol.*, vol. 5, pp. 31–47, 2014.
15. N. A. F. Mohd Hori, N. F. Mohd Nasir, N. A. Mohd, E. M. Cheng, and S. N. Sohaimi, "The fabrication and characterization of hydroxyapatite-ubi gadong starch based tissue engineering scaffolds," *IECBES 2016- IEEE-EMBS Conference on Biomedical Engineering Science*, pp. 220–225, 2016.

16. Z. A. A. Hamid, M. H. M. Idris, N. A. A. B. Arzami, and S. F. M. Ramle, "Investigation on the chemical composition of *Discorea hispida* Dennst. (ubi gadong)," *AIP Conf. Proc.*, vol. 2068, no. February,pp. 020041-1–020041-6, 2019.

17. S. Kumar, A. K. Parida, and P. K. Jena, "Ethno-medico-biology of Bān-Aālu (Dioscorea species): a neglected tuber crops of Odisha, India," *Int. J. Pharm. Life Sci.*, vol. 4, no. 12, pp. 3143–3150, 2013.

18. A. Cahyo Kumoro, D. Susetyo Retnowati, and C. Sri Budiyati, "Removal of cyanides from gadung (*Dioscorea hispida* Dennst.) tuber chips using leaching and steaming techniques," *J. Appl. Sci. Res.*, vol. 7, no. 12, pp. 2140–2146, 2011.

19. S. Theerasin and A. T. Baker, "Analysis and identification of phenolic compounds in *Dioscorea hispida* Dennst.," *As. J. Food Ag.-Ind.*, vol. 2, no. 204, pp. 547–560, 2009.

20. Rudito *et al.*, "Physical and chemical characteristics of fermented dayak wild yam (*Dioscorea hispida* Dennst.), purple yam (*Dioscorea alata* var. purpurea) and *Air Potato* (*Dioscorea bulbifera* L.) flour as food ingredient," *Int. J. ChemTech Res.*, vol. 11, no. 11, pp. 369–378, 2018.

21. J. Tattiyakul and T. Naksriarporn, "X-ray diffraction pattern and functional properties of *Dioscorea hispida* Dennst. starch hydrothermally modified at different temperatures," *Food Bioprocess Technol.*, vol. 5, pp. 964–971, 2012.

22. L. Burhenne, J. Messmer, T. Aicher, and M. P. Laborie, "The effect of the biomass components lignin, cellulose and hemicellulose on TGA and fixed bed pyrolysis," *J. Anal. Appl. Pyrolysis*, vol. 101, pp. 177–184, 2013.

23. M. R. Bhandari and J. Kawabata, "Bitterness and toxicity in wild yam (*Dioscorea* spp.) tubers of Nepal," *Plant Foods Hum. Nutr.*, vol. 60, no. 3, pp. 129–135, 2005.

24. A. C. Kumoro and R. Amalia, "Preparation and characterization of physicochemical properties of glacial acetic acid modified Gadung (*Diocorea hispida* Dennst.) flours," *J. Food Sci. Technol.*, vol. 52, no. 10, pp. 6615–6622, 2015.

25. N. Singh, D. Chawla, and J. Singh, "Influence of acetic anhydride on physicochemical, morphological and thermal properties of corn and potato starch," *Food Chem.*, vol. 86, no. 4, pp. 601–608, 2004.

26. K. Harju, "United States patent," *Geothermics*, vol. 14, no. 4, pp. 595–599, 1985.

27. R. M. Gonza, "Starch-based polymers for food packaging 19," In *Multifunctional and Nanoreinforced Polymers for food packaging*, 1st ed., J.M. Lagaron, Ed., pp. 527–570, 2011.

28. U. Jambi, "As Potential Source of Staple Food," *Department of Food Technology, Faculty of Agricultural Technology, Universiti Jambi, Indonesia*, pp. 1–4, 2013.

29. W. Musa, *Laporan Program Penyelidikan Ubi Gadong (Dioscorea hispida Dennst.)*, Terengganu, Malaysia: Unisza, pp. 215–245, 2010.

30. K. Maneenoon, P. Sirirugsa, and K. Sridith, "Ethnobotany of *Dioscorea* L. (Dioscoreaceae), a major food plant of the sakai tribe at Banthad Range, Peninsular Thailand," *Ethnobot. Res. Appl.*, vol. 6, pp. 385–393, 2008.

31. M. Nashriyah, T. Salmah, M. Y. NurAtiqah, O. Siti Nor Indah, A. W. MuhamadAzhar, S. Munirah, Y. Nornasuha, and A. A. Manaf, "Ethnobotany and distribution of *Dioscorea hispida* Dennst. (Dioscoreaceae) in Terengganu, Peninsular Malaysia," *World Acad. Sci. Eng. Technol.*, vol. 72, no. 12, pp. 1782–1785, 2012.

32. G.U. Stuart, Philippine Alternative Medicine. Manual of Some Philippine Medicinal Plants. 2013. http://www.stuartxchange.org/OtherHerbals.html

33. I. H. Burkill, W. Birtwistle, F. W. Foxworthy, J. B. Scrivenor, and J. G. Watson, *A Dictionary of the Economic Products of the Malay Peninsula*. Kuala Lumpur: Ministry of Agriculture Malaysia, Vol. 2, 2nd edition, 2444 pp, 1966.

34. A. A. Sulaini and S. F. Sabran, "Edible and medicinal plants sold at selected local markets in Batu Pahat, Johor, Malaysia," *AIP Conf. Proc.*, vol. 2002, no. 02006, pp. 020006-1–020006-10, 2018.

35. S. Roslan, M. H. H. Razali, and W. I. W. Ismail, "An overview for the application of multispectral device for determination of alkaloid level in *Dioscorea hispida*," *Sci. J. Rev.*, vol. 1, no. 3, pp. 53–57, 2012.

36. M. M. Miah, P. Das, Y. Ibrahim, M. S. Shajib, and M. A. Rashid, "In vitro antioxidant, antimicrobial, membrane stabilization and thrombolytic activities of *Dioscorea hispida* Dennst.," *Eur. J. Integr. Med.*, vol. 19, no. January, pp. 121–127, 2018.

37. E. Faradianna Lokman, H. Muhammad, N. Awang, M. Hasyima Omar, F. Mansor, and F. Saparuddin, "Gene expression profiling associated with hepatoxicity in pregnant rats treated with ubi gadong (*Dioscorea hispida*) extract," *Int. J. Biomed. Sci.*, vol. 13, no. 1, pp. 26–34, 2017.

38. Z. Xiong *et al.*, "Preparation and characterization of poly (lactic acid)/starch composites toughened with epoxidized soybean oil," *Carbohydr. Polym.*, vol. 92, no. 1, pp. 810–816, 2013.

39. P. González, C. Medina, L. Famá, and S. Goyanes, "Biodegradable and non-retrogradable eco-films based on starch – glycerol with citric acid as crosslinking agent," *Carbohydr. Polym.*, vol. 138, pp. 66–74, 2016.

40. A. Edhirej, S. M. Sapuan, M. Jawaid, and N. I. Zahari, "Cassava : its polymer, fiber, composite, and application," *Polymer Composites*, vol. 38, no. 3, pp. 555–570, 2015.

41. J. Sahari, S. M. Sapuan, E. S. Zainudin, and M. A. Maleque, "A new approach to use *Arenga pinnata* as sustainable biopolymer: effects of plasticizers on physical properties," *Procedia Chem.*, vol. 4, pp. 254–259, 2012.

42. S. Pérez, P. M. Baldwin, and D. J. Gallant, "Starch Granules I," In: *Starch: Chemistry and Technology.*, 3rd ed., Cambridge, MA: Elsevier, Inc., 2009.

43. J. Sahari, S. M. Sapuan, E. S. Zainudin, and M. A. Maleque, "Thermo-mechanical behaviors of thermoplastic starch derived from sugar palm tree (*Arenga pinnata*)," *Carbohydr. Polym.*, vol. 92, no. 2, pp. 1711–1716, 2013.

44. N. Le and S. Molina-boisseau, "Production of PVAc – starch composite materials by co-grinding — influence of the amylopectin to amylose ratio on the properties," *Powder Technol.*, vol. 255, pp. 36–43, 2014.

45. W. Zou *et al.*, "Effects of amylose/amylopectin ratio on starch-based superabsorbent polymers," *Carbohydr. Polym.*, vol. 87, no. 2, pp. 1583–1588, 2012.

46. A. Cano, A. Jiménez, M. Cháfer, C. Gónzalez, and A. Chiralt, "Effect of amylose : amylopectin ratio and rice bran addition on starch films properties," *Carbohydr. Polym.*, vol. 111, pp. 543–555, 2014.

47. M. D. Ninago *et al.*, "Enhancement of thermoplastic starch final properties by blending with poly (ε-caprolactone)," *Carbohydr. Polym.*, vol. 134, pp. 205–212, 2015.

48. T. Gurunathan, S. Mohanty, and S. K. Nayak, "A review of the recent developments in biocomposites based on natural fibres and their application perspectives," *Compos. Part A Appl. Sci. Manuf.*, vol. 77, pp. 1–25, 2015.

49. G. Koronis, A. Silva, and M. Fontul, "Green composites: a review of adequate materials for automotive applications," *Compos. Part B Eng.*, vol. 44, no. 1, pp. 120–127, 2013.

50. Y. Zhang, C. Rempel, Q. Liu, Y. Zhang, C. Rempel, and Q. Liu, "Thermoplastic starch processing and characteristics-A review," *Crit Rev Food Sci Nutr,*vol. 8398, no. February, pp. 1353–1370, 2016.

51. R. A. Ilyas, S. M. Sapuan, M. R. Ishak, and E. S. Zainudin, "Sugar palm nanocrystalline cellulose reinforced sugar palm starch composite: Degradation and water-barrier properties," *IOP Conf. Ser. Mater. Sci. Eng.*, vol. 368, no. 1, pp. 1–12, 2018.

52. P. Müller, K. Renner, J. Móczó, E. Fekete, and B. Pukánszky, "Thermoplastic starch/wood composites : interfacial interactions and functional properties," *Carbohydr. Polym.*, vol. 102, pp. 821–829, 2014.

53. L. S. Lai and J. L. Kokini, "Physicochemical changes and rheological properties of starch during extrusion (a review)," *Biotechnol. Prog.*, vol. 7, no. 3, pp. 251–266, 1991.

54. N. R. Savadekar and S. T. Mhaske, "Synthesis of nano cellulose fibers and effect on thermoplastics starch based films," *Carbohydr. Polym.*, vol. 89, no. 1, pp. 146–151, 2012.

55. C. H. Remsen and J. P. Clark, "A viscosity model for a cooking dough," *J. Food Process Eng.*, vol. 2, no. 1, pp. 39–64, 1978.

56. O. V. López *et al.*, "Thermoplastic starch plasticized with alginate – glycerol mixtures: melt-processing evaluation and film properties," *Carbohydr. Polym.*, vol. 126, pp. 83–90, 2015.

57. L. Averous and N. Boquillon, "Biocomposites based on plasticized starch: thermal and mechanical behaviours," *Carbohydr. Polym.*, vol. 56, pp. 111–122, 2004.

58. Z. Yang, H. Peng, W. Wang, and T. Liu, "Crystallization behavior of poly(ε-caprolactone)/layered double hydroxide nanocomposites," *J. Appl. Polym. Sci.*, vol. 116, no. 5, pp. 2658–2667, 2010.

59. M. Fazeli, J. P. Florez, and R. A. Simão, "Improvement in adhesion of cellulose fibers to the thermoplastic starch matrix by plasma treatment modification," *Compos. Part B Eng.*, vol. 163, no. October, pp. 207–216, 2019.

60. M. Bootklad and K. Kaewtatip, "Biodegradation of thermoplastic starch/eggshell powder composites," *Carbohydr. Polym.*, vol. 97, no. 2, pp. 315–320, 2013.

61. M. Babaee, M. Jonoobi, Y. Hamzeh, and A. Ashori, "Biodegradability and mechanical properties of reinforced starch nanocomposites using cellulose nanofibers," *Carbohydr. Polym.*, vol. 132, pp. 1–8, 2015.

62. M. L. Sanyang *et al.*, "Effect of plasticizer type and concentration on dynamic mechanical properties of sugar palm starch – based films effect of plasticizer type and concentration on dynamic mechanical properties of sugar palm star," Int. J. Polym. Anal. Ch.., vol. 20, no. March, pp. 627–636, 2015.

63. M. L. Sanyang, S. M. Sapuan, M. Jawaid, M. R. Ishak, and J. Sahari, "Effect of plasticizer type and concentration on tensile, thermal and barrier properties of biodegradable films based on sugar palm (*Arenga pinnata*) starch," *Polymers (Basel).*, vol. 7, no. 6, pp. 1106–1124, 2015.

64. J. Prachayawarakorn and W. Pomdage, "Effect of carrageenan on properties of biodegradable thermoplastic cassava starch/low-density polyethylene composites reinforced by cotton fibers," *J. Mater.*, vol. 61, pp. 264–269, 2014.

65. A. M. Nafchi, M. Moradpour, M. Saeidi, and A. K. Alias, "Thermoplastic starches: Properties, challenges, and prospects," *Starch/Staerke*, vol. 65, pp. 61–72, 2013.

66. N. L. V. Mlingi, Z. A. Bainbridge, N. H. Poulter, and H. Rosling, "Critical stages in cyanogen removal during cassava processing in southern Tanzania," *Food Chem.*, vol. 53, no. 1, pp. 29–33, 1995.

67. A. Ashri, M. S. M. Yusof, M. S. Jamil, A. Abdullah, S. F. M. Yusoff, and M. M. Nasir, "Physicochemical characterization of starch extracted from Malaysian wild yam (*Dioscorea hispida* Dennst.)," *Emir. J. Food Agric.*, vol. 26, no. 8, pp. 652–658, 2014.

68. A. Gunaratne and R. Hoover, "Effect of heat-moisture treatment on the structure and physicochemical properties of tuber and root starches," *Carbohydr. Polym.*, vol. 49, no. 4, pp. 425–437, 2002.

69. P. N. Alam, Mukhlishien, H. Husin, T. M. Asnawi, Santia, and A. Yustira, "The utilization of gadung (*Dioscorea hispida* Dennst.) starch for edible coating making and its tomato packaging," *IOP Conf. Ser. Mater. Sci. Eng.*, vol. 543, no. 1, pp. 012015-1–012015-6, 2019.

70. S. Takahashi and P. Seib, "Paste and gel properties of prime corn and wheat starches with and without native lipids," *Cereal Chem.*, vol. 65, no. 6, pp. 474–483, 1988.

71. J. G. Akpa and K. K. Dagde, "Modification of cassava starch for industrial uses," *Int. J. Eng. Technol.*, vol. 2, no. 6, pp. 913–919, 2012.

11 Processing and Modification of Starch into Thermoplastic Materials

R. Jumaidin and Z.A.S. Saidi
Universiti Teknikal Malaysia Melaka

S.M. Sapuan
Universiti Putra Malaysia

CONTENTS

11.1 INTRODUCTION

Population growth has dramatically increased the usage of different types of petroleum-based polymers for packaging materials which have prompted environmental problems due to the large quantities developed and disposal issues. Petroleum-based plastics do not biodegrade for long durations of time and has become waste management concern [1]. Another example is air pollution since plastics also produce a greenhouse gas called "methane" which can increase global temperatures. Apart from air pollution and global warming, the use of non-biodegradable packaging materials can cause flooding since they do not decompose; they can prevent proper drainage and clog the sewage system. The human race and other living organisms could be destroyed if these pollutions and environmental issues are not fully resolved.

In recent years, thermoplastic starch (TPS)-based composites have become the developing environmental-friendly product useful for packaging and material applications. Apart from its biodegradable characteristics, starch is a promising renewable resource that has the ability to change or replace petroleum-based plastic packaging. However, TPS has disadvantages or limitations in terms of high water and moisture absorption and low flexibility. To solve these issues, many fillers have been used to enhance TPS properties such as the addition of fibers [2]. Among them, natural fibers are the most promising options due to their favorable characteristics such as biodegradability, low density, low cost, recyclability, and abundance.

11.2 BIOPOLYMERS

Biopolymers are polymers produced by living organisms, i.e., plants. They can be considered as the most organic compounds in the ecosphere since they are biodegradable chemical compounds. They can be chemically synthesized from biological starting materials such as sugars, fats, oils, and starch and can also be created by biological systems such as creatures, microorganisms, plants, and animals. The use of plastics over metals and glass for packaging means has increased due to easy processability, low weight, low cost, good properties, and accessibility. Waste disposal problems can be overcome and solved to a considerable extent by biopolymer-based materials [3].

Currently, biopolymers such polylactic acid (PLA), silk, and chitosan have been extensively researched for medical applications. Biocompatibility and biodegradability are essential properties of biopolymers, providing great advantages and developing their potential use in implantable medical applications. Since synthetic materials do not fulfill the needs of the living system, these new materials carry significant importance in medicine and medical purposes. For bone applications, approximately several examples of chitosan-based matrices were studied. The use of mixed chitosan with other materials has increased in order to serve as osteoconductive matrices since chitosan matrices on their own are mechanically weak [4]. Table 11.1 shows the applications of chitosan in the medical field.

In general, there are three main types of biopolymers: starch, cellulose, and synthetic biopolymers.

11.2.1 STARCH

Starch is present in most green plants as energy storage. The carbohydrate is in greater portions in larger plants. It is a small and dense powder with individual segments known as granules. It is insoluble in cold water. Since it exists in granular form, it can be found in large concentrations in cereal grains (wheat, rice, maize, barley, rye, oats, millet, sorghum) [5]. It can be processed to form glucose by malting. This polymer can also be found in vegetables such as tapioca, corn, wheat, and

TABLE 11.1

Applications of Chitosan Implants in the Medical Field [4]

Field of Application	Implants for
Cardiology	Heart valves – electrospun gelatin – chitosan–polyurethane
Dermatology	Gelatin, chitosan, hyaluronan scaffold for skin generation
Surgery	Nerve regeneration – chitosan, gelatin scaffold
	Liver – chitosan, gelatin scaffold
Ophthalmology	Contact lens – chitosan, gelatin

potatoes and is not present in animal tissues. The resulting raw material is convenient for conventional plastic-forming processes such as extruding and injection molding. Starch is classified as one of the most highly used biopolymers in the manufacturing industry as it can be applied as filler. Starch-based materials exhibit fine oxygen barrier functions that contain high amylose contents [6].

11.2.2 CELLULOSE

Cellulose is a polymer obtained from natural resources such as cotton, wood, wheat, and corn, and it has a natural long chain which is an essential part in the indirect cycle of human food resource. This polymer has semi-synthetic derivatives that are largely used in cosmetic and pharmaceutical industries. Cellulose has two main groups, cellulose ethers and cellulose derivatives, with different mechanical properties and physicochemical features. These polymers are used in healthcare products and dose production. To give the bundled chains even more strength and stability, they can commonly pack in places to establish hard and stable crystalline regions [7].

11.2.3 SYNTHETIC BIOPOLYMERS

Materials made from chemicals are widely used since they are more durable and lower in cost compared to natural materials. An example of a material made from plastics would be polyethylene. Such materials are known as synthetic materials. Synthetic biopolymer-based nanocomposites can be applied in scaffolding for tissue engineering mainly due to their modifiable properties and biocompatibility of linear biodegradation. The saturated poly(α-hydroxyl esters), including polylactic acid (PLA), polyglycolic acid (PGA), polylactic acid-*co*-glycolic acid (PLGA), and poly caprolactone (PCL), are customarily applied as synthetic biopolymers in tissue engineering for three-dimensional scaffolds [8].

11.3 NATURAL FIBER COMPOSITES

Reinforcement is the composite part that provides steadiness, strength, and the ability to carry a load [9]. Natural fiber is an innovative creation of supplements and reinforcements for polymer-based materials as well as a renewable source. Natural fiber

TABLE 11.2

Natural Fiber Composite Applications [12]

Fiber	Application in Buildings, Constructions, etc.
Hemp fiber	Infrastructure products, cordage, flexible materials, manufacture of pipes, electrical, and furniture applications
Palm oil fiber	Door frames for construction materials, windows, siding, roofing, fencing, and structural insulated panel building system
Wood fiber	Window frame, decking, door shutters, panels, fencing, and railing system
Sisal fiber	Shutting plate for infrastructure industry, panels, doors, roofing sheets, and also the manufacture of paper
Kenaf fiber	Mobile cases, packing materials, bags, soilless mixes, materials that absorb oils, liquids, and animal bedding
Cotton fiber	The furniture industry, textile and yarn, and cordage
Jute fiber	Building panels, door shutters, roofing sheets, packaging, transport, chipboards, and geotextiles

is a capable chemical substance that has the ability to replace fabrics, can be applied in energy conservation applications, and is lower in weight. The purpose of reinforcing natural fiber with natural-based resins and polymer composites is to establish an application to substitute current synthetic polymer or glass fiber-reinforced materials in large quantities [10]. Jute, palm oil, sisal, kenaf, and bananas are examples of dissimilar types of natural fiber plants that have been actively developed for automotive and aircraft industries regarding interior parts. The lower processing of natural fiber composites with high specific characteristics is attractive for different applications [10].

Researchers have progressed in the last two decades regarding the use of natural fibers as reinforcements for polymer-based structural materials. The combination of straws and mud to construct walls in small villages has been practiced in China and Korea centuries ago [11]. Many automotive industries have already produced components using natural composites, mostly based on polymers such as polypropylene and fibers such as kenaf, sisal, and jute. Automobile manufacturers in Germany such as Porsche, Opel, Mercedes-Benz, and Audi have found alternatives of deriving composites from natural fibers to be used in interior and exterior applications. Door trim panels for the Audi A2 midrange car were built from the reinforcement of mixed flax and sisal fibers with polyurethane [10]. Table 11.2 presents the applications of natural fiber composites.

11.4 STARCH-BASED COMPOSITES

Starch has received considerable consideration from the scientific world due to its capacity to form bioplastics. Starch has been used as a starting point for a wide range of green biomaterials. Diverse methods have been applied to transform starch as a means of improving product properties and develop a wide range of applications.

Various types of starches deemed to be suitable as matrix components of composites are highlighted, along with their mechanical properties and characteristics. Starch-based plastics are very special due to their properties as biodegradable polymers acquired from renewable sources and raw materials. It is a common natural polymer which can be converted into items for processing equipment after plasticization [13]. Starch is obtained from tubers, seeds, and roots of plants, making it suitable for industrial use. Next, it is distilled through the processes of washing, sieving, wet grinding, and drying. The properties of starch can be enhanced by numerous chemical and physical modifications such as blending, derivation, and graft copolymerization.

Biodegradable renewable materials have been identified as the most suitable for replacing single-use packaging materials. To boost microstructure, different approaches have been studied with the objective of solving these problems such as chemical modification of starches, blending with various biodegradable polymers, merging of natural fibers, and the inclusion of nanofillers [5]. Reinforcing commercially available TPS matrix with rapeseed fibers allowed the introduction of fully biodegradable composite materials. Espigulé et al. (2013) mentioned that when high reinforcement content is applied, the capacity of the TPS matrix to crystallize is lowered due to the restrictions of the mobility of starch macromolecules caused by the reinforcement [14].

11.5   POLYMERIZATION OF THERMOPLASTIC STARCH

Starch can be changed into TPS when processing occurs at high temperature and shear from a plasticizer [15]. The role of the plasticizer is to disturb the starch granules by separating the starch bonds between the intermolecular and intramolecular hydrogen. The plasticizer atom can recover the starch interaction with the starch–plasticizer intuition after killing the interaction. The extent of the plasticizer and its chemical nature has a two-way effect on the physical properties of the processed starch: first, by controlling its de-structuring and depolymerization, and second, by influencing the final properties of the fabric, such as its glass transition temperature and modulus. As described by Da Róz et al. (2006), TPS with high plasticizer content can produce phase separation, and these compositions are more delicate in surrounding humidity as the plasticizer substance expands [16]. Processing technology of starch plastics is one of the most important issues in their production. Adding suitable amounts of plasticizer is a commonly used method when creating plasticizer starches. Plasticizers are usually developed to increase the flexibility of the film by reducing the bonding of intramolecular hydrogen with chains of polymers, expanding the intermolecular spacing. This development in expanded molecular mobility reduces the degree of crystallinity and brings down the glass transition temperature [17]. Generally, the amount and type of plasticizer does not influence the crystallinity of the processed tests [16]. Thermoplastic is a polymeric material that can maintain its properties while transforming into softened and solidified states several times. Apart from being renewable and biodegradable, its characteristics make this material recyclable, increasing its positive qualities.

There are several kinds of plasticizers for creating TPS such as water (which has mostly become a fundamental plasticizer), glycerol (the most commonly used

plasticizer after water), glycols, urea, acetamide, citric acid, sorbitol, and mannitol. Considering the physical properties of TPS, the selection of plasticizer is very
important depending on the essential factors of utilization such as biodegradability,
processing, long-term stability, and resistance against the environment [18]. It was
found that the role of sorbitol, which is a low power of plasticization, induces an
anti-plasticization effect that reduces the volume strain but increases the cohesion of
the material. For glycerol, which is a proficient plasticizer, the portability of polymer
chains is expanding, but any surplus of glycerol will result in a phase separation and
restrict the volume deformation. The characteristics of mannitol starch-based materials not only consist of brittle and breaks in deformation, but also exude and crystallize onto the surface of materials after several days of testing [19]. Recent studies
have been made on TPS derived from several types of plant sources to enhance
future usage and strengthen its properties. Its application will be of minor consequence on the environment due to its wholly biodegradable nature, making it suitable
for various applications.

11.6   THERMOPLASTIC CASSAVA STARCH

Manihot Esculenta is commonly called cassava and classified as the third most
generally used food supply in tropical areas and the fifth extensively produced crop
in the world. The amount of cassava can be promising, resulting in the improvement of rural economies through the process of extracting cassava starch (CS) [20].
Cassava starch is mostly used for food and other applications throughout nature due
to its capabilities and abundance in tropical areas. As a natural resource, it has the
potential of being used as a thermoplastic which led to the comprehensive research
of using cassava starch as grinded materials for the biodegradability of thermoplastics. From the previous study reported, chitosan-coated starch films were formed
by coating the chitosan solutions with cassava starch film with glycerol used as a
plasticizer [21]. In this study, the starch films consist of different weight percentages (wt%) of glycerol, coated by various solutions of chitosan ranging (from 1 to
4 wt%). Some characteristics of the film include physical and thermal properties
which had improved from the coating of chitosan solutions. The tensile strength,
Young's modulus, and percent of elongation at break are greatly affected by the
concentrations of glycerol. The tensile strength and Young's modulus have reduced
with increasing amounts of glycerol.

Natural fibers have already been studied and used as reinforcements for polymer matrix composites due to their good properties and features; for instance, they
are inexpensive, lightweight, and renewable. They solve the poor characteristics of
TPS which has low mechanical strength. However, attributable to the diverse nature
of hydrophilic fiber and hydrophobic matrix, their production for petroleum-based
polymers is restricted, inhibiting optimum surface adhesion and wetting between
the matrix and fiber. Thus, the natural fiber or filler reinforced with TPS is found
to be a more encouraging way for enhancing and upgrading the properties of TPS
while maintaining its biodegradable and renewable qualities. Optimum fiber–matrix
interaction can be achieved by this biocomposite since TPS has the same hydrophilic

nature as the natural fiber. Prachayawarakorn et al. (2013) mentioned that mechanical properties for thermoplastic cassava starch (TPCS) and jute or kapok fibers were substantially improved [22]. The authors reported that the expansion of fiber content seems to enhance the strength of TPCS with jute fiber as well as TPCS with kapok fiber.

From the thermogravimetric analysis (TGA), when adding jute fibers, the TPCS matrix increased; when combining kapok fibers, the thermal degradation temperature decreased. The beginning thermal degradation temperature of dissimilar TPCS with jute fiber composites was always seemingly identical, identifying the shape of starch decomposition. The weight loss was condensed, revealing enhanced thermal stability for TPCS with jute fiber composites. This may be due to lesser jute fiber compared to the hydrophobic attributes of kapok fibers. Thus, improvement in thermal stability for TPCS with kapok fiber composites was due to the evidence of reduced weight loss.

Another interesting study had utilized cotton fiber reinforcement for cassava starch biofoams [23]. Cotton fiber and concentrated natural rubber latex (CNRL) were included in the biofoam products to overcome their two main weaknesses, which is susceptibility to moisture and lack of flexibility. Natural rubber (NR) is a biodegradable polymer that presents decent impact properties, high elongation, and high tensile strength. There are numerous formulations for preparing cassava starch biofoam using a conventional compression molding machine. The homogenous batter was obtained after all additives were mixed using a kitchen aid mixer at room temperature for 5 min; this includes starch, cotton fiber, CNRL (pre-mixed with an appropriate amount of vulcanization ingredients), and distilled water. Next, the homogenous batter was placed in the compression mold and poured into a $300 \times 300 \times 3\,mm^3$ mold. The pressure mold was finalized at a pressure of 1,000 atm for 4 min and a temperature of 220°C. Biofoam displayed denser cell structure, less porosity with higher density, and improved hydrophobicity with increasing CNRL content. Increasing the hydrophobicity leads to better dimensional stability as well as a decrease in the moisture adsorption rate and moisture adsorption capacity of the biofoam. CNRL stimulates the thermal degradation of the cotton fiber even though the use of CNRL did not affect the thermal degradation behavior of the biofoam. The suitable addition of CNRL concentrations led to better flexural properties for the biofoam; this is related to its cross-linking structure. The density of biofoam increased after the use of CNRL. The modulus rupture of the biofoam was enhanced with a process of reinforcing either the cotton fiber or the rubber network formed. The water content for biofoam products had significantly decreased due to the addition of CNRL to the starch batter because of its inherent hydrophobicity. When the biofoam was in contact with water, its dimensional stability improved. The delayed biodegradation was caused by the improvement of the biofoam's hydrophobicity that occurred via hydrolysis.

In food industries, eggshells are produced in large quantities and have become a major waste product. They are considered as excellent candidates to be used as a bio-filler in the preparation of polymer composites since they are cheap, abundant, low in density, environment-friendly, and a renewable resource. A comprehensive study on TPS from cassava was conducted using compressed molding and chicken eggshells which functioned as a filler to investigate the outcome of eggshells on the properties of TPS and make a comparison with the utilization of calcium carbonate [2]. Cassava starch and glycerol were mixed together with calcium carbonate or egg-shells as a control at 10–50 wt% using polyethylene bags to dry the starch mixture. The preparation of mixed TPS and composites was accomplished before the use of the compression molding machine at 160°C in 10 min. Soil burial degradation test run in Hat Yai, Songkhla, Thailand, was carried out by burying the samples in natural soil at 7 cm depth. The samples were gathered after 15 and 30 days and then cleaned several times by using distilled water. They were then placed in the oven to dry at 105°C for 24 h to verify the weight lost. The test stated that after 15 days of burial, the original weight loss for the matrix TPS was 71.55%, which completely degraded after 30 days of burial. The weight loss of TPS was due to the microorganism attack in the soil and biodegradation process by moisture. It was found that the water absorption of TPS and eggshell composites can be reduced by incorporating calcium carbonate. This was attributed to the more hydrophobic nature of calcium carbonate than the neat TPS matrix.

Natural fiber has become a possible material for replacing petroleum-based products with either the combination of other materials or alone to produce green composites. In order to make a hybrid composite, cassava bagasse (CB) with sugar palm fiber (SPF) was prepared by applying the casting method with cassava starch (CS) acting as a plasticizer [20]. Samples of 3.0 g were dissolved in 25 mL of distilled water and kept in pre-weighed centrifuge tubes for 25 min at 3,000 rpm in order to conduct the water absorption test. The dispersal was combined and placed at room temperature for 1 h after being kept in a centrifuge. From the results, CB absorbed a higher amount of water which signifies that it is greatly hydrophilic compared to SPF (257.04%); SPF absorbed lower amounts of water (122.33%). The moisture content somewhat decreased when SPF and CB were included in the CS films. In comparison with the CS film, the film with higher amounts of SPF showed lower moisture content. Table 11.3 presents a summary of previous studies on TPCS. TPS has great potential for numerous types of environmental-friendly applications due to its favorable properties.

11.7 THERMOPLASTIC RICE STARCH

Rice has been extensively consumed as a basic food worldwide. Over 500 million tons of rice is harvested, which provides sustenance to countless people and to many countries around the world every year. Rice starch is a fundamental biopolymer. Amylose and amylopectin are great raw materials used in packaging materials. A study investigated rice and waxy rice used as biodegradable polymers with TPS. Cotton fiber was included as reinforcement in the TPS matrix to reduce water absorption and increase the mechanical properties of TPS [28]. The results showed that the TPRS sample had greater stress at maximum load and Young's modulus; however, the strain decreased at maximum load of reinforcement because of higher amylose content of rice starch which created sturdiness compared to the TPWRS starch sample. Water absorption increased at the primary stage of TPRS and TPWRS samples without cotton fiber

TABLE 11.3
Thermoplastic Cassava Starch Composites

Type of Starch	Type of Fiber	Type of Testing	Main Finding	Potential Application	References
Cassava	Cotton fiber, natural rubber latex	Soil burial test	Better moisture resistance and dimensional stability	Packaging	[23]
Cassava	Rice straw	SEM, TEM, TGA, mechanical test, water drop test	Increased flexural strength and modulus of four times compared to the control	Non-food packaging	[24]
Cassava	Carrageenan, low-density polyethylene (LDPE) composites, cotton fiber	IR spectroscopic, XRD, mechanical testing, water absorption, soil burial test	Increased the stress at maximum load and Young's modulus of the TPCS/low–LDPE composites	Packaging	[25]
Cassava	Sugar palm	Water absorption, water content, chemical composition, thickness swelling, XRD	Increased the suitability of cassava starch composite films as environmental-friendly food packaging	Bio-based product	[20]
Cassava	Eggshell	SEM, water absorption, biodegradation in soil, thermal stability	Improved the water resistance and thermal stability of the TPS	Biodegradable material	[2]
Cassava	Jute, kapok	FTIR spectroscopic, XRD, SEM, thermal properties, mechanical testing, water absorption	Increased the stress at maximum load and Young's modulus by the incorporation of both jute and kapok fibers	Packaging	[22]
Cassava	Green coconut fiber	SEM, mechanical testing, water absorption, water sensitivity, moisture absorption	Improved the tensile properties of cassava starch, interfacial bonding between the matrix and fibers, as well as the hindrance to absorption caused by the fiber	Packaging	[26]

(Continued)

TABLE 11.3 (*Continued*)
Thermoplastic Cassava Starch Composites

Type of Starch	Type of Fiber	Type of Testing	Main Finding	Potential Application	References
Cassava	Pectin particle, cotton fiber	IR spectroscopic, XRD, SEM, mechanical tests, water absorption, thermal properties	Increased stress at the highest load and Young's modulus of the TPCS/pectin particle and the TPCS/cotton fiber composites	Biodegradable polymer	[27]
Cassava	Chitosan coating	XRD, physical and mechanical property measurements, water absorption, WVTR, mechanical properties evaluation	Increased in tensile stress at maximum load and tensile modulus, and reduced the percent of elongation at break	Packaging	[21]

SEM, scanning electron microscope; TGA, thermogravimetric analysis; WVTR, water vapor transmission; XRD, X-ray diffraction.

reinforcement. Due to higher contents of the more branched structure of the amylopectin molecule, TPRS was found to have higher water absorption. The percentage of water absorption for TPRS gradually increases and achieves its limit within 7 days, while for the TPWSR sample, the limit is achieved within 4 days.

As an alternative choice of fiber reinforcement because of inadequate mechanical properties and elevated water absorption of TPS, thermoplastic rice starch (TPRS) was improved by using natural silk protein fibers as reinforcement [29]. The addition of silk fibers into the TPRS matrix brought about a significant increase in the stress at maximum load and Young's modulus. The interaction between the TPRS matrix and silk triggered the reinforcing effect of the silk fibers. Meanwhile, the stress at maximum load and Young's modulus managed to reduce the silk fibers since the content of lightweight silk fibers loaded into the TPRS matrix was too much. Table 11.4 presents a review of TPS composites reinforced with rice starch. Apart from biodegradable packaging, the potential of this material was also explored in other industries such as agricultural and biomedical fields.

11.8 THERMOPLASTIC CORNSTARCH

Corn or maize has been produced a lot more than wheat or rice and is regarded as a food staple in most parts of the world. Cornstarch is obtained from the corn grain, and it is very versatile and easily modified, as well as applicable in various industries such as food, pharmaceutical, and textile manufacturing. One of the ways to develop TPS thermoplastic properties was using cornstarch in most experiments; numerous modifications have been conducted. The modification of films based on thermoplastic cornstarch (TPCRS) with the incorporation of chitosan and chitin was found by melt-mixing and thermocompression [33]. The films' tensile strength substantially improved when 10 g of chitosan with 100 g of starch were mixed with the TPS matrix. The films' elongation at break was reduced by around 30% with the addition of 10 g of chitosan. The interactions between chitosan or chitin and cornstarch reduce the accessibility of hydrophilic groups. Hence, this decreased the water vapor transmission in the matrix.

A study characterized cornstarch by using X-ray powder diffraction and thermal analysis, as well as characterized and processed banana starch and sugarcane bagasse fiber composites [34]. The composites were made by compression molding of the matrix, with banana and bagasse fibers presenting more identical composites with crude glycerin of cornstarch and glycerol as a plasticizer. TGA observations have shown positive thermal stability for making the composites. Some cracks were found between the smooth and rough surfaces by fractographic studies of 70 wt% starch and 30% glycerol matrix; this proves that the sample was ductile. Strength properties of the matrix were found to be enhanced in the tensile test of the composites. Higher quality of homogeneous composites is produced from crude glycerin which has better properties compared to production materials produced with commercial glycerol.

Table 11.5 presents a brief introduction of TPCRS composites extracted from cornstarch. Various actions were conducted to resolve the setbacks created by plastic waste in order to produce materials that are environmentally friendly, especially

TABLE 11.4

Thermoplastic Rice Starch Composite

Type of Starch	Type of Fiber	Type of Testing	Main Finding	Potential Application	References
Rice	Natural fiber	MFI, Fourier-transform infrared spectroscopy, Fourier-transform, tensile test, enzymatic tests, SEM	Improved tensile properties	Biodegradable composite	[30]
Rice	Cotton	Tensile tests, SEM, water absorption, TGA, soil burial test	Development of tensile strength and Young's modulus	Biodegradable polymer	[31]
Rice	Silk protein fiber	IR spectroscopic, water absorption, SEM mechanical test	Increased the stress at maximum load and Young's modulus composite	Biodegradable material	[29]
Rice, waxy rice	Cotton	IR spectroscopic, XRD, water absorption, SEM, mechanical testing, soil burial test, TGA	Increased the stress at maximum load and Young's modulus of both thermoplastic rice starch (TPRS) and thermoplastic waxy rice starch (TPWRS)	Biodegradable polymer	[32]
Waxy rice	Agar, cotton fiber	SEM, tensile test, water absorption, color measurement, IR Spectroscopic, tensile test, soil burial test, TGA, water uptake	Improved the tensile properties, thermal degradation temperature and thermal stability of thermoplastic waxy rice starch (TPWRS)	Packaging, agricultural, medical application	(Prachayawarakorn et al., 2012)

MFI, melt flow index; SEM, scanning electron microscope; TGA, thermogravimetric analysis; XRD, X-ray diffraction.

TABLE 11.5
Thermoplastic Cornstarch Composites

Type of Starch	Type of Fiber	Type of Testing	Main Finding	Potential Application	References
Corn	Chitosan	XRD, DSC, SEM optical properties, WVP, tensile test, DMA	Increased tensile strength and elastic modulus, and decreased elongation at break	Packaging	[33]
Corn	Sisal	Tensile test, SEM, infrared spectrum analysis, biodegradability test, water absorption	Replacement for expandable polystyrene (EPS) as packing material, especially under large compression load (0.7–6 MPa)	Plastic packing materials	[35]
Corn	Date palm, flax fibers	Tensile test, SEM water uptake test, soil burial, TGA	Increased the composite static tensile and flexural mechanical properties (stiffness and strength), thermal stability, water uptake, and biodegradation also improved	Biodegradable polymer	[36]
Corn	Clay	SEM, XRD soil burial test, contact angle measurements	Improved properties in relation to thermoplastic cornstarch (TPS)	Biodegradable material	[37]
Corn	Banana, sugarcane bagasse	XRD, moisture content, thermal analysis, tensile test, SEM	Improvements in tensile properties	Biodegradable composite	[34]

DMA, dynamic mechanical analysis; DSC, differential scanning calorimetric; SEM, scanning electron microscope; TGA, thermogravimetric analysis; WVP, water vapor permeability; XRD, X-ray diffraction.

disposable commodity materials that can be used only once. Many researchers focused on substituting petro-based commodity plastics with biodegradable materials for competitive mechanical properties and cost-effectiveness.

11.9   THERMOPLASTIC POTATO STARCH

Starch grains (leucoplasts) were extracted from a cell of root tubers of potato plants. The potatoes must be crushed as a means of extracting starch. The crushing method will release starch grains from the crumpled cells. A study by Wei et al. (2015) investigated the biodegradability of poly(butylene adipate-*co*-terephthalate) (PBAT) with TPS composites [38] by using the synthesized reactive compatibilizer which is a styrene-maleic anhydride glycidyl methacrylate (SMG) and the commercial compatibilizer. To change the part of PBAT in the PBAT/TPS blends, the second approach was used to modify PBAT (M-PBAT) with more molecular weight in the existence of a chain extension by the reactive extrusion of PBAT. SEM images show an improvement in the dispersion of TPS when applying M-PBAT, leading to higher tensile strength and elongation at break of PBAT/TPS blends. Even though the tensile strength maintained the same level or greater, in the existence of the compatibilizer (SMG), the elongation at break decreased. Generally, the tensile and elongation results explain the suitability of use for numerous applications, as well as the disposal of packaging and agricultural mulching films.

Table 11.6 summarizes the application of TPS composites derived from potato starch. Potato starch is promising as a renewable material with potential application in packaging and agricultural industries.

11.10   THERMOPLASTICS FROM VARIOUS SOURCES

The production of TPS was obtained through various types of starch. Table 11.7 presents a summary of TPS gained from many sources apart from plants, as discussed above, including mung bean, pea, sugar palm, and wheat. Several analyses of thermoplastics have been made on various starches since they possess good properties and characteristics that make them suitable for various applications besides improving the overall product.

11.11   CONCLUSION

In general, TPS is considered to be a potential replacement for conventional synthetic plastics. Although TPS has advantages such as low cost, biodegradable, renewable, and sustainable, it is not completely free of problems. A serious problem of TPS is its hydrophilic behavior which tends to absorb moisture from the environment, hence affecting the mechanical properties and dimensional stability of the material. Several modifications have been conducted on TPS including blending with other polymers, reinforced with natural fibers, and treatments to improve the functional characteristics of this material. The findings show that blending with other polymers, incorporating with fibers, and treatments of fibers can reduce the hydrophilic behavior of the material while improving mechanical properties as well. Various

TABLE 11.6
Thermoplastic Potato Starch

Type of Starch	Type of Fiber or Polymer	Type of Testing	Main Finding	Potential Application	References
Potato, sweet potato, corn	Chitosan	DSC, TGA, SEM	Improved the thermal properties	Biodegradable composite	[39]
Potato	Clay	XRD, TEM, tensile test, thermal properties	Improved the mechanical properties by tensile and impact test	Packaging application	[40]
Potato	Agar	FT-IR, SEM, water absorption, tensile test, WVP	Improved microstructure of starch film, meliorated mechanical properties, and WVP at high moisture environment	Food packaging	[41]
Potato, sweet potato, corn	Natural fibers (sisal, jute, and cabuya)	Tensile test, impact test, the fracture surface	Improved mechanical properties by tensile and impact tests	Biodegradable composite	[42]
Potato	PBAT	PBAT/TPS blends film, PBAT/TPS/ SMG blends films, PBAT/M-PBAT/ TPS blends films	Improved mechanical properties by tensile strength and elongation	Disposable packaging, agricultural mulching film	[38]
Potato	Corn, poly(vinyl alcohol)	TGA, SEM, trays testing	Improved tensile properties	Biodegradable materials	[43]

DSC, differential scanning calorimetry; FT-IR, Fourier-transform infrared spectroscopy; SEM, scanning electron microscope; TEM, transmission electron microscopy; TGA, thermogravimetric analysis; WVP, water vapor permeability; XRD, X-ray diffraction.

TABLE 11.7

Thermoplastic Starch from Various Sources

Type of Starch	Type of Fiber	Type of Testing	Main Finding	Potential Application	References
Mung bean	Cotton, polyethylene	SEM, tensile test, water absorption, thermal properties, soil burial test	Increased the mechanical strength and rigidity of the films but water vapor permeability of the films decreased	Hard capsules for pharmaceutical applications	[44]
Sugar palm	Seaweed	FT-IR, SEM, tensile test, flexural testing, impact test, TGA, density determination balance, moisture content, water absorption, thickness swelling, moisture absorption, water solubility, soil burial test, statistical analysis	Improved tensile and flexural properties accompanied by lower impact resistance	Biodegradable product	[45]
Wheat	Poly(lactic acid), methylenediphenyl diisocyanate	DMA, tensile test, water absorption	Improved the tensile properties, microstructure, and water absorption	Short-term disposable applications	[46]
Pea	Citric acid	TGA, DMTA, WVP, mechanical testing	Improved the storage modulus, the glass transition temperature, the tensile strength, and the water vapor barrier, but decreased thermal stability CARS/TPS	Biodegradable plastics	[47]

DMA, dynamic mechanical analysis; DMTA, dynamic mechanical thermal analysis; FT-IR, Fourier transform infrared; TEM, transmission electron microscopy; WVP, water vapor permeability.

modifications and treatments have achieved various levels of success in improving the properties of TPS. In conclusion, TPS is a promising material for replacing the non-biodegradable polymer in the market, provided that proper modifications are carried out to achieve the desired characteristics of the final product.

ACKNOWLEDGMENT

The authors would like to thank Universiti Teknikal Malaysia Melaka for the financial support provided through Journal Publication Incentive Grant (JURNAL/2018/FTK/Q00004).

REFERENCES

1. R. Jumaidin, S. M. Sapuan, M. Jawaid, and M. R. Ishak, "Effect of Agar on Flexural, Impact, and Thermogravimetric Properties of Thermoplastic Sugar Palm Starch," *Curr. Org. Synth.*, vol. 14, no. 2, pp. 200–205, 2017.
2. M. Bootklad and K. Kaewtatip, "Biodegradation of thermoplastic starch/eggshell powder composites," *Carbohydr. Polym.*, vol. 97, no. 2, pp. 315–320, 2013.
3. F. Garavand, M. Rouhi, S. H. Razavi, I. Cacciotti, and R. Mohammadi, "Improving the integrity of natural biopolymer films used in food packaging by crosslinking approach: a review," *Int. J. Biol. Macromol.*, vol. 104, pp. 687–707, 2017.
4. R. Rebelo, M. Fernandes, and R. Fangueiro, "Biopolymers in medical implants: a brief review," *Procedia Eng.*, vol. 200, pp. 236–243, 2017.
5. N. Soykeabkaew, C. Thanomsilp, and O. Suwantong, *A Review: Starch-Based Composite Foams*, vol. 78, Elsevier Ltd, 2015.
6. A. E. Wiacek, "Effect of surface modification on starch biopolymer wettability," *Food Hydrocoll.*, vol. 48, pp. 228–237, 2015.
7. D. R. Bogati, "Cellulose based biochemicals and their applications," *Degree Program. Pap. Technol.*, vol. Bachelor, p. 37, 2011.
8. M. Okamoto and B. John, "Synthetic biopolymer nanocomposites for tissue engineering scaffolds," *Prog. Polym. Sci.*, vol. 38, no. 10–11, pp. 1487–1503, 2013.
9. A. Chauhan and P. Chauhan, "Natural fiber reinforced composite: a concise review article," *J. Chem. Eng. Process Technol.*, vol. 03, no. 02, pp. 2–4, 2012.
10. M. R. Sanjay, G. R. Arpitha, L. L. Naik, K. Gopalakrishna, and B. Yogesha, "Applications of natural fibers and its composites: an overview," *Nat. Resour.*, vol. 07, no. 03, pp. 108–114, 2016.
11. K. Lau, P. Hung, M. Zhu, and D. Hui, "Properties of natural fi bre composites for structural engineering applications," *Compos. Part B*, vol. 136, no. September, pp. 222–233, 2018.
12. L. Mohammed, M. N. M. Ansari, G. Pua, M. Jawaid, and M. S. Islam, "A review on natural fiber reinforced polymer composite and its applications," *Int. J. Polym. Sci.*, vol. 2015, pp. 1–15, 2015.
13. K. M. Gupta, "CHAP 8 starch based composites for packaging applications," *Handbook of Bioplastics and Biocomposites Engineering Applications*, pp. 189–266, 2011.
14. E. Espigulé, X. Puigvert, F. Vilaseca, J. A. Mendez, P. Mutjé, and J. Girones, "Thermoplastic starch-based composites reinforced with rape fibers: water uptake and thermomechanical properties," *BioResources*, vol. 8, no. 2, pp. 2620–2630, 2013.
15. R. Jumaidin et al., "Water Transport and Physical Properties of Sugarcane Bagasse Fibre Reinforced Thermoplastic Potato Starch Biocomposite," *J. Adv. Res. Fluid Mech. Therm. Sci.*, vol. 61, no. 2, pp. 273–281, 2019.

16. A. L. Da Roz, A. Carvalho, A. Gandini, and A. Curvelo, "The effect of plasticizers on thermoplastic starch compositions obtained by melt processing," *Carbohydr. Polym.*, vol. 63, pp. 417–424, 2006.

17. Y. Zuo, J. Gu, H. Tan, and Y. Zhang, "Thermoplastic starch prepared with different plasticizers: relation between degree of plasticization and properties," *J. Wuhan Univ. Technol. Sci. Ed.*, vol. 30, no. 2, pp. 423–428, 2015.

18. F. Ivanič, D. Jochec-Mošková, I. Janigová, and I. Chodák, "Physical properties of starch plasticized by a mixture of plasticizers," *Eur. Polym. J.*, vol. 93, pp. 843–849, 2017.

19. P. Y. Mikus *et al.*, "Deformation mechanisms of plasticized starch materials," *Carbohydr. Polym.*, vol. 114, pp. 450–457, 2014.

20. A. Edhirej, S. M. Sapuan, M. Jawaid, and N. I. Zahari, "Cassava/sugar palm fiber reinforced cassava starch hybrid composites: physical, thermal and structural properties," *Int. J. Biol. Macromol.*, vol. 101, pp. 75–83, 2017.

21. C. Bangyekan, D. Aht-Ong, and K. Srikulkit, "Preparation and properties evaluation of chitosan-coated cassava starch films," *Carbohydr. Polym.*, vol. 63, pp. 61–71, 2006.

22. J. Prachayawarakorn, S. Chaiwatyothin, S. Mueangta, and A. Hanchana, "Effect of jute and kapok fibers on properties of thermoplastic cassava starch composites," *Mater. Des.*, vol. 47, pp. 309–315, 2013.

23. W. Sanhawong, P. Banhalee, S. Boonsang, and S. Kaewpirom, "Effect of concentrated natural rubber latex on the properties and degradation behavior of cotton-fiber-reinforced cassava starch biofoam," *Ind. Crops Prod.*, vol. 108, no. July, pp. 756–766, 2017.

24. S. Narkchamnan and C. Sakdaronnarong, "Thermo-molded biocomposite from cassava starch, natural fibers and lignin associated by laccase-mediator system.," *Carbohydr. Polym.*, vol. 96, no. 1, pp. 109–117, 2013.

25. J. Prachayawarakorn and W. Pomdage, "Effect of carrageenan on properties of biodegradable thermoplastic cassava starch/low-density polyethylene composites reinforced by cotton fibers," *Mater. Des.*, vol. 61, pp. 264–269, 2014.

26. M. G. Lomelí Ramírez, K. G. Satyanarayana, S. Iwakiri, G. B. De Muniz, V. Tanobe, and T. S. Flores-Sahagun, "Study of the properties of biocomposites. Part I. Cassava starch-green coir fibers from Brazil," *Carbohydr. Polym.*, vol. 86, pp. 1712–1722, 2011.

27. J. Prachayawarakorn and W. Pattanasin, "Effect of pectin particles and cotton fibers on properties of thermoplastic cassava starch composites," *IOP Conf. Ser. Mater. Sci. Eng.*, vol. 38, no. 2, pp. 129–136, 2016.

28. J. Prachayawarakorn, P. Ruttanabus, and P. Boonsom, "Effect of cotton fiber contents and lengths on properties of thermoplastic starch composites prepared from rice and waxy rice starches," *J. Polym. Environ.*, vol. 19, pp. 274–282, 2011.

29. J. Prachayawarakorn and W. Hwansanoet, "Effect of silk protein fibers on properties of thermoplastic rice starch," *Fibers Polym.*, vol. 13, no. 5, pp. 606–612, 2012.

30. G. H. Yew, A. M. Mohd Yusof, Z. A. Mohd Ishak, and U. S. Ishiaku, "Water absorption and enzymatic degradation of poly(lactic acid)/rice starch composites," *Polym. Degrad. Stab.*, vol. 90, no. 3, pp. 488–500, 2005.

31. J. Prachayawarakorn, P. Sangnitidej, and P. Boonpasith, "Properties of thermoplastic rice starch composites reinforced by cotton fiber or low-density polyethylene," *Carbohydr. Polym.*, vol. 81, no. 2, pp. 425–433, 2010.

32. J. Prachayawarakorn, N. Limsiriwong, R. Kongjindamunee, and S. Surakit, "Effect of agar and cotton fiber on properties of thermoplastic waxy rice starch composites," *J. Polym. Environ.*, vol. 20, no. 1, pp. 88–95, 2011.

33. O. Lopez, M. a. Garcia, M. a. Villar, a. Gentili, M. S. Rodriguez, and L. Albertengo, "Thermo-compression of biodegradable thermoplastic corn starch films containing chitin and chitosan," *LWT - Food Sci. Technol.*, vol. 57, no. 1, pp. 106–115, 2014.

34. J. L. Guimarães, F. Wypych, C. K. Saul, L. P. Ramos, and K. G. Satyanarayana, "Studies of the processing and characterization of corn starch and its composites with banana and sugarcane fibers from Brazil," *Carbohydr. Polym.*, vol. 80, no. 1, pp. 130–138, 2010.

35. Q. Xie *et al.*, "A new biodegradable sisal fiber–starch packing composite with nest structure," *Carbohydr. Polym.*, vol. 189, pp. 56–64, 2018.

36. H. Ibrahim, M. Farag, H. Megahed, and S. Mehanny, "Characteristics of starch-based biodegradable composites reinforced with date palm and flax fibers," *Carbohydr. Polym.*, vol. 101, pp. 11–19, 2014.

37. N. F. Magalhães and C. T. Andrade, "Thermoplastic corn starch/clay hybrids: effect of clay type and content on physical properties," *Carbohydr. Polym.*, vol. 75, no. 4, pp. 712–718, 2009.

38. D. Wei, H. Wang, H. Xiao, A. Zheng, and Y. Yang, "Morphology and mechanical properties of poly(butylene adipate-co-terephthalate)/potato starch blends in the presence of synthesized reactive compatibilizer or modified poly(butylene adipate-co-terephthalate)," *Carbohydr. Polym.*, vol. 123, pp. 275–282, 2015.

39. C. Gómez, F. G. Torres, J. Nakamatsu, and O. H. Arroyo, "Thermal and structural analysis of natural fiber reinforced starch-based biocomposites," *Int. J. Polym. Mater.*, vol. 55, no. 11, pp. 893–907, 2006.

40. H. M. Park, X. Li, C. Z. Jin, C. Y. Park, W. J. Cho, and C. S. Ha, "Preparation and properties of biodegradable thermoplastic starch/clay hybrids," *Macromol. Mater. Eng.*, vol. 287, no. 8, pp. 553–558, 2002.

41. Y. Wu, F. Geng, P. R. Chang, J. Yu, and X. Ma, "Effect of agar on the microstructure and performance of potato starch film," *Carbohydr. Polym.*, vol. 76, no. 2, pp. 299–304, Mar. 2009.

42. F. G. Torres, O. H. Arroyo, and C. Gomez, "Processing and mechanical properties of natural fiber reinforced thermoplastic starch biocomposites," *J. Thermoplast. Compos. Mater.*, vol. 20, no. 2, pp. 207–223, 2007.

43. P. Cinelli, E. Chiellini, J. W. Lawton, and S. H. Imam, "Foamed articles based on potato starch, corn fibers and poly(vinyl alcohol)," *Polym. Degrad. Stab.*, vol. 91, no. 5, pp. 1147–1155, 2006.

44. J. Prachayawarakorn, L. Hommanee, D. Phosee, and P. Chairapaksatien, "Property improvement of thermoplastic mung bean starch using cotton fiber and low-density polyethylene," *Starch/Staerke*, vol. 62, no. 8, pp. 435–443, 2010.

45. R. Jumaidin, S. M. Sapuan, M. Jawaid, M. R. Ishak, and J. Sahari, "Effect of seaweed on mechanical, thermal, and biodegradation properties of thermoplastic sugar palm starch/agar composites," *Int. J. Biol. Macromol.*, vol. 99, pp. 265–273, 2017.

46. H. Wang, X. Sun, and P. Seib, "Mechanical properties of poly(lactic acid) and wheat starch blends with methylenediphenyl diisocyanate," *J. Appl. Polym. Sci.*, vol. 84, no. 6, pp. 1257–1262, 2002.

47. X. Ma, P. R. Chang, J. Yu, and M. Stumborg, "Properties of biodegradable citric acid-modified granular starch/thermoplastic pea starch composites," *Carbohydr. Polym.*, vol. 75, pp. 1–8, 2009.

12 Modification of Thermoplastic Starch with Natural Fiber

R. Jumaidin and N.W. Adam
Universiti Teknikal Malaysia Melaka

S.M. Sapuan
Universiti Putra Malaysia

CONTENTS

12.1 INTRODUCTION

In recent years, plastics have been widely used in packaging, electric and electronic devices, and automotive industries. However, plastics derived from non-biodegradable materials produced from petroleum-based fuel gave rise to critical issues for the environment and human beings [1]. The development of biodegradable products based on agricultural materials has seen great progress [2]. Hence, to tackle environmental and sustainable issues, this century focuses on greater achievements in green technology products through the development of bio-composites. Nowadays, there is increased attention on natural resources for creating biodegradable polymers as alternatives in the replacement of oil-based polymers [3].

Starch has been one of the most promising biopolymers due to being abundant, low in cost, biodegradable, and renewable. Starch can be a thermoplastic material by adding plasticizer under high temperature. Plasticizer agents are significant materials in modifying the properties of starch to increase its flexibility and processability in thermoplastic materials [1]. The uses of natural reinforcing fibers such as flax, ramie, and hemp into the biopolymer matrix derived from starch have been the new development in bio-composites [4].

Incorporation of these materials can create environmentally friendly conditions and reduce the dependency on conventional fibers (such as glass fibers) which carry potential hazards in manufacturing. Thus, the development of environmentally friendly materials is the best way for industries to overcome ecological issues for both disposal problems and production of synthetic polymers.

12.2   BIOPOLYMERS

Biopolymers are polymers synthesized from living organisms such as cellulose, protein, and starch. When natural reinforced fibers such as ramie, jute, or flax are embedded into the biopolymer matrix that comes from the derivatives of starch, chitosan, cellulose, or lactic acid, this new fiber-reinforced material is known as bio-composites [4]. Bio-composites consist of biodegradable polymers as the matrix material and usually produce reinforced materials called bio-fibers. This composite is biodegradable [4]. According to Mohanty et al. [4], when the enzyme reacts with living organisms such as bacteria and fungi and secretion products, this reaction is called biodegradability. There are also abiotic reactions such as photodegradation, oxidation, and hydrolysis which take place during or instead of the biodegradation process due to environmental factors that cause change in the polymer.

Biodegradable polymers can be categorized as biosynthetic, semi-biosynthetic, and chemosynthetic types [4]. Most biosynthetic polymers that come from renewable resources are biodegradable. In addition, semisynthetic and chemosynthetic types can also be biodegradable if they have a chemical bond in a structure that occurs in the natural compound. Thus, biodegradability also results from natural compounds. There are three main types of biopolymers: starch-based, synthetic-based, and cellulose-based biopolymers. Biopolymers can be produced either naturally or through polymerization of bio-based monomers [5].

As shown in Figure 12.1, bio-based materials can be categorized into different origins. In current polymerization processes, the polymer is suited to be stiff, soft, rubbery, a conductive insulator, optical, permeable, and non-permeable, as well as biodegradable. The most prominent examples of the application of polymers would be in food and medical packaging materials, automotive and aerospace industries, lightweight engineering plastic, electronic components, construction materials, composite materials, and biomedical applications such as wound dressing and water purification [5].

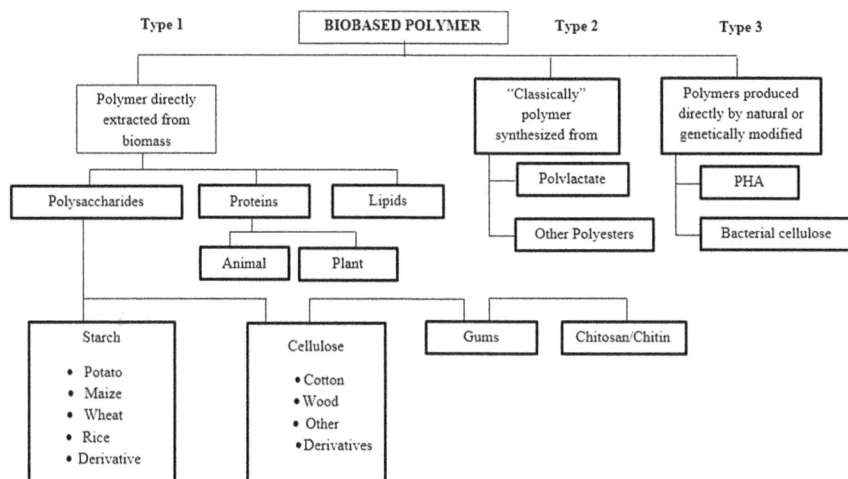

FIGURE 12.1   Origin of different categories of biologically based materials [6,7].

12.2.1   STARCH BIOPOLYMER

Mülhaupt [5] revealed that starch is formed as granules and stored as energy in most green plants. Its characteristics include high water solubility and easy degradation, and it contains a high mixture of amylose and has a branch of amylopectin. The structural unit of starch consists of amylose (20%–30%) and amylopectin (70%–80%). Both contain a branch of polymer a-D-glucose unit, and in amylose, there is a linkage which has oxygen atom on the same side. Amylopectin consists of linkage that forms branch points. This link depends on the sources of starch. Starch is found in potato, maize, wheat, and other plants and living organisms. Most starches are utilized in the food industry and non-food applications such as cosmetics, textile, adhesives, and dextrose sugar for biotechnology [5].

In contrast to cellulose, amylose can exhibit thermoplastic behavior by adding water and glycerol which act as plasticizers [5,8]. The limitations of starch-based foams are weakness and water sensitivity. To overcome the limitations and improve the mechanical, thermal, and water resistance, moldability, and weight properties of starch, various techniques of chemical modifications (such as blending starch with biodegradable polymer or adding natural fibers and fillers) have been accomplished in previous works [9]. Generally, by adding plasticizers such as glycerol, sorbitol, and water under high temperature, the granular structure of starch should be partially destroyed and turned into amylose or amylopectin semi-crystalline matrix. The addition of plasticizer into starch can improve the flexibility of starch and enable it to melt at lower decomposition temperatures, leading to the improvement of processing starch into thermoplastic.

When starch is blended with a biodegradable polymer such as polylactic acid (PLA), it can enhance its compatibility [9]. This improvement can be made by using the method of gelatinization of starch in the presence of water and another plasticizer.

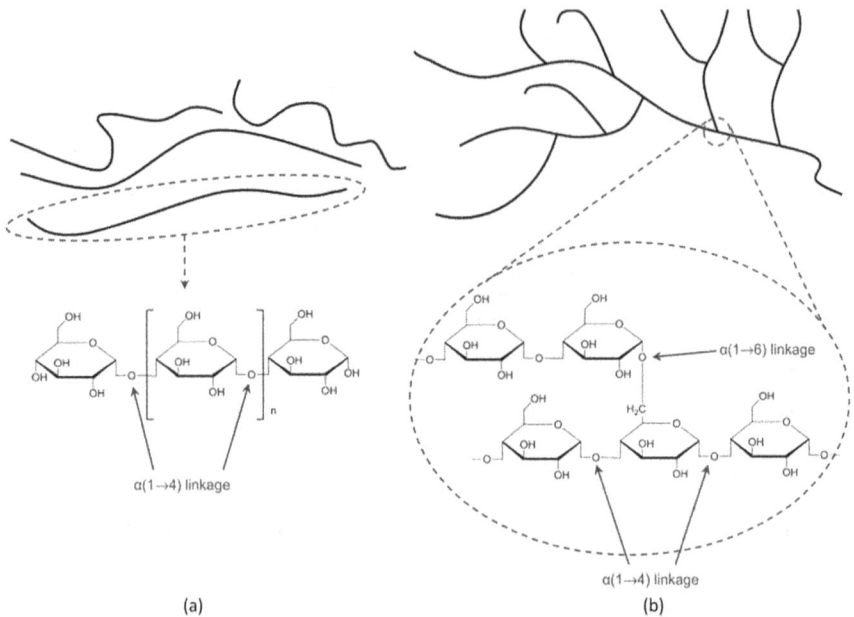

FIGURE 12.2   Starch's molecular structure (a) amylose (b) amylopectin [12].

Glass transition temperature and mechanical properties of starch are dependent on the plasticizer content [9].

A previous study also proved that adding natural fibers as reinforcement materials which have lignocellulose (such as jute, corn, wood, kenaf, and nano-fillers like clay) resulted in the high performance of starch-based composites. There is a drawback when thermoplastic starch (TPS) is produced with biodegradable films where poor mechanical properties are present compared to other synthetic polymers. This is due to the hydrophilic nature that occurs in the component [10]. The advantage of this natural polymer is that starch is low in cost and an extremely abundant biopolymer [11]. Figure 12.2 shows the molecular structure of starch.

12.3   COMPOSITES

Due to environmental and sustainability issues, this century has sought awareness in the accomplishment of a green economy in the field of engineering materials through the development of composites [13]. For example, potential utilization of bamboo composite materials for structural applications was examined by using durability and toughness test [14]. To put it simply, composites are materials composed from two or more materials where one has different properties. Incorporation of reinforcement into matrix will improve the strength of the material such as sugarcane bagasse into thermoplastic composites [15]. Most composites are based on natural materials that are biological.

Natural composites can be found in plants or animals. For example, wood is a composite made up of long cellulose fiber surrounded by lignin. Composites are also found in the bone of the human body. Bones have a combination of strength with the blend of calcium phosphate and collagen called "protein." Nowadays, composites are in high demand in countries because they provide numerous advantages; they are lightweight, have high strength, and are corrosion resistant. They are produced to optimize the mechanical, physical, and optical properties of materials.

The opportunity of creating an advanced composite from natural fiber will help develop a sustainable product for automotive industries, sound proofing, as well as for structural applications [16]. Currently, composites are widely used in civil engineering fields (e.g., bridge structure) and also in biomedical products such as prosthetic devices [17]. The challenges of composites are related to the automotive composite design. In engineering materials, the matrix can be polymers, metals, and ceramics. Thus, these combinations are known as matrix materials categorized into three types: polymer matrix composite (PMC), metal matrix composite (MMC), and ceramic matrix composite (CMC).

12.3.1   POLYMER MATRIX COMPOSITE (PMC)

PMC is made up of various types of organic polymers which consist of short or continuous fibers as the reinforcement materials, making it possible to improve properties such as fracture toughness, strength, and stiffness [18]. The most popular composite material now is PMC. The mechanical loads applied to PMC are supported by fibers. The function of the polymer matrix is to promote the fibers as well as transfer loads between them. Thus, the aim of the PMC design is to transfer mechanical loads subjected to the structures which are supported by reinforcement materials [18]. Fibers in PMC have low structural value where stiffness is along the linear direction but there is no cross-stiffness of strength. For the polymer matrix, it is less strong and less stiff, but tougher and often more chemically inactive than fibers. PMC reinforced with fibers has higher resistance and good interfacial bonding between the matrix and fibers, maintaining mechanical and chemical properties.

Production of PMC is simpler and easier compared to CMCs or MMCs in terms of fabrication. In engineering applications, thermoplastic polymers are more commonly used due to their good chemical resistance, good mechanical properties, and lower cost [18]. However, there is a drawback in using polymers: they are non-biodegradable after their end of life phase.

Thus, to overcome this drawback, composites are made by incorporating polymer matrix with natural fibers. This composite has superior advantages such as being lightweight, rigid, and dynamic reinforcement [15].

12.3.2   METAL MATRIX COMPOSITE (MMC)

MMCs can be formed by other metals as reinforcement. In engineering materials, MMCs are developed by combining metal and alloy with other materials to increase their performance [15]. This type of composite consists of several different and distinct materials according to their base metal, such as copper, aluminum, and titanium. Processing MMC has several reasons. Making composites from the metal matrix is one approach for significant metal oxide or carbide [19]. As reinforcement, iron is frequently used with carbide and oxide because carbon and oxygen are dissolved in molten metal. Second, the purpose of this composite is to produce metallic materials by considering the specific modulus of the metal, elastic modulus, and transverse elasticity for the component, as well as to know the deformation limit for structural applications [19].

12.3.3 CERAMIC MATRIX COMPOSITE (CMC)

Recently, the demand for high performance of materials has been increasing in aerospace and biomedical applications. To achieve product quality with fracture toughness properties, CMC is made to fulfill the demand. CMC is the combination of ceramic matrix reinforced with intractable fibers such as silicon carbide and silicon fiber [15].

The natural ceramic matrix contains bitumen concrete, dental composite, mastic bitumen, and pure pearl [15]. In order to achieve high strength and fracture toughness, the elements to prevent break from spreading were added in ceramic materials. The research and theory have been discovered to evaluate the factor of elevated temperature effects on the mechanical behavior of reinforced glass and ceramic [20].

12.4   NATURAL FIBER-REINFORCED COMPOSITES

Fibers are commonly categorized according to their origin, whether derived from plants, animals, or minerals. Generally, plant fibers have higher properties, with high strength and stiffness, compared to animal fibers that are readily available. Thus, this makes the plant fiber more suitable for use in composites. Natural fiber-reinforced composites (NFRCs) are composite materials consisting of matrix material (polymer) implanted with natural fibers such as jute, palm oil, kenaf, and flax. Basically, matrices that are used in NFRC are from polymeric sources. Polymers can be categorized into two types: thermoset and thermoplastic [21].

Thermoset polymers can be defined as polymers with high cross-linked structures that are preserved using heat, pressure, or light radiation, whereas thermoplastic materials consist of two-dimensional molecular structures [22]. The structure of thermoset provides good properties for the polymer such as high flexibility, increased strength, and great modulus. Polyurethane (PU), polypropylene (PP), polyvinyl chloride (PVC), and epoxy are examples of synthetic polymers [22].

Thermoplastic is the most useful polymer that plays a crucial role of increasing the performance of polymer composites. In thermoset composites, several materials are involved in its formulation such as base resin, curing agents, following agents, catalysts, and hardness. The structure of this composite is highly cross-linked and is three-dimensional for the work structure. It has characteristics of high solvent resistance, toughness, and creep resistance. High fiber content (80%) can cause the high fiber loading to enhance the properties of the composite [23].

Thermoplastics provide numerous advantages over thermosets. The advantages of thermoplastics are low processing cost, flexibility in design, and ease for molding complex parts. To produce and process these composites, the simple methods of extrusion and injection molding are used [23]. In the development of NFRC, the properties of fiber and the ratio of fiber and matrix contribute to the properties of the composite. Moreover, the performance of composites can be evaluated by the transmission of stress from the matrix to fiber that usually comes from the surface adhesion between the fiber and matrix. Other factors that influence the properties of NFRCs are the composition of fiber, the ratio of fiber, fiber matrix adhesion, and orientation [23].

The polymer used as matrix materials in NFRC are natural polymers such as starch, cellulose, silk, and wool, and the reinforcement material used is natural fibers [18]. Usually, the hydrophilic behavior present in natural fiber and fiber loading provides impact to composite properties [13]. It was noticed that high fiber content displays improvement in tensile properties of the composite. Another factor that affects the properties of the composite involves parameters where suitable process parameters should be chosen to obtain the best characteristics of the composite. The characteristics of the composite also depend on the chemical composition of natural fibers represented by the percentage of cellulose, hemicellulose, lignin, and waxes.

NFRCs do have some limitations such as poor compatibility between fiber and matrix and high water absorption. Therefore, to improve the adhesion of fiber with different matrices, fiber surface properties must be modified. Strength and stiffness can be achieved with brittle surface that has strong interface and stands crack propagation between fibers and matrix. The stress transfer from the matrix can be more efficient when the interface becomes weaker [13]. Poor compatibility of NFRC has opened a new research gap for researchers to tackle these issues. Most studies on natural fiber composites involved the effect between fiber content and mechanical properties, affected by physical and chemical treatments of fiber [24].

Goulart et al. [25] developed PP-reinforced sugarcane bagasse to examine the mechanical behavior of NFRC. Sugarcane bagasse was pretreated with 10% of sulfuric acid solution at 120°C in 10 min, and then, 1% of sodium hydroxide solution (NaOH) was dissolved in the extracted lignocellulosic fraction. The cellulose fiber was obtained by solvent mixtures. These fibers were mixed with polypropylene, and the range of fiber used was 5–20 wt%. It was declared that the chemical treatment affected the mechanical properties of the fiber matrix adhesion compared to pure polymer PP. The result illustrated that tensile strength of the composite increased by 16% and the tensile modulus increased by 51% compared to pure PP. The increase in tensile strength and modulus was due to the high strength of Young's modulus of fibers and good interfacial bonding between the fibers and matrix. In addition, the flexural properties are higher compared to PP. The composites showed an increase of 45% value of impact strength compared to pure PP. This was attributed to the matrix interface and test condition. Figure 12.3 presents the fracture of PP/F SB (20%) composite by using SEM testing. This literature proved that the modification of fibers from sugarcane bagasse did enhance the tensile, flexural, and impact strength compared to pure polymer.

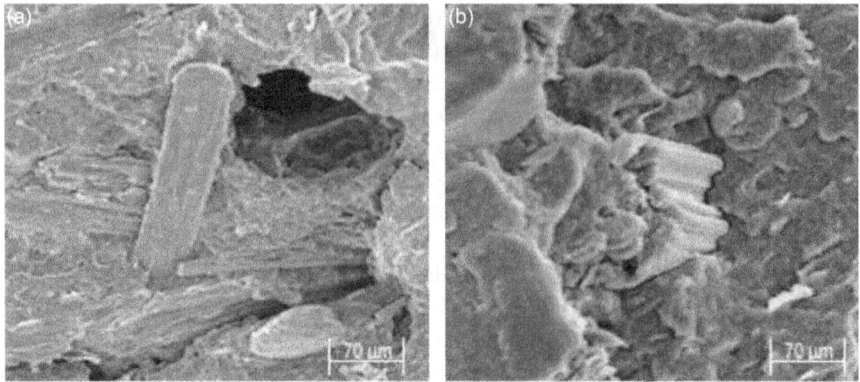

FIGURE 12.3   SEM of fracture of PP/F SB 20% composite [25].

12.5   THERMOPLASTIC STARCH

TPS is one of the bio-based materials available in the production of biodegradable plastics [11]. This material is important for environmental consciousness to avoid environmental problems due to petrochemical synthetic polymers and the high cost of petroleum-based materials. Starch is a natural polymer that has biodegradable characteristics. The renewable raw materials are abundant in the natural agricultural environment. Thermoplastic materials derived from starch have numerous advantages such as low cost and biodegradable characteristics [26]. Previous researchers are now focused on modifying the structure of TPS by chemical modification during the extrusion process. The aims of this research are to reduce its hydrophobic nature and improve mechanical properties [2].

To develop TPS, the granular structure of the starch must be either totally or partially destroyed and changed into amylose/amylopectin semi-crystalline structure under high temperature, and apply plasticizers such as glycerol and sorbitol. The biodegradable composite produced from pure TPS has poor mechanical properties and is water sensitive compared to synthetic polymers. This is due to the hydrophobic nature of starch [27]. During fabrication of thermoplastics, glycerol plays an important role as a plasticizer agent and forms hydrogen bonds within starch. Starch has multi hydroxyl groups with three hydroxyl branches of monomers. When starch is incorporated with a plasticizer, the original bonds of hydroxyl groups are destroyed, allowing starch to be plasticized [28].

12.5.1   THERMOPLASTIC CORN STARCH

Corn is one of the main sources of starch in the agricultural environment. The countries that provide the best condition for the growth of corn are the United States, China, and Brazil. The large production of corn starch can be found in Brazil which could be largely expanded to non-food applications [29]. TPS is one of the bio-based materials for biodegradable plastic [30]. In fact, the composition of amylose in corn starch is about 28 wt% higher than cassava starch with 17 wt%. The mechanical properties, processing condition, and barrier all depend on the ratio of amylose and amylopectin. In general, increasing the amount of amylose can improve mechanical properties [31]. Starch can be flexible with the addition of suitable plasticizers, such as glycerol, as the chemical agent. Glycerol can be obtained by chemical production or in the production of bio-sel [29]. Due to the awareness of global concerns for recycling natural resources such as corn starch and the waste of glycerol, a new approach has been made by applying modifications on native corn starch which could improve the poor properties of TPS derived from corn starch. To tackle this issue, the previous research has opened new research gaps.

Ma et al. [28] utilized the properties of composites derived from winceyette fiber and corn starch. Thermoplastic corn starch (TPCS) reinforced with winceyette fiber composites was prepared by incorporating winceyette into TPCS which was premixed with urea and form-amide as plasticizers (200 rpm in 2 min) and mixed together. The results showed that the incorporation of winceyette fiber has improved thermal stability, water resistance, and tensile strength. The fiber with a low percentage (5%) also improved the thermal stability. The mass loss onset temperature for composites and TPCS-matrix was 15% and 10%, respectively. The result in mass loss for pure TPCS came from moisture content and plasticizer [2]. The result of higher mass loss in the composite was due to good adhesion between the fiber and TPCS-matrix. As the fiber content increased, the mass loss and thermal stability of the composites increased. The modification of TPS derived from corn starch achieved better water resistance. This improvement was ascribed to the less hydrophobic character of the fiber. The tensile strength also rose as fiber content increased at 20%. The result further showed that by incorporating fiber into TPCS, Young's modulus of TPCS composite displayed good behavior. Obviously, the increase of tensile strength was attributed to good compatibility between the matrix and fiber. This was further due to the good adhesion between the fiber and matrix where starch has good interface with the cellulose fiber.

Guimarães et al. [29] used sugarcane bagasse and banana fiber as a reinforcement material for TPCS. The sugarcane and banana fiber were prepared and mixed with the corn starch and glycerol mixture.

The TPCS/sugarcane and banana fiber were mixed using a mixer and subjected to thermal molding at temperatures between 110°C and 170°C. The author reported significant findings of good thermal stability of TPCS reinforced with sugarcane and banana fiber (Figure 12.4). It was found that the tensile strength and Young's modulus also increased after adding banana fiber (30 wt%). It was declared that crude glycerin provides good adhesion between TPCS and fiber. Increasing tensile strength also depends on the processing method of the type of glycerol to improve adhesion. Table 12.1 presents a summary of TPCS composites from previous researches.

12.5.2 Thermoplastic Cassava Starch

Cassava starch (*Manihot esculenta*) is a starch material that contains proteins, lipids, lignocellulosic fiber, and sugar. This starch has been developed to a great extent in tropical countries such as Brazil. Cassava bagasse can be obtained by undergoing the extraction of cassava starch which includes the elimination of dissolvable sugars and

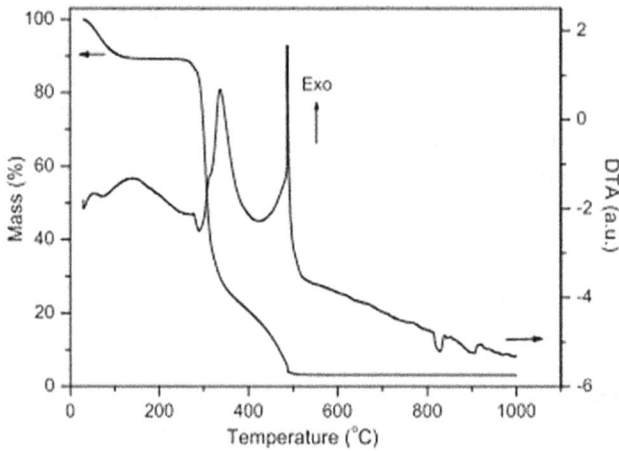

FIGURE 12.4   Thermal stability of TPS/sugarcane and banana fiber [29].

TABLE 12.1
Thermoplastic Corn Starch Composites

Starch	Fiber	Types of Testing	Main Finding	References
Corn	Banana and sugarcane	Thermo-gravimetric analysis (TGA), tensile test, X-ray diffraction (XRD), and water absorption test	Good thermal stability, ductile properties of the composite, tensile strength, and Young's modulus increased	[29]
Corn	Winceyette	Scanning electron microscopy (SEM), XRD, and thermal analysis (TA)	Water content, thermal stability, and water resistance have improved and tensile strength increased	[28]
Corn	Lignocellulosic	TGA, water permeability, film deterioration in soil, soil burial, and Fourier transform infrared spectroscopy (FTIR)	Thermal stability decreased, strong water permeability, and strongly adherence to the film surface	[32]

the separation of fiber to obtain pure starch and solid waste. This bagasse is predominantly made out of water (70–80 wt%), remaining starch, and cellulose fibers. The cellulose fiber content ranges between 15 and 50 wt% of solid waste (dry weight), and the rest is starch [33].

Extensive studies have been conducted on cassava starch as a biodegradable thermoplastic. In this study, the preparation of foam was made by mixing cassava starch, glycerol, water, sugarcane fiber, and polyvinyl alcohol. It was found that this mixture undergoes an extrusion process at an elevated temperature of glass transition (T_g) of

TPCS. The result indicated that by adding polyvinyl alcohol, the expansion index increased, leading to the reduction of water absorption of starch foam and finally enhancing foam properties. The main finding also showed that high addition of fiber content (90%) resulted in improved compression strength of TPCS. This study concludes that the addition of fibers can improve expansion properties, reducing water uptake, improving strength, and enhancing starch molecules [33].

Another study [34] concurred that TPCS/luffa fiber affects the chemical structure, fracture surface morphology, thermal stability, and mechanical properties of the composite. The TPCS/luffa fiber composite was prepared by incorporating the luffa fiber at 0, 5, 10, 15, and 20 wt% into TPCS which was premixed with glycerol (15 wt%). FTIR observation shows the good formation of hydrogen bonds between the luffa fiber and TPCS matrix compared to pure TPCS. The SEM micrograph in Figure 12.5 shows that fracture surface of TPCS composite has a smooth surface.

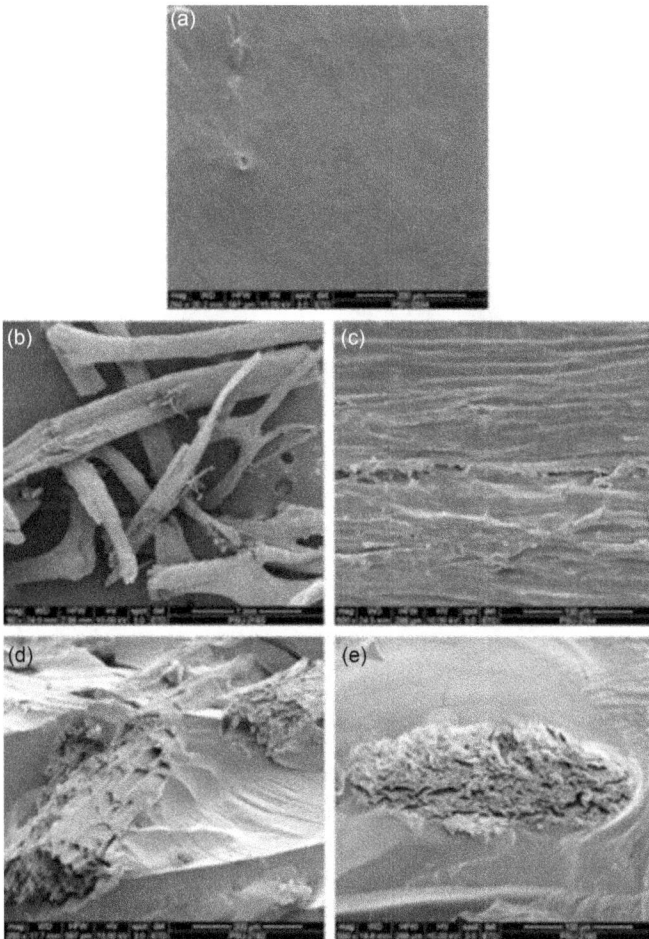

FIGURE 12.5   (a–e) SEM micrograph of TPS luffa fiber [34].

The coarse surface resulted in good adhesion between the luffa fiber and TPCS matrix. The composites also displayed a reduction in water absorption of materials (Figure 12.6). This was due to the less hydrophilic nature of the fiber compared to the starch matrix where starch is more hydrophilic than cellulose in fiber. Table 12.2 displays a summary of thermoplastic cassava starch composites.

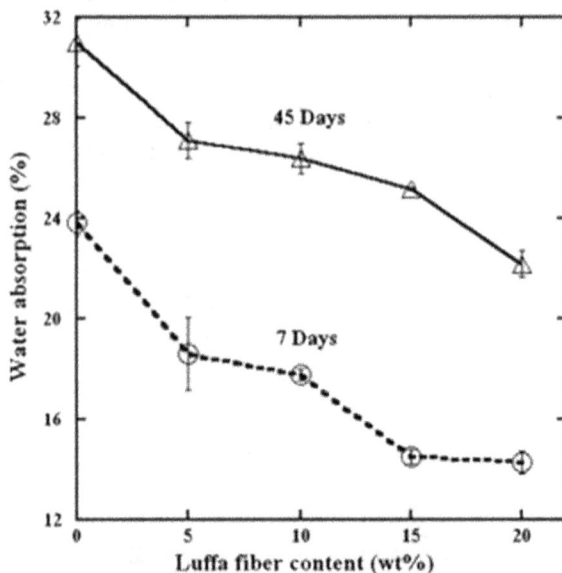

FIGURE 12.6   Water absorption (%) of luffa fiber content (%) for TPS and TPS luffa fiber composite [34].

TABLE 12.2
Thermoplastic Cassava Starch Composites

Starch	Fiber	Types of Testing	Main Finding	References
Cassava	Luffa	Tensile test, TGA, water absorption test, SEM, and FTIR	Tensile strength and thermal stability increased, and water absorption decreased	[34]
Cassava	Wheat fiber	Tension test, XRD, gravimetric, static method (moisture absorption), and SEM	Tensile strength did not alter, and rigidity films increased	[35]
Cassava	Coir (green coconut)	TGA, SEM, tensile test, water absorption, and swelling test	Tensile properties, improved and better interfacial bond was present	[36]
Cassava	Sugarcane	SEM and water absorption	Improved compression strength, foam properties, and water resistance	[37]

12.5.3 THERMOPLASTIC POTATO STARCH

Cellulose is an abundant biopolymer in the agriculture environment and a sustainable source for living organisms such as plants, animals, and bacteria. Potato is one of the current agricultural sources used for human consumption and fulfills demands for food products. Potato peel wastes are usually discarded and have negligible value because most potato peels are small in quantity and can only be sold at low prices for animal feed. The previous study proved that extraction of potato pulp produces cellulose microfibrils that can be reinforcing additives [38,39]. Chen et al. [39] investigated the effect of cellulose nanocrystals on the properties of thermoplastic potato starch (TPPS). Nanocrystals from potato peels were treated with alkali and hydroxide solution. Glycerol was added at 30 wt% to the potato starch as a plasticizer. The author reported that extracted cellulose of potato peel waste takes a longer time to obtain the fiber from microfibrils during acid hydrolysis compared to cellulose nanocrystal (CNC). By incorporating CNC into the polymer matrix, the experiment yielded good improvement of mechanical and barrier properties.

The addition of clay as a filler is an alternative method used to enhance mechanical and physical properties of TPPS/clay nanocomposite films [40]. In this study, it was found that the preparation of nanocomposites using natural filler is the best way to improve the properties of biodegradable polymer. By adding clay into the polymer, it will produce nanocomposites that exhibit an improvement in thermal, barrier, and oxidative properties compared to traditional composites. The addition of Na-montmorillonite (MMT) clay into the polymer blend of TPS showed higher hydrophilic character of starch. This was due to the high polarity of MMT. The incorporation of MMT (clay) presented remarkable results of great thermal stability.

This finding was associated with the reduction of water during the degradation process at varied temperatures. The increase of thermal stability was due to the heat barrier of the clay. Since the clay is an inorganic material, it has better chemistry compared to organic materials. Thus, the combination of inorganic materials enhances the thermal stability of composites.

Table 12.3 summarizes the previous studies on TPPS composites. In general, it can be seen that TPS potato starch has promising mechanical, thermal, and physical properties, improving the drawback of pure TPS.

12.6 APPLICATION OF THERMOPLASTIC STARCH

A wide range of TPS from biopolymers has been produced by utilizing current levels of innovation. The enrolment of various activities and advancements in bioplastics has aligned TPS towards appealing applications and market expansion. TPS has various applications in the industry by undergoing the manufacturing process, such as extrusion, to produce packaging material, biomedical instrument, agriculture, and textile. TPS is the best material to be used in packaging due to its moderate strength and easy degradation when disposed in the environment [42]. It can replace petroleum-based polymers which cannot degrade or are difficult to decompose, leading to environmental pollution and an unstable ecosystem. Increasing awareness on

TABLE 12.3

Thermoplastic Potato Starch Composites

Starch	Filler/Polymer	Types of Testing	Main Finding	References
Potato	CNCs and polyvinyl alcohol	FTIR, XRD, tensile testing, and water vapor transmission testing	Tensile modulus increased, marginal reduction of water permeability, and mechanical and barrier properties improved	[39]
Potato	Nano-clay	XRD, TGA, FTIR	Thermal resistance improved, Young's modulus increased and, water absorption decreased	[40]
Potato	Fluoro-elastomer	XRD, moisture absorption test, and tensile test	Mechanical properties improved, and moisture content decreased	[41]

sustainability, eco-efficiency, and green economy has driven researchers to recognize the potential use of TPS as a packaging material [43].

TPS can be used as biodegradable packaging material such as packaging bags by undergoing the blown films process [42]. Mechanical behavior of TPS films and thermosealing capacity were studied to develop packaging bags for food products [44]. It was found that the addition of talc nanoparticles into TPS films resulted in good thermo-seal, proving the good quality of the seal. This was attributed by the vanishing interface between both individual layers because of their high atomic collaboration, permitting the new shape of the homogeneous layer. Thermo-sealed tightness and tear resistance for packaging material are important properties to ensure in industrial products. These properties can be evaluated by inherent characteristics and tearing behavior of packaging materials. This study shows that the tightness of the packaging bag was improved by the addition of 3 wt% of talc due to the factor of increasing water films and oxygen barrier properties. From this study, it can be concluded that the incorporation of TPS with mineral filler like talc is suitable as a bio-nanocomposite for packaging since their functional properties are enhanced.

Apart from this, TPS was found to be a potential application in biomedical usage. The previous study has shown that materials from blended TPS with synthetic polymers can be used in biomedical applications such as scaffolds for bone tissue, bone cement, and filler for bone defects and cartilage [45]. As shown in previous works, starch-based polymers are cheaper. The thermal properties of TPS bio-based materials are suitable for making products for biomedical applications to prevent unpredictable physical changes when PLA reacts in the human body [45]. TPS also exhibited degradation, good adhesion, and excellent biocompatibility features.

12.7 SUMMARY

From the reviews conducted on natural fiber, biopolymers, composites, NFRCs, and TPS, it can be concluded that

i. The natural fiber structure's presence along the length of the fiber, its hydrophilic nature, and other linkages provide strength and stiffness suitable as a reinforcement material. The hydrophilic nature of natural fibers leads to some adhesion problems with the hydrophobic polymer matrix.

ii. Modification of NFRCs through physical and chemical treatments is another promising technique for improving the compatibility between fiber and matrix. Good adhesion of fiber with different matrices does enhance fiber surface properties.

iii. Starch has good properties when turned into TPS. Previous studies have proved that by incorporating starch with filler or fiber, improvements are made for this bio-based polymer material.

iv. The drawback of TPS can be improved through the addition of fiber. Biodegradable polymers derived from natural fibers can maintain their biodegradation characteristics compared to non-biodegradable polymers.

v. Sugarcane fiber is an abundant natural fiber presently utilized as a common renewable fiber for the development of composite materials. Biopolymers derived from this fiber have good mechanical and thermal properties which show their potential for making a rigid composite.

vi. The potential of potato starch blended with filler and fiber has been proven in the literature. However, TPPS reinforced with sugarcane fiber has not been studied yet.

ACKNOWLEDGMENT

The authors would like to thank Universiti Teknikal Malaysia Melaka for the financial support provided through Journal Publication Incentive Grant (JURNAL/2018/FTK/Q00004).

REFERENCES

1. J. Sahari, S. M. Sapuan, E. S. Zainudin, and M. A. Maleque, "Physico-chemical and thermal properties of starch derived from sugar palm tree (*Arenga pinnata*)," *Asian J. Chem.*, vol. 26, no. 4, pp. 955–959, 2014.

2. J. A. M. Curvelo, A. A. S., de Carvalho, A. J. F., & Agnelli, "Thermoplastic starch-cellulosic fibers composites," *Prelim. Results Carbohydr. Polym.*, vol. 45, no. 2, pp. 183–188, 2001.

3. L. Avérous and N. Boquillon, "Biocomposites based on plasticized starch: thermal and mechanical behaviour," *Carbohydr. Polym.*, vol. 56, pp. 111–122, 2004.

4. A. K. Mohanty, M. Misra, and G. Hinrichsen, "Biofibres, biodegradable polymers and biocomposites: an overview," *Macromol. Mater. Eng.*, vol. 276–277, pp. 1–24, 2000.

5. R. Mülhaupt, "Green polymer chemistry and bio-based plastics: dreams and reality," *Macromol. Chem. Phys.*, vol. 214, no. 2, pp. 159–174, 2013.

6. R. Van Tuil, P. Fowler, M. Lawther, and C. J. Weber, "Properties of biobased packaging materials," In *Production of Biobased Packaging Materials for the Food Industry*. Landskab og Planlægning/Københavns Universite: Center for Skov, p. 8±33, 2000.

7. C. J. Weber, V. Haugaard, R. Festersen, and G. Bertelsen, "Production and applications of biobased packaging materials for the food industry," *Food Addit. Contam.*, vol. 19, no. January, pp. 172–177, 2002.

8. L. P. B. M. Janssen and L. Moscicki. "Scaling-Up of Thermoplastic Starch Extrusion." In L. P. B. M. Janssen, & L. Moscicki (Eds.), *Thermoplastic Starch: A Green Material for Various Industries.* Weinheim, Germany: Wiley, pp. 219–229, 2009.

9. N. Soykeabkaew, C. Thanomsilp, and O. Suwantong, *A Review: Starch-Based Composite Foams, Compos. Part A Appl. Sci. Manuf.,* vol. 78, pp. 246–263, 2015.

10. P. Kampeerapappun, D. Aht-ong, D. Pentrakoon, and K. Srikulkit, "Preparation of cassava starch/montmorillonite composite film," *Carbohydr. Polym.,* vol. 67, pp. 155–163, 2007.

11. L. Yu, L., Dean, K., and Li, "Polymer blends and composites from renewable resources," *Prog. Polym. Sci.,* vol. 31, pp. 576–602, 2006.

12. F. Xie, E. Pollet, P. J. Halley, and L. Avérous, "Starch-based nano-biocomposites," *Prog. Polym. Sci.,* vol. 38, no. 10–11, pp. 1590–1628, 2013.

13. O. Faruk, A. K. Bledzki, H. P. Fink, and M. Sain, "Biocomposites reinforced with natural fibers: 2000–2010," *Prog. Polym. Sci.,* vol. 37, no. 11, pp. 1552–1596, 2012.

14. N. Rahman *et al.,* "Enhanced bamboo composite with protective coating for structural concrete application," *Energy Procedia,* vol. 143, pp. 167–172, 2017.

15. R. Jumaidin et al., "Water Transport and Physical Properties of Sugarcane Bagasse Fibre Reinforced Thermoplastic Potato Starch Biocomposite," *J. Adv. Res. Fluid Mech. Therm. Sci.,* vol. 61, no. 2, pp. 273–281, 2019.

16. I. Naghmouchi, F. X. Espinach, P. Mutjé, and S. Boufi, "Polypropylene composites based on lignocellulosic fillers: How the filler morphology affects the composite properties," *Mater. Des.,* vol. 65, pp. 454–461, 2015.

17. S. Kalpakjian and S. R. Schmid, *Manufacturing Engineering and Technology,* Prentice Hall, Pearson Education South Asia, vol. 6, 2010.

18. B. Yogesha, "Polymer matrix-natural fiber composites : an overview," *Cogent Eng.,* vol. 5, no. 1, Article no. 1446667, 2018.

19. A. Mortensen and J. Llorca, "Metal matrix composites," *Annu. Rev. Mater. Res.,* vol. 40, no. 1, pp. 243–270, 2010.

20. I. W. Donald and P. W. McMillan, "Ceramic-matrix composites," *J. Mater. Sci.,* vol. 11, no. 5, pp. 949–972, 1976.

21. H. Ku, H. Wang, N. Pattarachaiyakoop, and M. Trada, "A review on the tensile properties of natural fiber reinforced polymer composites," *Compos. Part B Eng.,* vol. 42, pp. 856–873, 2011.

22. L. Mohammed, M. N. M. Ansari, G. Pua, M. Jawaid, and M. S. Islam, "A review on natural fiber reinforced polymer composite and its applications," *Int. J. Polym. Sci.,* vol. 2015, pp. 1–15, 2015.

23. N. Saheb and J. Jog, "Natural fiber polymer composites : a review," *Adv. Polym. Technol.,* vol. 2329, no. July, pp. 351–363, 2015.

24. K. L. Pickering, M. G. A. Efendy, and T. M. Le, "A review of recent developments in natural fibre composites and their mechanical performance," *Compos. Part A Appl. Sci. Manuf.,* vol. 83, pp. 98–112, 2016.

25. S. A. S. Goulart, T. A. Oliveira, A. Teixeira, P. C. Miléo, and D. R. Mulinari, "Mechanical behaviour of polypropylene reinforced palm fibers composites," *Procedia Eng.,* vol. 10, pp. 2034–2039, 2011.

26. A. L. Da Róz, A. M. Ferreira, F. M. Yamaji, and A. J. F. Carvalho, "Compatible blends of thermoplastic starch and hydrolyzed ethylene-vinyl acetate copolymers," *Carbohydr. Polym.,* vol. 90, no. 1, pp. 34–40, 2012.

27. K. Kampeerapappum, P., Aht-ong, D., Pentrakoon, D., and Srikulkit, "Preparation of cassava starch: montmorillonite composite films," *Carbohydr. Polym.,* vol. 67, pp. 155–163, 2007.

28. X. Ma, J. Yu, and J. F. Kennedy, "Studies on the properties of natural fibers-reinforced thermoplastic starch composites," *Carbohydr. Polym.,* vol. 62, no. 1, pp. 19–24, 2005.

29. J. L. Guimarães, F. Wypych, C. K. Saul, L. P. Ramos, and K. G. Satyanarayana, "Studies of the processing and characterization of corn starch and its composites with banana and sugarcane fibers from Brazil," *Carbohydr. Polym.*, vol. 80, no. 1, pp. 130–138, 2010.

30. A. L. Da Róz, M. D. Zambon, A. A. S. Curvelo, and A. J. F. Carvalho, "Thermoplastic starch modified during melt processing with organic acids: the effect of molar mass on thermal and mechanical properties," *Ind. Crops Prod.*, vol. 33, no. 1, pp. 152–157, 2011.

31. J. F. Mendes *et al.*, "Biodegradable polymer blends based on corn starch and thermoplastic chitosan processed by extrusion," *Carbohydr. Polym.*, vol. 137, pp. 452–458, 2016.

32. S. H. Imam, P. Cinelli, S. H. Gordon, and E. Chiellini, "Characterization of biodegradable composite films prepared from blends of poly(vinyl alcohol), cornstarch, and lignocellulosic fiber," *J. Polym. Environ.*, vol. 13, no. 1, pp. 47–55, 2005.

33. E. D. M. Teixeira, D. Pasquini, A. a S. Curvelo, E. Corradini, M. N. Belgacem, and A. Dufresne, "Cassava bagasse cellulose nanofibrils reinforced thermoplastic cassava starch," *Carbohydr. Polym.*, vol. 78, no. 3, pp. 422–431, 2009.

34. K. Kaewtatip and J. Thongmee, "Studies on the structure and properties of thermoplastic starch/luffa fiber composites," *Mater. Des.*, vol. 40, pp. 314–318, 2012.

35. C. M. O. Müller, J. B. Laurindo, and F. Yamashita, "Composites of thermoplastic starch and nanoclays produced by extrusion and thermopressing," *Carbohydr. Polym.*, vol. 89, no. 2, pp. 504–510, 2012.

36. M. G. Lomelí Ramírez, K. G. Satyanarayana, S. Iwakiri, G. B. De Muniz, V. Tanobe, and T. S. Flores-Sahagun, "Study of the properties of biocomposites. Part I. Cassava starch-green coir fibers from Brazil," *Carbohydr. Polym.*, vol. 86, pp. 1712–1722, 2011.

37. F. Debiagi, M. V. E. Grossmann, and F. Yamashita, "Starch, sugarcane bagasse fibre, and polyvinyl alcohol effects on extruded foam properties : a mixture design approach," *Ind. Crop. Prod.*, vol. 32, no. 3, pp. 353–359, 2010.

38. M. J. H. Keijbets, "Potato processing for the consumer: developments and future challenges," *Potato Res.*, vol. 51, pp. 271–281, 2008.

39. D. Chen, D. Lawton, M. R. Thompson, and Q. Liu, "Biocomposites reinforced with cellulose nanocrystals derived from potato peel waste," *Carbohydr. Polym.*, vol. 90, no. 1, pp. 709–716, 2012.

40. V. P. Cyras, L. B. Manfredi, M.-T. Ton-That, and A. Vázquez, "Physical and mechanical properties of thermoplastic starch/montmorillonite nanocomposite films," *Carbohydr. Polym.*, vol. 73, pp. 55–63, 2008.

41. M. Thuwall, A. Boldizar, and M. Rigdahl, "Extrusion processing of high amylose potato starch materials," *Carbohydr. Polym.*, vol. 65, pp. 441–446, 2006.

42. J. Lörcks, "Properties and applications of compostable starch-based plastic material," *Polym. Degrad. Stab.*, vol. 59, pp. 245–249, 1998.

43. L. Mościcki, M. Mitrus, A. Wójtowicz, T. Oniszczuk, A. Rejak, and L. Janssen, "Application of extrusion-cooking for processing of thermoplastic starch (TPS)," *Food Res. Int.*, vol. 47, no. 2, pp. 291–299, 2012.

44. O. V. López, L. a. Castillo, M. A. García, M. a. Villar, and S. E. Barbosa, "Food packaging bags based on thermoplastic corn starch reinforced with talc nanoparticles," *Food Hydrocoll.*, vol. 43, pp. 18–24, 2014.

45. J. F. Mano, D. Koniarova, and R. L. Reis, "Thermal properties of thermoplastic starch/synthetic polymer blends with potential biomedical applicability," *J. Mater. Sci. Mater. Med.*, vol. 14, no. i, pp. 127–135, 2003.

13 Biocomposite Materials in Design for Sustainability

H.N. Salwa and S.M. Sapuan
Universiti Putra Malaysia

M.T. Mastura
Universiti Teknikal Malaysia Melaka

M.Y.M. Zuhri
Universiti Putra Malaysia

CONTENTS

13.1 INTRODUCTION OF DESIGN FOR SUSTAINABILITY

The definition of sustainable development is "development that meets the needs of the present without compromising the ability of the future generations to meet their own needs" was reported by Brundtland Report in 1987. This definition is a popular explanation of the term "sustainable development" and frequently cited [1]. Sustainability incorporates the need to deal with the economic development, social equity, and protection to the environment [2]. United Nations Member States in 2015 launched the global goals to end poverty, protect the planet, and ensure that all people enjoy peace and prosperity by 2030 for sustainable development which incorporate people, planet, and profit (Figure 13.1).

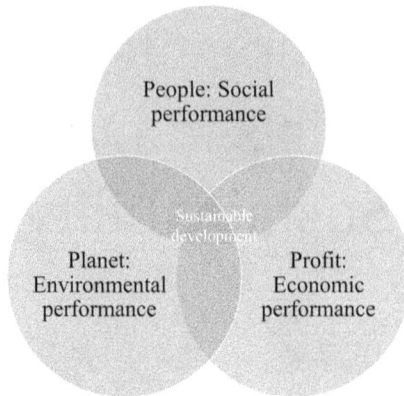

FIGURE 13.1   Elements of sustainable developments.

Reported by Arnette et al. [3], the term "triple bottom line" in design is referring to the three E's: ecology (environmental protection), equity (social equity), and economy (economic growth). Further, Ceschin and Gaziulusoy [1] clarify the contemporary understanding of sustainability which is a system property and not a property of individual elements of systems.

Design for sustainability (DFS) has evolved from emphasizing technical and product-centric towards a focus on large-scale complex system, and sustainability is concluded as a socio-technical challenge [1]. The focus on sustainable development is now shifting from minimizing negative impacts (eco-efficiency) towards optimizing positive impacts (eco-effectiveness) [4]. In the 1980s, the term "design for" was used to a study on the aspects of assembly considerations regarding its constraints and costs during design phase of a product. A series of guidelines was then developed to facilitate the assembly process which then increased the efficiency. The term "design for assembly" (DFA) was then created. Later, it was spreading in all related aspects of product design. Manufacturing, supply chain, and environment are among the focused topics in applying the "design for" techniques aspiring the term "design for X" (DFX) [3]. *X* is stand for a specific attempt, feature, or objective that is being considered in the phase of product design.

The "design for environment" (DFE) technique was concerning about the product and focus on environmentally friendly practices over the course of the product's life cycle. However, Arnette et al. [3] highlighted that environment is only one element in sustainable development. DFS approaches were initially associated with the product innovation and termed "green design," ecodesign, and biomimicry which principally demand technical knowledge on materials, production processes, and renewable energies. Later the focus of DFS is extended from single products to complex systems with elevated interest towards "people-centered" aspects of sustainability. The aspects of people, such as labor conditions, poverty alleviation, integration of weak and marginalized people, social cohesion, democratic empowerment of citizens, and in general quality of life, have been progressively embraced into the DFS approaches. Designers now also demanded to be equipped with a different set of

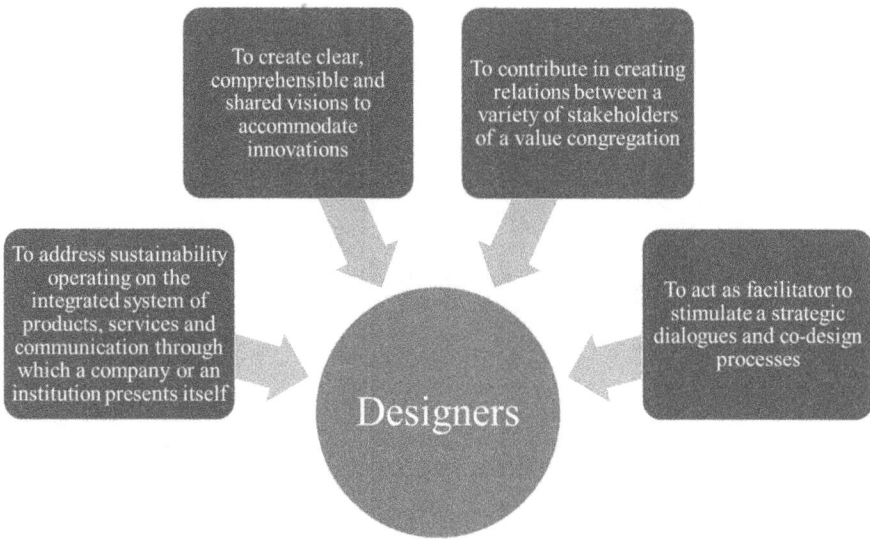

FIGURE 13.2   Designers' expected capabilities [1].

expertise particularly in human-centered design skills such as emotionally durable design and design for sustainable behavior [1].

DFS approaches and the concept of circular economy are considered as a potential solution to cultivate protection to the environment without limiting economic growth. It is described as "an industrial economy that is restorative or regenerative by intention and design" with the fundamental of 3R principles – reduction of resources, reuse, and recycling [1]. It is also a recognition of material closed-loop flow systems. DFS moved from product thinking to system thinking and to become more strategic. Expectation of designers's capability today is summarized in Figure 13.2.

Ceschin and Gaziulusoy [1] proposed four different innovation levels of DFS as shown in Figure 13.3. The first level is only focusing on improving existing or developing completely new products. The next level is going beyond individual products and towards integrated combinations of products and services. DFS in the next higher level would look into the context of innovation on human settlements and the spatio-social conditions of their communities, from neighborhoods to cities scales, and ultimately, at the highest level of DFS approach would concentrating on promoting radical changes on how communal needs, such as nutrition and transport/mobility, are fulfilled and consequently supporting transitions to new socio-technical systems [1].

13.2   RELATED AREA OF STUDIES IN DFS

The role of sustainability in business has grown, and it cannot be denied that product design plays a major role in helping to achieve sustainability. Abundance of studies have been done covering varieties of topics to look at sustainability and product

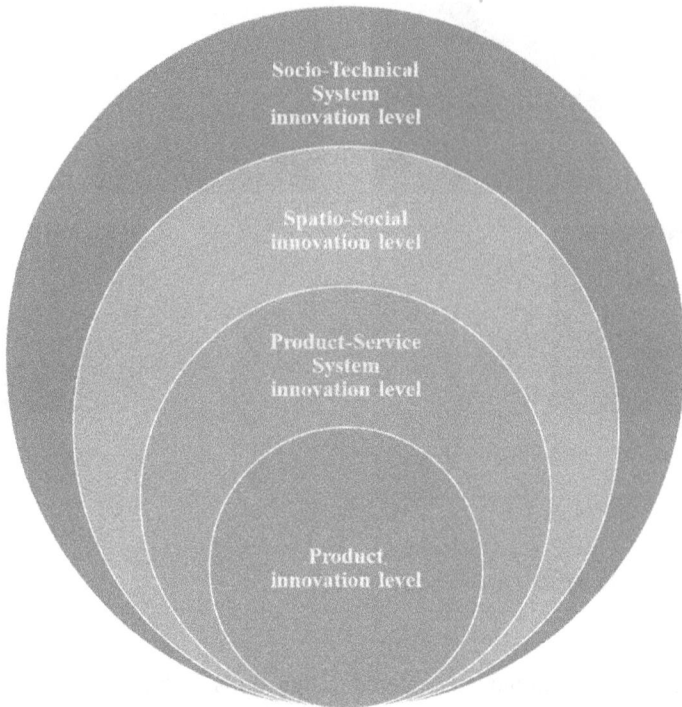

FIGURE 13.3   Four different innovation levels of DFS approaches [1].

design from different perspectives. "Environment" has always been an important dimension in DFS topic of study. It covers a broad discussion area about a product and process design and development with issues associated with environmental safety and health over product's full life cycle including waste reduction, material selection, and energy use to reduce environmental impacts. These can be detail narrowed down to raw material acquisition, processing, manufacturing and assembly, product use and service, and ending with product retirement and following by recycling, reuse, remanufacture, or disposal. Further sub-categories under the environment dimension of DFS are chronic risk reduction, accident prevention, waste recovery and reuse, and contaminant avoidance [3]. A study on the environmental performance of packaging materials derived from packaged food waste using life cycle assessment (LCA) attained negative results in all LCA categories, meaning the materials tested have beneficial effect on the environment [5]. Ironically, de Medeiros et al. [6] concluded in their report that environmental issues are still not a dominant topic in product development practices. They suggested that these practices shall be included in distinct phases of the PDP, mainly in the phases before concept generation and in the post-launch phases.

"Quality" is another important aspect of sustainable product design where the main concern is to reduce or eliminate defects in production processes. Additionally, the objective of study could be on fulfilling customer requirements and ensuring a robust product for manufacturing and use [3]. Besides, in terms of quality in DFS, matters of techniques or methods that could increase product life, reduce costs, and ultimately increase customer satisfaction always have attract researchers' interests. For example, Yalcinkaya et al. [7] carried out a study on modified conventional vacuum-assisted transfer molding (VARTM) for manufacturing to produce higher quality composites. They discovered that that pressurized VARTM led to laminates with less than 1% void content, and fiber volume fraction and flexural strength were increased.

"Manufacturing" in DFS is another subject often studied by researchers which includes efforts to eliminate expensive manufacturing processes and materials, and to ensure process feasibility. This area of research also involves topics on production efficiency increment, quality, flexibility, reliability, innovation, and approaches on reduction of production costs. "Supply chain" research topics in DFS also attract researcher's attention. The topics discuss about designing products for efficiency within the supply chain where the design can be used to improve supply chain performance. An example of research on this topic is done by [8] which studied four second-generation biomass supply chain scenarios and each are evaluated and compared on economic and environmental comparison of mobile versus fixed pyrolysis scenarios. A significant aspect of manufacturing is "assembly" which focuses on the assembly stage of a production. The topics relevant to DFS are usually on design to reduce the number of parts, tasks, and motions during assembly process; design to combine functionality of parts; design to reduce difficulty of assembly processes, increasing production efficiency; and suggestions on decreasing production costs and time get product into market. Meanwhile, "remanufacture" mostly investigates potential of environmentally friendly outcomes for a product that has reached end-of-life. DFS efforts would enable disassembly, assembly, cleaning, testing, repair, and replacement. The new product enters markets of remanufactured goods and products sold at greater profits with reduced costs, raw materials, and energy consumption [3].

Another area of interest is the "reuse" of a product. This includes "as-is" product or harvesting working parts and components for reuse, and they are often in the form of repairs and replacements. Studies on this facet focus on design to standardize components across the age of product models and to enhance durability of reuse-targeted components. Researchers might aim to improve the recovery of products and parts, as well as increasing components recovery. Reduction of cost, energy, and raw material consumption in designing products too are important topics discussed in this area of research, whereas, in the subject of "recycle," researches focus on recyclability of inputs and outputs materials. Efforts in minimizing material variety too are another noteworthy area of study. Issues discussed are on percentage of recyclability increment; more recycled and recyclable inputs, increased recycling efficiency, and also raw materials consumption reduction [3].

"Logistics" is also another interesting area of study in DFS. In this subject, research studies discuss the design to reduce packaging while protecting the product and be able to withstand transportation environmental factors. Additionally, studies also investigate on solutions to reduce product size for storage and transportation, and on methods or techniques to ensure compatibility with material handling equipment. They also study on solutions for decreased packaging and shipping costs; decreased transit and storage damage; and decreased logistics lead times [3].

Recently, the subject of "social responsibility" in designing products is getting more attention too. Studies discussed the design of a product that is produced in good conditions and does not hurt any communities. Research studies in this subject also support humanity where studies on designs have great potential to enable linkages with society and considering non-traditional markets. Some studies proposed theories on eliminating social problems by innovative efforts in product development process (PDP). Research studies also examine deeper into the issues of increased worker retention rates, increased value for society, and change in societal/user behavior [3].

13.3 GREENER MATERIALS IN SUSTAINABLE PRODUCT DEVELOPMENT

PDP provides a process road map to designers and deliverables required in designing, developing, and manufacturing a product. The main objectives of a PDP are to minimize the life cycle cost and maximize product quality, as well as maximize customers' satisfaction, maximize flexibility, and minimize lead time [9,10]. Earlier, [11] simply outlined the main objective of PDP is to create a product that an individual consumer or a manufacturing company or a service organization will buy. Activities and main component as described are illustrated in Figure 13.4. Another author briefly defined PDP as the implementation of steps that move the product from the concept to its launch which involves a sequence of practices in which information is processed through the decomposition of project steps into smaller sub-tasks [6].

PDP is considered one of the critical processes for companies to grow and sustain themselves in the market. Companies need to focus on the success of new product

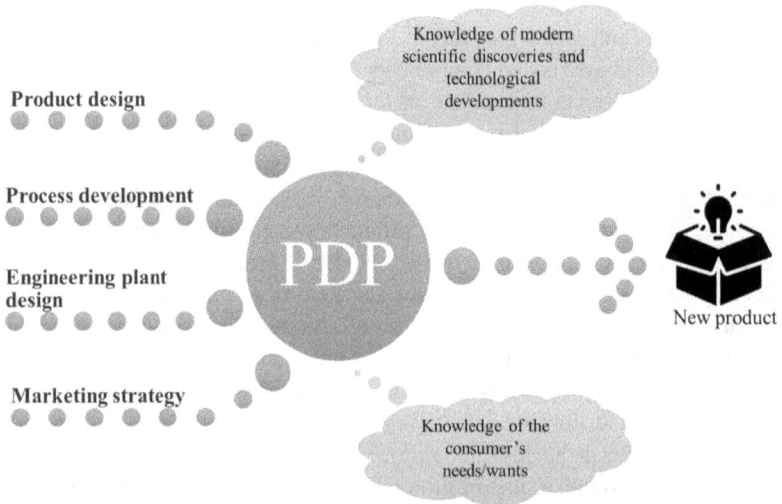

FIGURE 13.4 Activities and main components of PDP. (Adapted from [11].)

development because of the reduction of the product life cycle and the increasing pressure in the global competitive market. At the same time, companies must deal with the new regulations and social demands for environmental awareness on a global scale. According to [12], these challenges have an impact on the entire product life cycle, urging companies to invest in the concept of operations management and reducing the complexity and quantity of materials in a new PDP. In addition, the stages of the PDP have a direct influence on the final product and it is the most critical decision in relation to selection of materials. Prendeville et al. [13] stated in their report that "materials are at the heart of innovation and sustainability while also being pivotal to economic" and supported by a report from the industrial arm of the Research and Innovation Department of the European Commission in 2011 saying that approximately 70% of all new product innovation is based on materials with new or improved properties. Materials and their characteristics are significant not only for determining the mechanical properties of a product, but also to explore products' metaphysical features as basis in customer satisfaction attributes [14].

Many traditional materials that have been utilized in various engineering applications are now being replaced by new green materials to facilitate meeting the growing demand of environmental issues and weight reduction, as well as performance enhancement [15–17]. Polymer composite products are mainly custom-made design. In the environment of concurrent engineering (CE), in the process of a product development, selecting the materials and manufacturing process are also studied at the very early stage of the development process [18]. There are many research activities using CE, and material selection is one of the most significant activities which is carried out simultaneously with other tasks in conceptual design stage.

The research and development of biodegradable or bio-based polymer materials has been growing rapidly due to environmental awareness and new rules and regulations. Consumer demand for good and safe food products also have directed the interests of researchers to find potential of exploiting available natural feedstocks to produce a fully bio-based materials [19–21]. There are many reports about impurities in petroleum-based food contact packaging and consumables which tend to migrate from food packaging into food and create potential risks to human health [22]. Environmental elements of sustainability are usually considered through ecodesign examples. Designers are key decision-makers with power to foster preferable material applications and to portray creative, future visions of material use [13].

13.3.1 SELECTION OF THE RIGHT MATERIALS IN SUSTAINABLE PRODUCT DESIGN

Material selection is the process of picking the best materials for a specific product design [23]. It is a significant step in developing new products or enhancing existing ones [24, 25]. It contributes a crucial part in the development of products that are efficient, safe, quality, and satisfactory to consumers. Material selection also helps to eradicate materials with potential environmental hazards from early stages of product development. Maleque and Salit [24] recommended that one should know what product is to be developed or manufactured, what it does, and how it does it. Providing appropriate answers to these questions at the initial stages of product development will enable effective product design and proper material selection.

In order to resolve the issue of material selection for better product performance, many computerized material selection systems have been developed to select the most fitting materials for a certain application.

A study by [26] developed a seven-ring circular chain with important links in the sustainability chain which includes material, economy, design, market, equity, technology, and ecology. They then evaluated the sustainability of six different types of materials in order to explore the role that material selection plays in sustainable product development and the concept of "triple bottom line" was incorporated. Wrong selection of materials frequently leads to huge cost contribution and eventually pushes a company towards early product failure. Therefore, the designers and stakeholders need to identify and select proper materials with specific functionalities in order to get the desired output with minimum cost involvement. But, the selection of proper materials for engineering applications is not an easy task to perform. Designers must consider many material selection-attributes and various alternative materials with complex relationships between various attributes.

A systematic, efficient, and easy approach for material selection is thus required to help the manufacturing organizations for selecting the best material for an application [27]. By choosing an appropriate combination of matrix and reinforcement material, a new material can be made that exactly meets the requirements of an application. Abundance of studies have been carried out on material selection in the process of PDP. Recent studies on materials selection process and selection system are summarized in the Table 13.1.

13.3.2 BIO-FILLER-REINFORCED POLYMER COMPOSITE (BFRP) MATERIALS

The term "bioplastic" means a plastic produced from a biological source (short carbon cycle), whereas the term "biodegradable" refers to a material that can be degraded relatively rapidly by microbes in a bio-active environment under suitable conditions [21]. Bioplastics can be defined as plastics based on renewable resources (bio-based) or as plastics which are biodegradable and/or compostable. Natural bio-fiber composites are emerging as a viable alternative to glass-fiber-reinforced plastics [36]. Biocomposite material is a composite material in which at least one of the components is derived from natural resources and can be categorized into three group described as shown in Figure 13.5. Biocomposites in which both matrix and filler/fiber derived from natural resources are called "green biocomposites" which are more environmentally friendly [37]. Plant natural fibers are used as reinforcing material (fillers) for polymer-based matrices to produce a fully bio-based polymer composite. Among them are sisal, hemp, kenaf, coir, jute, flax, and sugar palm due to encouraging emphasis on environmental regulations as well as sustainability concepts, ecological, social, and economical awareness [18, 37].

Bio-filler-reinforced polymer composite (BFRP) materials from local and renewable resources provide significant sustainability; industrial ecology, eco-efficiency, and green chemistry are guiding the development of the next generation of materials, products, and processes. Other advantages of BFRP compared to synthetic fiber composites are durability, easy maintenance, renewable, bio-degradable, combustible, and cost-effective [38].

TABLE 13.1

Studies on Material Selection in PDP (2015–2018)

No	Product Application	Materials Considered	Tools Utilized	References
1	Packaging fruits, dry food, and dairy products	Thirty types of bio-based polymer materials (PLA, PHA, PHB, PHBHV, etc.)	Exsys Corvid expert system	[28]
2	Composite eco-design	GF-reinforced plastic composite (GFRP) and flax fiber-reinforced composites (FFRP)	Comparison of the environmental performance using Life Cycle Assessment (LCA)	[29]
3	Plastic pipe design	Poly(vinyl chloride) (PVC), PP, and polyethylene (PE)	Grey relational analysis (GRA) with an analytic hierarchy process (AHP)	[30]
3	Refillable glass bottles	Glass, PET, and aluminum	Environmental accounting based on energy	[31]
3	Spur gear	Twelve types of carbon steels and low alloy steels	Ashby's method	[32]
	Hybrid biocomposite material in the design for an automotive anti-roll bar	Nine natural fibers (sugar palm, kenaf, oil palm, sisal, jute, hemp, flax, pineapple, and coir)	Combination of AHP and quality function deployment for environment (QFDE)	[33]
	Automotive parts – engine block and intake manifold	Grey cast iron, aluminum cast, GF-reinforced nylon, magnesium, and CGI	Entropy and fuzzy TOPSIS	[34]
4	Automotive anti-roll bar	Thermoplastic polyurethane, high-density polyethylene (HDPE), low-density polyethylene (LDPE), polystyrene, and PP	QFDE – voice of customers and the voice of the environment, and AHP-based software	[35]

13.4 ENVIRONMENTAL ASSESSMENT OF BIO-FILLER-REINFORCED POLYMER MATERIAL PRODUCTS

Evaluation of environmental impacts of a product during its entire life cycle starts with raw material acquisition, all the way through processing, manufacturing, and assembly, on to product use and service, and ending with product retirement and subsequent recycling, reuse, remanufacture, or disposal. Innovative design efforts with environmental concerns incorporated in the process would minimize long-term ecological harm. Reduction of energy usage is anticipated throughout supply chain and product's useful life. Less materials used in supply chain and product' useful life give advantages to the environment as well as minimizing waste generated from the

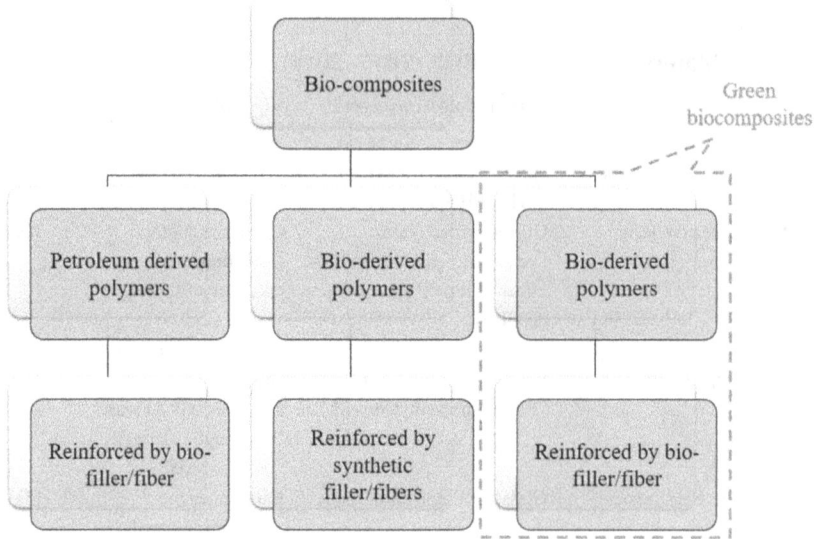

FIGURE 13.5 Biocomposites categorization.

product. Further benefits are achieved if waste is made recoverable (for recycling, energy creation, etc.).

The LCA methodology is the most widespread technique for assessing the environmental impacts associated with material products focused on studying the whole product system. LCA comprises the material extraction, manufacturing and waste production, packaging, transportation, product use, and product disposal. To calculate the amount of emissions and waste generated during the life cycle of a product, detailed fundamental information on manufacturing processes, materials, and energy use is required. A life cycle inventory (LCI) compiles record of the emissions and resources consumed that can be attributed to a specific product. Applying LCA to newly developed materials can be tough, especially if the materials are still in research and development or in pilot scale [39]. Shortage of data about process parameters, material formulation, and unavailability of material properties may cause some uncertainties when developing the model. Commonly in those cases, a preliminary LCA with the available data were performed and to be developed when more data is available in the future.

Korol et al. [40] studied environmental assessment of a plastic pallet produced from composites based on polypropylene (PP), reinforced with glass fibers (GFs), and natural fibers, i.e., cotton fibers (CF), jute fibers (JFs), and kenaf fibers (KFs). Table 13.2 summarizes the results of environmental assessment obtained for pallet production of composites and bio-filler-reinforced composites in [40].

From the results obtained in [40], all composite materials and PP have about the same values of impact for climate change although PP is just a bit higher. PPGF and PPCF are having similar values and only very slight lower value than PP, whereas both bio-filler-reinforced composites of PPKF and PPJF have much lower values

TABLE 13.2

Results of Environmental Assessment Obtained for Pallet Production of Composites and Bio-Filler-Reinforced Composites and PP in [40]

Impact Category	Unit	PPKF	PPJF	PPGF	PPCF	PP
Climate change	kg CO_2 eq/FU	58.2287	58.3897	69.8297	69.4565	70.9783
Ozone depletion	kg CFC-11 eq/FU	0.0001	0.0001	0.0001	0.0001	0.0001
Terrestrial acidification	kg SO_2 eq/FU	0.6090	0.6140	0.7240	0.7010	0.7730
Freshwater eutrophication	kg P eq/FU	0.0140	0.0143	0.0139	0.0292	0.0124
Marine eutrophication	kg N eq/FU	0.0402	0.0445	0.0333	0.0645	0.0355
Human toxicity	kg 1,4-DB eq/FU	13.2159	13.2806	17.8975	16.8541	15.1013
Photochemical oxidant formation	kg NMVOC/FU	1.0820	1.0780	1.3837	1.1365	1.5160
Particulate matter formation	kg PM_{10} eq/FU	0.1946	0.1944	0.2408	0.2215	0.2569
Terrestrial ecotoxicity	kg 1,4-DB eq/FU	0.0069	0.0071	0.0076	0.7176	0.0080
Freshwater ecotoxicity	kg 1,4-DB eq/FU	0.2417	0.2421	0.2728	0.4683	0.2502
Marine ecotoxicity	kg 1,4-DB eq/FU	0.2863	0.2867	0.3325	0.4076	0.3156
Ionizing radiation	kBq U235 eq/FU	22.2561	22.2629	26.8822	24.2952	27.5044
Agriculture land occupation	m^2a/FU	9.6216	10.8802	1.9974	53.2596	1.9624
Urban land occupation	m^2a/FU	0.1011	0.0984	0.1104	0.8793	0.0934
Natural land transformation	m^2/FU	0.0035	0.0034	0.0039	0.0054	0.0032
Water depletion	m^2/FU	2.8292	5.5525	0.7294	7.2581	0.7477
Metal depletion	kg Fe eq/FU	0.7895	0.7659	1.0549	2.0189	0.8414
Fossil depletion	kg oil eq/FU	29.0026	28.9560	36.1989	31.8024	38.1517

where both have similar values with approximately only 0.2 difference. As for human toxicity category, the highest value goes to PPGF followed by PPCF and afterward PP, whereas PPKF and PPJF have lower values. Roughly similar findings are shown for ionizing radiation and fossil depletion categories. Nonetheless, for agriculture land occupation category and water depletion category, all bio-filler-reinforced PP composites, i.e., PPCF, PPKF, and PPJF, have higher values than PPGF and PP and this is understood since the fillers are originated from plantation and the technique to produce the fibers. In addition, PPCF shows significantly much higher value than the other two biocomposites. It also has remarkably higher score at terrestrial ecotoxicity, metal depletion, and urban land occupation category. Notably, ozone depletion and natural land transformation categories have minimal values compared to the other categories and have similar values among all materials.

These results were then further analyzed by the researchers based on the eco-efficiency assessment which include the cost of raw materials and particular impact categories and generated results shown in Figure 13.6. They found that the chosen impact categories determine eco-efficiency results. In the case of using total damage category, human toxicity, fossil fuel depletion, particulate matter formation, and climate change as impact categories, the highest eco-efficiency was indicated for the PPKF and PPJF scenarios. For the human toxicity impact category, the lowest eco-efficiency was indicated for the PPGF scenario. The lowest eco-efficiency value for agriculture land occupation impact category was indicated for the PPCF

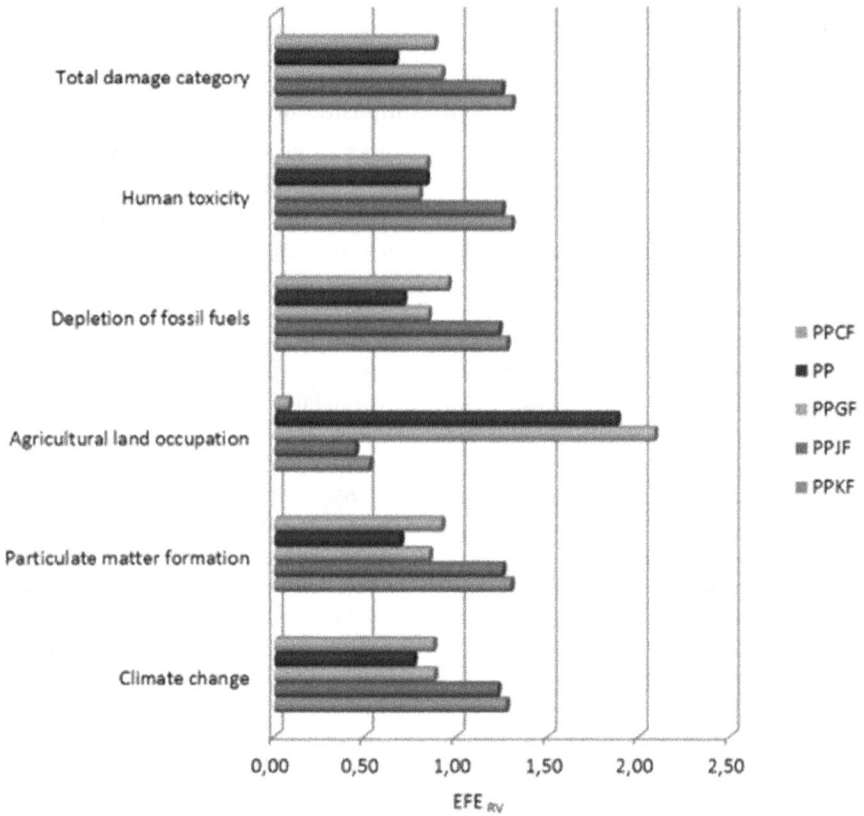

FIGURE 13.6 Eco-efficiency analysis results in [40].

scenario. In the case of using GHG emissions as an impact category, the highest eco-efficiency was noted for pallets produced from PPKF and PPJF. It was then concluded that PP composites reinforced by KFs and JFs had similarly high levels of eco-efficiency for all impact categories except for agriculture land occupation. Designers and decision-makers would understand more on how environmental impact categories affect the eco-efficiency of biocomposites and hence help them to select materials cautiously.

13.5 IMPACTS OF BIO-FILLER-REINFORCED POLYMER MATERIALS ONTO THE ECONOMY

Efficiency within the supply chain is the key factor to create favorable impacts to the economy. Aspect of logistics generates costs which includes distribution stage where products must be designed so that it can be shipped effectively. Production in efficient manner in the manufacturing stage would also give benefits to the economy. The ease of product maintenance so that the life of the product can be extended is another important aspect of successful and sustainable product in market. Manufacturers adopting sustainable design practices including utilizing green materials will bring positive economy impacts to a community and region in a long run. For example, the sustainable manufacturing industry could foster a market for recycled materials and energy-efficient systems, creating enough economies of scale to reduce the price of these types of products to be more competitive with conventional products. Furthermore, the surrounding communities and the region may experience economic development which include job creation through growth of businesses that make biocomposites materials, produce energy-efficient technologies, and provide sustainable design. A region inclined to practicing sustainable design and manufacturing is an attractive place for these companies to establish and therefore potentially generating additional local jobs and income. A government sustainable design projects can be proposed to be a "seed" for growth of sustainable communities and regions.

13.6 BIO-FILLER-REINFORCED POLYMER MATERIALS AFFECTING SOCIAL DEVELOPMENT

Producing products in good conditions and not imposing harm on communities as well as support humanity is the fundamental principle of social performance in sustainability concept. Design development in harmony with the environment can make a person's access to better health where people get a better living environment. The destruction of the living environment of the current design, in the long run, affects the health and safety of life of future generations. The people and the natural environment are interdependent or described as a symbiotic relationship; hence, products should be designed to protect the environment to create a good and harmonious to the ecological environment [41]. Green materials can contribute to quality of life in many ways. As described in the previous section, green materials industry could create local jobs and income, and therefore, occupants who experience increased job satisfaction, health, and productivity will carry these experiences back to their families and friends in the community, thus influencing overall well-being. Benefits can potentially disseminate beyond the factory and promoting more use of the sustainable design practices and environmental behavioral change in the bigger scale of the community. These changes include increased recycling, purchasing green products, and investing in energy-efficient technologies. Environmentally conscious manufacturing practices will tend to generate lower amounts of dust, pollution, noise, traffic congestion, and other community annoyances, and this would contribute to a better public health, safety, and well-being. Manufacturing practices and production operation practices that foster recycling and reduce waste generation, energy use, and water consumption will eventually reduce the demand for new landfills, electric utility plants, and wastewater treatment facilities, and will decrease the public nuisance associated with them. Locally produced and manufactured products strengthen the local economy and provide jobs in the community as well as reducing energy use and emissions caused by long-distance transportation. Such development would empower rural communities to tackle their poverty and creates the opportunity for local villagers to learn advanced techniques and skills for proper natural fiber processing methods. This then could lead to the development of locally manufactured biocomposites. Alternatively, the prepared natural fibers could be supplied to high-technology engineering companies for manufacturing commercial biocomposite components or products [42].

13.7 CONCLUSION

New product development should be able to achieve new levels of performance that bring products to customers that add value in ways that were not previously conceived or expected. This chapter presented the concept of DFS covering the three key dimensions of sustainability. DFS aims to improve human welfare and business profitability without compromising the utilization of the most appropriate materials, suitable technology, and manufacturing processes for the development of products.

ACKNOWLEDGMENT

The authors wish to express the highest appreciation to the Public Service Department (JPA), Malaysia, for the study sponsorship to the main author and financial support from the Ministry of Education Malaysia through Universiti Putra Malaysia Grant Scheme HICoE (6369107).

REFERENCES

1. F. Ceschin and I. Gaziulusoy, "Evolution of design for sustainability: from product design to design for system innovations and transitions," *Des. Stud.*, vol. 47, pp. 118–163, 2016.
2. S. M. Sapuan, *Composites Materials: Concurrent Engineering Approach.* Oxford: Butterworth-Heinemann Elsevier Inc., 2017.
3. A. N. Arnette, B. L. Brewer, and T. Choal, "Design for sustainability (DFS): the intersection of supply chain and environment," *J. Clean. Prod.*, vol. 83, pp. 374–390, 2014.
4. B. de Koeijer, R. Wever, and J. Henseler, "Realizing product-packaging combinations in circular systems: shaping the research agenda," *Packag. Technol. Sci.*, vol. 30, no. 8, pp. 443–460, 2017.
5. G. Vitale, D. Mosna, E. Bottani, R. Montanari, and G. Vignali, "Environmental impact of a new industrial process for the recovery and valorisation of packaging materials derived from packaged food waste," *Sustain. Prod. Consum.*, vol. 14, pp. 105–121, 2018.
6. J. F. de Medeiros, N. C. Lago, C. Colling, J. L. D. Ribeiro, and A. Marcon, "Proposal of a novel reference system for the green product development process (GPDP)," *J. Clean. Prod.*, vol. 187, pp. 984–995, 2018.
7. M. A. Yalcinkaya, E. M. Sozer, and M. C. Altan, "Fabrication of high quality composite laminates by pressurized and heated-VARTM," *Compos. Part A Appl. Sci. Manuf.*, vol. 102, pp. 336–346, 2017.
8. D. M. Yazan, I. van Duren, M. Mes, S. Kersten, J. Clancy, and H. Zijm, "Design of sustainable second-generation biomass supply chains," *Biomass Bioenerg.*, vol. 94, pp. 173–186, 2016.
9. B. C. Mitra, "Environment friendly composite materials: biocomposites and green composites," *Def. Sci. J.*, vol. 64, no. 3, pp. 244–261, 2014.
10. M. S. Salit, *Tropical Natural Fibre Composites: Properties, Manufacture and Application.* Singapore: Springer Singapore, 2014.

11. M. Earle, R. Earle, and A. Anderson, "The product development process," *Food Product Development*, Cambridge, UK: Woodhead Publishing Limited, 2001, pp. 95–148.

12. O. Canciglieri Junior, A. Y. U. Reche, and C. C. A. Estorilio, "How can green supply chain management contribute to the product development process," *Adv. Transdiscipl. Eng.*, vol. 7, pp. 1175–1183, 2018.

13. S. Prendeville, F. O. Connor, S. O. Rafferty, and L. Palmer, "Material selection and design for sustainable material innovation," *Proc. Eur. Acad. Des. Conf. Crafting Futur.*, pp. 1–12, 2012.

14. F. M. Al-Oqla, A. Almagableh, and M. A. Omari, "Design and fabrication of green biocomposites," M. Jawaid, M.S. Salit and O.Y. Alothman (Eds.), *Green Biocompopsites: Design and Applications*. Cham: Springer International Publishing, 2017, pp. 45–67.

15. F. M. Al-Oqla, M. S. Salit, M. R. Ishak, and N. A. Aziz, "Selecting natural fibers for bio-based materials with conflicting criteria," *Am. J. Appl. Sci.*, vol. 12, no. 1, pp. 64–71, 2015.

16. F. M. AL-Oqla and M. S. Salit, *Materials Selection of Natural Fibre Composites*. Duxford, UK: Woodhead Publishing, pp. 235–272, 2017.

17. M. Ramesh, K. Palanikumar, and K. H. Reddy, "Plant fibre based bio-composites: sustainable and renewable green materials," *Renew. Sustain. Energy Rev.*, vol. 79, no. February, pp. 558–584, 2017.

18. M. S. Sapuan, *Tropical Natural Fibre Composites: Properties, Manufacture and Applications*. Singapore: Springer, 2014.

19. A. P. Ananda *et al.*, "A relook at food packaging for cost effective by incorporation of novel technologies," *J. Packag. Technol. Res.*, vol. 1, no. 2, pp. 67–85, 2017.

20. N. Peelman, "Application of bioplastics for food packaging," *Trends Food Sci. Technol.*, vol. 32, no. August, pp. 128–141, 2013.

21. A. Soroudi and I. Jakubowicz, "Recycling of bioplastics, their blends and biocomposites: a review," *Eur. Polym. J.*, vol. 49, no. 10, pp. 2839–2858, 2013.

22. S. Q. Li, H. G. Ni, and H. Zeng, "PAHs in polystyrene food contact materials: an unintended consequence," *Sci. Total Environ.*, vol. 609, pp. 1126–1131, 2017.

23. N. Saba, M. Jawaid, M. T. H. Sultan, and O. Y. Alothman, "Green biocomposites for structural applications," M. Jawaid, M. S. Salit, and O. Y. Alothman, Eds., *Green Biocomposites*. Cham: Springer, 2017, pp. 1–27.

24. M. A. Maleque and M. Salit, "Mechanical failure of materials," *Mater. Sel. Des.*, vol. 20, pp. 17–38, 2013.

25. A. Hambali, S. M. Sapuan, N. Ismail, and Y. Nukman, "Material selection of polymeric composite automotive bumper beam using analytical hierarchy process," *J. Cent. South Univ. Technol.*, vol. 17, no. 2, pp. 244–256, Apr. 2010.

26. L. Y. Ljungberg, "Materials selection and design for development of sustainable products," *Mater. Des.*, vol. 28, no. 2, pp. 466–479, 2007.

27. D. Rai, G. Kumar Jha, P. Chatterjee, and S. Chakraborty, "Material selection in manufacturing environment using compromise ranking and regret theory-based compromise ranking methods: a comparative study," *Univers. J. Mater. Sci.*, vol. 1, no. 2, pp. 69–77, 2013.

28. M. L. Sanyang and S. M. Sapuan, "Development of expert system for biobased polymer material selection: food packaging application," *J. Food Sci. Technol.*, vol. 52, no. 10, pp. 6445–6454, 2015.

29. A. Corona, B. Madsen, M. Z. Hauschild, and M. Birkved, "Natural fibre selection for composite eco-design," *CIRP Ann. Manuf. Technol.*, vol. 65, no. 1, pp. 13–16, 2016.

30. R. Zhao, H. Su, X. Chen, and Y. Yu, "Commercially available materials selection in sustainable design: an integrated multi-attribute decision making approach," *Sustainability*, vol. 8, p.79, 2016.

31. C. M. V. B. Almeida, A. J. M. Rodrigues, F. Agostinho, and B. F. Giannetti, "Material selection for environmental responsibility: the case of soft drinks packaging in Brazil," *J. Clean. Prod.*, vol. 142, pp. 173–179, 2017.

32. H. Delibaş, Ç. Uzay, and N. Geren, "Advanced material selection technique for high strength and lightweight spur gear design," *Eur. Mech. Sci.*, vol. 1, no. 4, pp. 133–140, 2017.

33. M. T. Mastura, S. M. Sapuan, M. R. Mansor, and A. A. Nuraini, "Environmentally conscious hybrid bio-composite material selection for automotive anti-roll bar," *Int. J. Adv. Manuf. Technol.*, vol. 89, no. 5–8, pp. 2203–2219, 2017.

34. S. S. Yang, N. Nasr, S. K. Ong, and A. Y. C. Nee, "Designing automotive products for remanufacturing from material selection perspective," *J. Clean. Prod.*, vol. 153, pp. 570–579, 2017.

35. M. T. Mastura, S. M. Sapuan, M. R. Mansor, and A. A. Nuraini, "Materials selection of thermoplastic matrices for 'green' natural fibre composites for automotive anti-roll bar with particular emphasis on the environment," *Int. J. Precis. Eng. Manuf. Green Technol.*, vol. 5, no. 1, pp. 111–119, 2018.

36. N. Jabeen, I. Majid, and G. A. Nayik, "Bioplastics and food packaging: a review," *Cogent Food Agric.*, vol. 1, no. 1, Article: 1117749, 2015.

37. F. M. AL-Oqla and M. A. Omari, "Challenges, potential and barriers for development," M. Jawaid, S.M. Sapuan and M.Y. Alothmani (Eds.), *Green Biocomposites: Manufacturing and Properties.* Cham: Springer International Publishing AG, 2017, pp. 13–29.

38. N. Jauharia, R. Mishrab, and H. Thakur, "Natural fibre reinforced composite laminates – a review," *Mater. Today Proceed.*, vol. 2, no. 4–5, pp. 2868–2877, 2015.

39. D. Civancik-Uslu, L. Ferrer, R. Puig, and P. Fullana-i-Palmer, "Are functional fillers improving environmental behavior of plastics? A review on LCA studies," *Sci. Total Environ.*, vol. 626, pp. 927–940, 2018.

40. J. Korol, D. Burchart-Korol, and M. Pichlak, "Expansion of environmental impact assessment for eco-efficiency evaluation of biocomposites for industrial application," *J. Clean. Prod.*, vol. 113, pp. 144–152, 2016.

41. Y. Gu, "Study on harmonious development of design and environment," *Adv. Mater. Res.*, vol. 889–890, pp. 1609–1612, 2014.

42. M. S. Salit, *Tropical Natural Fibre Composites.* Singapore: Springer, 2014, pp. 103–118.

14 Tensile Properties of Sugar Palm Fiber-Reinforced Polymer Composites
A Comprehensive Review

R.A. Ilyas, S.M. Sapuan, and M.S.N. Atikah
Universiti Putra Malaysia

R. Ibrahim
Forest Research Institute Malaysia

*R. Syafiq, M.D. Hazrol, A. Nazrin,
and M.I.J. Ibrahim*
Universiti Putra Malaysia

CONTENTS

14.1 INTRODUCTION

The recent increase in petroleum prices and the introduction of strict environmental regulations in developed and developing nations have led to the recovery of natural fibers to replace petroleum-based fibers. Natural fiber is seen as a major alternative to petroleum-based fibers. This might be attributed to their cost-effectiveness, non-toxicity appreciable mechanical property, abundant availability, less abrasion, biodegradability, corrosion resistance, less tool wear, and safer processing. Natural fibers reinforced with polymers are now attaining huge attention in various applications such as building, packaging, medical, automobile, sporting, and electronics fields [1–5]. Although these natural fibers do not have much strength as carbon, glass, rayon, nylon, polyester (PE), acrylic, spandex, etc., scientists are identifying ways to make at par with petroleum-based materials. Besides strength, other disadvantages of natural fibers include incompatibility with resin, poor resistance to high temperature, variability of fiber properties (plant age, insect attack, part of the plant, etc.), hydrophilic character, and non-uniform mechanical properties. In general, the properties of natural fibers diverge depending on their growing conditions, method of fiber preparations, species and geographical location, and many other factors [6,7].

Natural fibers are also known as lignocellulosic fibers. This is due to the composition of natural fibers that are mostly cellulose fibrils embedded in the lignin matrix [8–10]. The mechanical properties of natural fiber-reinforced polymer composites are affected by several factors such as cellulose content, their spatial distribution in the composite system, fiber anatomy, interfacial\adhesion between the hydrophilic fibers and hydrophobic matrices, crystallinity index of the fibers, and weight or volume ratio of the fibers in the composites. They are improved by transferring the stress applied on the natural fibers in the composite system [11–14].

Sugar palm tree also known as *Arenga pinnata ((Wurmb.) Merr)* is a popular multipurpose tree dominantly found in tropical regions. It belongs to the Palmae family which has about 181 genera with around 2,600 known species [15]. It is a tall and large palm tree with a single unbranched stem. It can grow up to 65 cm in diameter and 20 m high. Besides the trunk, stem also acts as storage for starch. Sugar palm starch (SPS) can be converted into sugars at the commencement of flowering, for the production of palm juice, vinegar, sweet sap, and sugar block. *Arenga pinnata ((Wurmb.) Merr)* is one of the oldest cultivated plants and probably a source of plant sugar long before sugarcane was cultivated for that purpose. SPS can be extracted and applied for other purposes when the tree is unfertile in terms of fruits and sugar. Besides that, sugar palm tree's trunk is covered with long black sugar palm fibers (SPFs) and the bases of broken leaves [16,17].

In addition, this tree can also be found growing productively in many tropical countries from Australia continental such as North Australia, to Asia continental such as Taiwan, Thailand, Vietnam, Papua New Guinea, Philippines, India, Indonesia, Malaysia, and Burma, as far as African continental such as Senegal, The Gambia, Guinea Bissau, and other West African countries. Sugar palm tree can be found in various altitudes ranging from sea level up to 1,500 m. It grows at more humid parts of the Asian tropics in areas with a rain fall of 500–1,200 cm³ [18].

Currently in Malaysia, this tree can widely be seen along the bushes and rivers at the rural areas such as Raub, Pahang; Maran, Pahang; Kuala Pilah, Negeri Sembilan; Bruas-Parit, Perak; and Jasin, Melaka. In terms of proper plantation management, it was found that approximately around 809 ha of sugar palm plantation was planted in Kebun Rimau Sdn.Bhd which are located at Tawau, Sabah (East Malaysia). SPFs like most other natural fibers are lignocellulosic, where the cellulose and hemicellulose are reinforced in a lignin matrix. These SPFs comprise 43.88%

cellulose, 33.24% lignin, 7.24% hemicellulose, 2.7% extractive, and 1.01% ash. In the past decades, SPFs are being used in various biocomposite applications. Therefore, for SPFs to attain comparable results as a potential alternative material to petroleum-based fibers in composites, the basic mechanical properties (i.e., tensile, flexural, and impact properties) need to be determined and possible enhancement by processing via mechanical and chemical treatments needs to be determined [19–24].

14.2 SUGAR PALM FIBERS

Figure 14.1 shows the photographs of the sugar palm tree. This plant belongs to the sub-family of *Arecoideae* and the tribe of Caryoteae [27]. It was earlier given a number of taxonomic names such as *Saguerus rumphii* and *Arenga saccharifera Labill*. However, in 1917, during the International Congress of Botany in Vienna, it was officially renamed as *Arenga pinnata ((Wurmb.) Merr)*. Sugar palm is a natural forest species that originates from the Palmae family. It is known as a fast-growing palm that is able to reach maturity within 10 years [28].

Field emission scanning electron microscope (FESEM) pictures of SPFs were taken to investigate the structure of SPFs to reveal their homogeneity and micrometric dimensions, and are also shown in Figure 14.1. Microscopic examination of the

FIGURE 14.1 (a–d) Photographs of the sugar palm tree and SPF [19].

TABLE 14.1

Chemical Composition of Sugar Palm Fibers at Different Stages of Treatment [19,25, 26]

Sample	Cellulose (%)	Hemicellulose (%)	Holocellulose (%)	Lignin (%)	Extractive (%)	Ash (%)
Sugar palm fibers	43.88	7.24	51.12	33.24	2.73	1.01
Sugar palm frond	66.5	14.7	81.2	18.9	2.5	3.1
Sugar palm bunch	61.8	10.0	71.8	23.5	2.2	3.4
Sugar palm trunk	40.6	20.5	61.1	46.4	6.3	2.4

cross section of SPFs was depicted in Figure 14.1c and d. As shown in Figure 14.1c and d, the view from the outer to the inner part showed that SPFs consist of a middle lamella, a primary cell wall, a secondary cell wall, and a tertiary cell wall, buildup around an opening, the lumen [19,21]. The middle lamella as seen in Figure 14.1d, which surrounds the cell wall is mainly composed of pectins (macromolecules of galacturonic acid) that hold fibers together into a bundle, with a size of around $1.98 \pm 0.15\,\mu m$. The interior of the SPFs consists of primary cell wall, secondary cell wall, and lumen. The primary cell wall is made of cellulose (a polymer based on glucose units) fibrils in an organic matrix of amorphous hemicelluloses and lignin, proteins, and low methyl-esterified pectins, with an average diameter of around $10.38 \pm 0.57\,\mu m$. The secondary cell wall consists of three layers of cellulose fibrils with different axial orientations that are bound by lignin. Primary cell wall and secondary cell wall also provide mechanical support to the plant. Figure 14.1d shows a lumen with a thickness of around $3.72 \pm 0.15\,\mu m$.

The chemical compositions of the SPFs at different stages of treatment are shown in Table 14.1. From the table, it can be seen that the raw SPFs consist of 43.88% cellulose, 7.24% hemicellulose, 33.24% lignin, 2.73% extractive, and 1.01% ash.

14.3 SUGAR PALM FIBER-REINFORCED POLYMER COMPOSITES

It is a well-known fact that the development of biocomposites would produce environmental-friendly materials as alternatives to synthetic polymers for many applications. This leads this issue to become an attractive study to be carried out among materials scientists in recent years. In the past decades, SPF had been reinforced with various polymers such as epoxy, unsaturated polyester (UP), PE resin, phenol formaldehyde (PF), high-impact polystyrene, polyurethane, phenolic, SPS, vinyl ester (VE), polyurethane, sago starch, and cassava starch (Table 14.2). In addition, researchers such as Ilyas et al. [23,29,30], Sahari et al. [31,32], Sanyang et al. [33–35], Jumaidin et al. [36–38], Atikah et al. [20], Adawiyah et al. [39], Jatmiko et al. [40,41], Poeloengasih et al. [42], Apriyana et al. [43], and Mansor et al. [44] (Table 14.3) also studied various types of natural fiber-reinforced SPS-based biopolymers. Another potential product from sugar palm is biopolymer. Since sugar palm produces starch from its trunk, the

TABLE 14.2
Sugar Palm Fibre-Reinforced Various Polymer Composites

Fibre	Matrix	Authors
SPFs	Epoxy	Leman et al. [59]
		Safri et al. [60]
		Munawar et al. [61]
		Chandrasekar et al. [62]
	UP	Sahari et al. [63]
		Misri et al. [64]
		Norizan et al. [65]
		Ishak et al. [66]
	PE resin	Ticoalu et al. [67]
		Oumer and Bachtiar [68]
		Munawar et al. [61]
	Polypropylene	Hadi et al. [69]
	PF	Ishak et al. [15]
	High impact polystyrene	Bachtiar et al. [70]
	Polyurethane	Mohammed et al. [71]
		Radzi et al. [72]
		Atiqah et al. [5]
	Phenolic	Agrebi et al. [73]
	SPS	Sahari et al. [74]
	PVDF	Alaaeddin et al. [75]
	Vinyl ester	Razali et al. [76]
		Huzaifah et al. [77]
		Ammar et al. [78]
		Razali et al. [76]
		Munawar et al. [61]
	Polyurethane	Atiqah et al. [5,79]
		Radzi et al. [80,81]
	Sago starch	Halimatul et al. [51,82]
	Cassava starch	Edhirej et al. [83]
Sugar palm cellulose	SPS	Sanyang et al. [34]
Sugar palm nanocrystalline cellulose	SPS	Ilyas et al. [23,84]
Sugar palm nanofibrillated cellulose	SPS	Ilyas et al. [24,85]
		Hazrol et al. [86,87]

starch can be used to make biodegradable polymer. The existence of the starch to develop biopolymer is proven by many researchers [29,30, 45–58].

The differences in the mechanical properties of SPF-reinforced polymer composites as those of petroleum-based composite materials are much influenced by the method of fabrication, the composition of fibre, ratio of fibre to matrix, and the type resin used. Moreover, the mechanical properties of SPF-reinforced polymer composites are also affected by the treatments on natural fibers (i.e., retting process,

TABLE 14.3
Sugar Palm Starch-Reinforced Various Natural Fibers

Matrix	Natural Fibre	Authors
SPS	Seaweed (*Eucheuma cottonii*)	Jumaidin et al. [37]
	Seaweed (*Eucheuma cottonii*)/SPF	Jumaidin et al. [36]
	SPF	Sahari et al. [74]
	Sugar palm cellulose	Sanyang et al. [34]
	Sugar palm nanocrystalline cellulose	Ilyas et al. [23]
	Sugar palm nanofibrillated cellulose	Ilyas et al. [24]
		Hazrol et al. [86]

and chemical and mechanical treatments) and cultivation of the SPF, as well as the moisture content property of the fibers. Table 14.4 shows the tensile properties of SPF-reinforced polymer composites. As shown in the table, the tensile properties of various SPF-reinforced composites are almost half of the properties of the synthetic fibre composites, whereas experimental studies conducted by Ishak et al. [66] showed that SPF-reinforced UP using vacuum resin impregnation shows an increase in tensile strength which is double the value of synthetic fibre composites. Therefore, given an appropriate strength design, SPF-reinforced polymer composites can be a potential substitute to synthetic fibers.

Currently, there are numerous studies published on the use of SPFs (macro- to nanosize) as reinforcing fibers in fibre composites. The following studies on sugar-palm fibers/various polymer composites have been carried out.

14.3.1 Macro-Size Sugar Palm Fibre-Reinforced Polymer Composites

Radzi et al. [80] studied the effect of alkaline treatment on thermal, mechanical, and physical properties of hybrid roselle (RF)/SPF-incorporated thermoplastic polyurethane composites. RF/SPF hybrid composites were manufactured at different concentrations of NaOH (3%, 6%, and 9%) through the process of melt mixing and compression molding. The result showed that the treated RF/SPF hybrid composites had improved the thermal, mechanical, and physical properties accompanied by lower impact resistance. The fiber treated with 6% NaOH concentration on RF/SPF hybrid composites shows the highest tensile, flexural strength, and impact strength that are 14.26 MPa, 14.05 MPa, and 23.76 kJ/m^2, respectively.

A study by Afzaluddin et al. [91] also discussed the effect of the ratio of sugar palm (SP) and glass (G) on the physical and mechanical properties of sugar palm/glass fiber-reinforced thermoplastic polyurethane hybrid composites. The fabrication of hybrid composites was carried out at a constant weight fraction of total fiber loading at 40 wt% using the melt compounding method. The mechanical property such as tensile property of the hybrid composites was increased with the increase in of SPF content (30/10 SP/G) (21.34 MPa) as compared to glass fiber-reinforced composites (0/40 SP/G) (17.86 MPa). This might be attributed to the excellent hybrid performance of the both synthetic and natural fibers.

TABLE 14.4
Tensile Properties of Sugar Palm Fibre-Reinforced Polymer Composites

Fibre	Matrix	Processing	Samples/fiber/polymer ratio	Tensile strength	Tensile modulus	E (%)	References
SPFs (macro-sized)	Epoxy	Hand lay-up	Neat epoxy untreated	37.50 MPa	–	–	Sastra et al. [88]
			Chopped random 10 wt%	32.03 MPa	1.08 GPa	–	
			Long random 10 wt%	44.33 MPa	1.16 GPa	–	
			Woven 10 wt%	51.73 MPa	1.23 GPa	–	
			Chopped random 15 wt%	32.53 MPa	1.15 GPa	–	
			Chopped random 20 wt%	30.49 MPa	1.06 GPa	–	
			Long random 20 wt%	30.84 MPa	1.03 GPa	–	
	Epoxy	Hand lay-up	Chopped random 10 wt%	33.76 MPa	1.07 GPa	–	Suriani et al. [89]
			Long random 10 wt%	50.39 MPa	1.04 GPa	–	
			Woven 10 wt%	51.72 MPa	1.01 GPa	–	
			Chopped random 15 wt%	31.58 MPa	1.16 GPa	–	
			Chopped random 20 wt%	30.49 MPa	1.25 GPa	–	
			Chopped random 20 wt%	30.89 MPa	1.20 GPa	–	
	Epoxy	Hand lay-up	Freshwater treatment	13,776.12 kPa	–	–	Leman et al. [59]
			Day 0	17,306.91 kPa	–	–	
			Day 6	18,458.35 kPa	–	–	
			Day 12	20,817.45 kPa	–	–	
			Day 18	18,877.22 kPa	–	–	
			Day 24	21,266.5 kPa	–	–	
			Day 30				
			Freshwater treatment	13,776.12 kPa	–	–	
			Day 0	16,715.14 kPa	–	–	
			Day 6	17,159.09 kPa	–	–	
			Day 12	18,392.64 kPa	–	–	
			Day 18	21,148.25 kPa	–	–	
			Day 24	23,042.48 kPa	–	–	
			Day 30				
	Epoxy	Vacuum impregnation process	SPF/control	75 MPa	20 GPa	11 MPa	Munawar et al. [61]
			SPF/epoxy	112 MPa	31 GPa	10 MPa	

(Continued)

TABLE 14.4 (Continued)
Tensile Properties of Sugar Palm Fibre-Reinforced Polymer Composites

Fibre	Matrix	Processing	Samples/fiber/polymer ratio	Tensile strength	Tensile modulus	E (%)	References
	Epoxy	Hot compression	All flax	122.27 ± 10.35 MPa	2.03 ± 0.59 GPa	8.04 ± 0.54	Chandrasekar et al. [62]
			All SPF	33.58 ± 4.00 MPa	2.61 ± 0.11 GPa	2.16 ± 0.52	
			Flax (outer)/SPF (core)	93.79 ± 6.08 MPa	2.94 ± 0.41 GPa	5.57 ± 0.61	
			SPF (outer)/Flax (core)	81.54 ± 7.83 MPa	3.09 ± 0.14 GPa	4.88 ± 0.39	
	UP	Hand lay-up	Sugar palm fibre	19.20 MPa	1,586.50 MPa	1.64	Misri et al. [64]
			Hybrid fiber	30.58 MPa	1,840.63 MPa	2.76	
	Unsaturated PE	Hand lay-up	30 wt% untreated	41.52	5.29 GPa	1.28	Norizan et al. [65]
			70/30	56.14	6.88 GPa	1.74	
			60/40	64.91	6.70 GPa	1.82	
			50/50				
			40 wt% untreated	63.74	7.24 GPa	1.83	
			70/30	77.00	7.52 GPa	2.32	
			60/40	83.04	8.12 GPa	2.56	
			50/50				
			30 wt% treated	53.41	6.42 GPa	1.41	
			70/30	59.26	6.70 GPa	1.36	
			60/40	83.823	7.78 GPa	1.30	
			50/50				
			40 wt% treated	67.64	7.49 GPa	1.79	
			70/30	80.13	7.52 GPa	1.72	
			60/40	90.26	8.28 GPa	1.65	
			50/50				
	UP	Vacuum resin impregnation	Control	241.93 MPa	3.07 GPa	25.16	Ishak et al. [66]
			5	283.75 MPa	3.11 GPa	23.09	
			10	280.98 MPa	3.12 GPa	22.21	
			15	285.06 MPa	3.13 GPa	20.62	
			20	288.17 MPa	3.11 GPa	21.82	
			25	290.36 MPa	3.11 GPa	21.67	

(Continued)

TABLE 14.4 (Continued)
Tensile Properties of Sugar Palm Fibre-Reinforced Polymer Composites

Fibre	Matrix	Processing	Samples/fiber/polymer ratio	Tensile strength	Tensile modulus	E (%)	References
	PE resin	Hand lay-up	Form of fibers	15.40 ± 8.64 MPa	–	–	Ticoalu et al. [67]
			Random original	14.52 ± 1.20 MPa	–	–	
			Chopped	24.49 ± 6.30 MPa	–	–	
			Unidirectional	9.24 ± 1.86 MPa	–	–	
			Woven mat				
	High-impact polystyrene	Mixing and hot compression	0%	29.9 MPa	1.27 GPa	–	Oumer and Bachtiar [68]
			8%	26.2 MPa	1.52 GPa	–	
			17%	23.2 MPa	1.62 GPa	–	
			26%	19.3 MPa	1.71 GPa	–	
			35%	24.3 MPa	1.66 GPa	–	
			45%	28.47 MPa	1.65 GPa	–	
	PE resin	Vacuum impregnation process	SPF/control	75 MPa	20 GPa	11	Munawar et al. [61]
			SPF/PE	47 MPa	13 GPa	5	
	Polypropylene	Mixing and hot compression	Neat PP	22.89 MPa	798 MPa	–	Hadi et al. [69]
			10% SPF	19.88 MPa	948 MPa	–	
			20% SFF	20.07 MPa	1,025 MPa	–	
			30% SPF	21.73 MPa	1,275 MPa	–	
	Polypropylene	Mixing and hot compression	Untreated	21.73 MPa	1,276 MPa	–	Hadi et al. [69]
			2% alkali	15.02 MPa	933 MPa	–	
			4% alkali	21.83 MPa	1,006 MPa	–	
			6% alkali	19.8 MPa	993 MPa	–	
	Polypropylene	Mixing and hot compression	Untreated	21.73 MPa	1,276 MPa	–	Hadi et al. [69]
			2% MAPP	24.06 MPa	1,296 MPa	–	
			4% MAPP	23.09 MPa	1,190 MPa	–	
			6% MAPP	20.14 MPa	1,040 MPa	–	

(Continued)

TABLE 14.4 (Continued)
Tensile Properties of Sugar Palm Fibre-Reinforced Polymer Composites

Fibre	Matrix	Processing	Samples/fiber/polymer ratio	Tensile strength	Tensile modulus	E (%)	References
	PF	Vacuum resin impregnation	Control	241.93 MPa	3.07 GPa	25.16	Ishak et al. [15]
			5	254.85 MPa	3.13 GPa	19.37	
			10	256.30 MPa	3.17 GPa	17.29	
			15	248.51 MPa	3.17 GPa	18.71	
			20	254.67 MPa	3.18 GPa	15.44	
			25	257.14 MPa	3.17 GPa	16.25	
	High-impact polystyrene	Mixing and hot compression	Neat HIPS	29.92 MPa	1.270 MPa	–	Bachtiar et al. [70]
			Untreated	24.34 MPa	1.662 MPa	–	
			2% MAH	26.04 MPa	907 MPa	–	
			3% MAH	28.91 MPa	760 MPa	–	
			4% NaOH	32.94 MPa	1.354 MPa	–	
			6% NaOH	30.34 MPa	1.316 MPa	–	
	Polyurethane	Extruder and hot compression	Untreated TPU/SPF	9.90 MPa	95.31 MPa	23.43	Mohammed et al. [71]
			Treated with 6% NaOH	5.49 MPa	30 MPa	48.01	
			Treated with 70°C	18.42 MPa	1,307.56 MPa	2.49	
			Treated with 80°C	17.13 MPa	1,063.63 MPa	3.21	
			Treated with 90°C	18.17 MPa	1,067.20 MPa	3.43	
	Polyurethane	Melt mixing and hot compression	Untreated RF/SPF	13.76 MPa	270.92 MPa	–	Radzi et al. [72]
			RF/SPF-3% NaOH	13.54 MPa	143.79 MPa	–	
			RF/SPF-6% NaOH	14.20 MPa	127.17 MPa	–	
			RF/SPF-9% NaOH	13.43 MPa	118.84 MPa	–	
	SPS	Hot press	SPS	2.48 MPa	80.59 MPa	8.03	Sahari et al. [74]
			SPF10	3.23 MPa	306.59 MPa	6.76	
			SPF20	4.04 MPa	441.73 MPa	4.1	
			SPF30	5.34 MPa	542.44 MPa	3.32	
	PVDF	Melt melting and hot compression	PVD/FSPF 70:30	23.06 MPa	2,243.80 MPa	–	Alaaeddin et al. [75]

(Continued)

TABLE 14.4 (Continued)
Tensile Properties of Sugar Palm Fibre-Reinforced Polymer Composites

Fibre	Matrix	Processing	Samples/fiber/polymer ratio	Tensile strength	Tensile modulus	E (%)	References
	Vinyl ester	Hand lay-up process	Neat VE	20.00 MPa	2,505.00 MPa	–	Huzaifah et al. [77]
			SPF Jempol/VE	24.06 MPa	2,525.04 MPa	–	
			SPF Indonesia/VE	17.37 MPa	2,487.45 MPa	–	
			SPF Tawau/VE	19.74 MPa	2,549.54 MPa	–	
	Vinyl ester	Hand lay-up process	Neat VE	38.84 MPa	1,413 MPa	–	Ammar et al. [78]
			Unidirectional (0°)	15.41 MPa	2,501 MPa	–	
			0°/90° woven	11.65 MPa	2,413 MPa	–	
			±45° woven	15.67 MPa	2,369 MPa	–	
	Vinyl ester	Hand lay-up process	Neat VE	15.57 MPa	2,223.64 MPa	–	Razali et al. [76]
			RF	24.81 MPa	2,201.01 MPa	–	
			SPF	21.35 MPa	1,519.93 MPa	–	
			RF/SP 25:75	19.76 MPa	900.66 MPa	–	
			RF/SP 50:50	26.34 MPa	2,336.37 MPa	–	
			RF/SP 75:25	24.945 MPa	1,024.48 MPa	–	
	Vinyl ester	Vacuum impregnation process	SPF/control	75 MPa	20 GPa	11	Munawar et al. [61]
			SPF/VE	85 MPa	17 GPa	10	
	Polyurethane	Melt melting and hot compression	30/10 SP/G	21.34 MPa	742.34 MPa	–	Afzaluddin et al. [90]
			20/20 SP/G	17.26 MPa	655.29 MPa	–	
			10/30 SP/G	16.35 MPa	586.35 MPa	–	
			0/40 SP/G	17.86 MPa	724.28 MPa	–	
	Polyurethane	Melt melting and hot compression	RF/SPF 100:0	14.24 MPa	330 MPa	–	Radzi et al. [80,81]
			RF/SPF 75:25	13.4 MPa	117 MPa	–	
			RF/SPF 50:50	12.5 MPa	99 MPa	–	
			RF/SPF 25:75	11.1 MPa	106 MPa	–	
			RF/SPF 0:100	10.1 MPa	96 MPa	–	

(Continued)

TABLE 14.4 (Continued)
Tensile Properties of Sugar Palm Fibre-Reinforced Polymer Composites

Fibre	Matrix	Processing	Samples/fiber/polymer ratio	Tensile strength	Tensile modulus	E (%)	References
	Cassava starch	Solution casting	CS film	5.28 MPa	66.8 MPa	17.63	Edhirej et al. [91]
			CS/CB	10.81 MPa	575.1 MPa	10.64	
			CS-CB/SPF2	16.03 MPa	827.88 MPa	5.56	
			CS-CB/SPF4	17.86 MPa	839.43 MPa	5.08	
			CS-CB/SPF6	20.72 MPa	1,115 MPa	4.46	
			CS-CB/SPF8	17.10 MPa	790 MPa	4.11	
Sugar palm cellulose (micro-sized)	SPS	Solution casting	Neat SPS	7.79 MPa	20.11 MPa	46.00	Sanyang et al. [34]
			SPC-C1	10.5 MPa	31.38 MPa	40.99	
			SPC-C3	12.3 MPa	41.33 MPa	38.49	
			SPC-C5	15.3 MPa	69.30 MPa	34.0	
			SPC-C10	19.68 MPa	92.33 MPa	32.8	
Sugar palm nanocrystalline cellulose	SPS	Solution casting	Neat SPS	4.80 ± 0.41 MPa	53.97 ± 8.74 MPa	38.10 ± 1.16	Ilyas et al. [23]
			SPNCCs-0.1 wt%	6.60 ± 0.47 MPa	98.10 ± 9.54 MPa	35.52 ± 0.43	
			SPNCCs-0.2 wt%	7.19 ± 0.25 MPa	107.98 ± 10.52 MPa	32.06 ± 0.93	
			SPNCCs-0.3 wt%	8.15 ± 0.43 MPa	122.93 ± 10.4 MPa	28.30 ± 1.24	
			SPNCCs-0.4 wt%	8.60 ± 0.48 MPa	133.94 ± 10.82 MPa	26.08 ± 0.32	
			SPNCCs-0.5 wt%	11.47 ± 0.25 MPa	178.83 ± 4.48 MPa	24.42 ± 0.96	
			SPNCCs-1.0 wt%	7.78 ± 0.46 MPa	117.19 ± 8.45 MPa	24.02 ± 1.15	
Sugar palm nanofibrillated cellulose	SPS	Solution casting	Neat SPS	4.80 ± 0.41 MPa	53.97 ± 8.74 MPa	38.10 ± 1.16	Ilyas et al. [24]
			SPNFCs-0.1 wt%	6.80 ± 0.25 MPa	59.07 ± 2.10 MPa	37.70 ± 0.66	
			SPNFCs-0.2 wt%	7.30 ± 0.11 MPa	63.63 ± 2.09 MPa	36.21 ± 0.75	
			SPNFCs-0.3 wt%	7.55 ± 0.07 MPa	72.46 ± 3.51 MPa	34.24 ± 0.55	
			SPNFCs-0.4 wt%	8.10 ± 0.23 MPa	84.46 ± 4.21 MPa	30.74 ± 0.67	
			SPNFCs-0.5 wt%	8.53 ± 0.13 MPa	99.52 ± 7.20 MPa	27.54 ± 0.76	
			SPNFCs-1.0 wt%	10.68 ± 0.67 MPa	121.26 ± 5.69 MPa	25.38 ± 0.50	

Munawar et al. [61] carried out an experiment on the effect of various thermosetting resins (epoxy, VE, and PE) on the tensile properties of vacuum resin-impregnated SPF biocomposites. The result shows that the impregnation of SPF with epoxy resin, VE, and PE increased the tensile strength of SPF by 112, 85, and 47 MPa, respectively. According to Ishak et al. [66], the enhancement of the strength of the biocomposites might be attributed to the resin implanted inside the fiber cell lumen and cell wall, which reinforces the structure of the fibers.

Razali et al. [76] conducted an experiment on morphological and mechanical properties of treated roselle fiber (RF)/SPF-incorporated VE hybrid composites. RF was undergone water retting and 6% NaOH pretreatments before being blended with the matrix to form the composites via hand lay-up technique. However, for SPF, no chemical treatment was applied to the fiber. The result indicated that the RF, SPF, and roselle/sugar palm hybrid composites showed an improvement in tensile strength and modulus compared to neat VE. The optimum tensile strength can be found at fiber percentages of sample E hybrid composites (50% roselle and 50% sugar palm). This phenomenon occurred because fiber acts as a load carrier in the matrix and anti-crack propagation, as well as good interfacial bonding between fiber/matrix and uniform dispersion of the fiber in the matrix.

Ammar et al. [92] studied mechanical properties of SPF-reinforced VE biocomposites at different fiber arrangements. Hand lay-up technique was used to fabricate the biocomposites. The result showed that unidirectional fibre biocomposites show good performance in flexural strength, impact strength, tensile modulus, and flexural modulus with a value of 93.08 MPa, 33.66 kJ/m^2, 2,501 and 3,328 MPa, respectively, compared with other fibre arrangements such as 0°/90° woven fibers and ±45° woven fibers.

Huzaifah et al. [77] conducted a study on the thermal, physical, and mechanical properties of SPF-incorporated VE composites that were obtained from three different geographical locations: Tawau (West Malaysia), Kuala Jempol (Peninsular Malaysia), and Tasikmalaya (Indonesia). The VE composites were prepared using a wet hand lay-up compression molding method. The result showed that the highest tensile strength of biocomposite is SPF Kuala Jempol/VE biocomposites, followed by SPF Tawau/VE biocomposites, and SPF Indonesia/VE biocomposites, with a value of 24.06, 19.74, and 17.37 MPa, respectively. According to Huzaifah et al. [77], the mechanical properties of the biocomposite mainly depend on the strength of the natural fibre. Besides that, the strength properties of natural fiber are influenced by cellulose content.

Alaaeddin et al. [75] investigated the physical, mechanical, and nanomechanical properties of polyvinylidene fluoride (PVDF)-reinforced short sugar palm fiber (SSPF) biocomposites. The result shows that the mechanical properties of the PVDF biopolymer improved when the SSPF were reinforced with PDVF. The tensile strength, stress at maximum load, the average of Young's modulus, and modulus (Aut.) were 23.059, 23.874, 2,243.799, and 2,204.015 MPa, respectively. The improvement of mechanical properties of the biocomposite is due to the good compatibility characteristics of both the PVDF matrix and SSPF fiber.

Radzi et al. [72] studied the effect of alkaline treatment on the mechanical, physical, and thermal properties of RF/SPF-reinforced thermoplastic polyurethane hybrid composites. Melt mixing and compression molding technique were used to fabricate RF/SPF hybrid biocomposites at different NaOH concentrations (3%, 6%, and 9%). The result showed that the effect of NaOH treatment on the surface of the fibre improved the physical and thermal properties. The highest tensile, flexural strength, and impact strength (14.26 MPa, 14.05 MPa, and 23.76 kJ/m^2, respectively)were obtained from the fiber treated with 6% NaOH concentration on RF/SPF hybrid composites. In general, the surface treatment on RF/SPF hybrid composite significantly improved the composite properties.

Mohammed et al. [71] conducted an experiment on the effect of microwave treatment on the tensile properties of treated SPF with 6% NaOH-reinforced thermoplastic polyurethane composites. The SPFs and polyurethane resin composite were fabricated using extruder and hot press machine. The highest tensile strength is 18.42 MPa with 6% alkali pretreatment and microwave temperature at 70°C. This might be due to the treatment, which cleaned the fiber surface, modified the chemistry on the surface, as well as lower the moisture up take, and increased the surface roughness of fibers.

Bachtiar et al. [70] studied on the effects of alkaline treatment and a compatibilizing agent on the tensile properties of SPF-reinforced high-impact polystyrene (HIPS) composites. Two concentrations of an alkali solution (4% and 6%) and two percentages of a compatibilizing agent (2% and 3%) were used. The fibers were blended with HIPS using melt mixer at a temperature of 165°C. The result shows that the tensile strength of treated fiber composite is higher compared to that of untreated composites. It can be summarized that 4% alkali treatment is the best treatment for improving the tensile properties of SPF-HIPS composites. This might be due to the improvement in the adhesion between the surfaces of the fiber and the HIPS matrixes.

Sahari et al. [32] studied the effect of fiber content on mechanical properties, water absorption behavior, and thermal properties of SPF-reinforced plasticized SPS biocomposites. The composites were fabricated with various loading (i.e., 10%, 20%, and 30% by weight percent and coded as SPS, SPF10, SPF20, and SPF30, respectively) by using glycerol as plasticizer for the starch. SPS 70 wt% and glycerol 30 wt% were mixed using mechanical stirrer and later hot pressed at 130°C for 30 min under the load of 10 tonnes. Besides that, from this experiment conducted by Sahari et al. [32], it is found that the mechanical properties of SPS biopolymer matrix improved with the reinforcement of SPF. The tensile strength and modulus of SPF/SPS biocomposites showed increasing trend with increasing SPF loading, whereas the addition of the SPF made the elongation fall from 8.03% to 3.32%. According to Sahari et al. [32], this phenomena might be attributed to the outstanding intrinsic adhesion of the fiber–matrix interface caused by the chemical similarity of starch and the cellulose fiber. However, elongation data showed decreasing trend when increasing fiber loading. This is due to a normal consequence of the increase in fiber weight percentage which is having a low strain compared to rubbery SPS materials. Besides that, the flexural strength and modulus of SPF/SPS biocomposites revealed increasing trend with increasing fiber loading. Similar trend also was displayed for impact strength of biocomposites, in which the impact strength increases with increasing fiber loading. As conclusion to their work, mechanical properties of plasticized SPS are highly affected by the incorporation of SPF.

Ishak et al. [31] studied the effects of impregnation modification via vacuum resin impregnation on physical and mechanical properties of sugar palm (*Arenga pinnata*) fibers. The fiber was evacuated at a constant impregnation pressure of 1,000 mmHg impregnation times (0, 5, 10, 15, 20 and 25 min) with two different impregnation agents: PF and UP. Result showed that the mechanical properties of SPF were be enhanced by impregnating the fiber with thermosetting polymer (PF and UP) for 5 min. This was due to the resin that was enclosed and locked in the fiber cell wall and lumen.

Hadi et al. [69] studied on the mechanical properties of SPF-reinforced polypropylene (PP) composites. Internal mixer and hot press machine were used to fabricate composites with different weight of sugar palm, i.e., 10, 20, and 30 wt%. In order to improve the adhesion between PP matrix and fibers, two treatments are performed: mercerization and add compatibilizing (maleic anhydride polypropylene (MAPP)) with 30 wt% of SPF. The result shows that the tensile strength and the flexural strength increase with the increase of SPF loading in PP matrix. However, the impact strength decreases with increase of SPF loading in PP matrix. Besides that, the effect of both treatments also displayed the enhancement for every mechanical properties of the biocomposites. Mercerization and MAPP treatments improved the mechanical properties of biocomposite. This might be due to the improvement of bonding between PP matrix and SPF fiber.

Oumer and Bachtiar [68] conducted experiment on the tensile properties of short sugar palm fiber-reinforced high-impact polystyrene (SPF-HIPS) composites obtained by means of statistical approach. Hot compression technique was used to fabricate composites. The experimental results presented that the tensile strength of the composite reduced due to the addition of SPF, whereas the elastic modulus increased by a factor of up to 1.34. This was due to the poor interface bonding between HIPS matrix and SPF.

Nurazzi et al. [93] conducted experiment on the effect of fiber loadings and alkaline treatment on the mechanical properties of sugar palm yarn/woven glass fiber-reinforced UP composites. The composites fabricated at the ratio of sugar palm yarn fiber to glass fiber was selected at 70:30, 60:40, and 50:50. The results showed that the mechanical properties of the hybrid composites were increased with an increase of glass fiber loading for both 30 and 40 wt% reinforcement content. The mechanical properties showed enhancement after the treatment of alkaline of sugar palm yarn fibers compared with the untreated hybrid composites. This phenomenon was due to a better compatibility and balance ratio of the pack arrangement between the sugar palm yarn fiber and glass fiber with the UP matrix.

Sahari et al. [31] studied the effect of water absorption on mechanical properties of SPF-reinforced sugar palm starch (SPF/SPS) biocomposites. The experiment was conducted due to concern of moisture absorption in biocomposites for further improvement potential application of biocomposite. Therefore, the effect of water absorption on the mechanical properties of palm fibre-reinforced sugar palm starch (SPF/SPS) biocomposite has been studied. For the fabrication of SPF/SPS biocomposites, SPS and glycerol with a ratio of 70:30 wt% were mixed together using mechanical stirrer for 30 min and later was cured by hot pressing in a Carver hydraulic hot press at 130°C for 30 min under the load of 10 tonnes. The loading of SPFs was varied to 0, 10, 20, and 30 wt% coded as SPS, SPF10, SPF20, and SPF30, respectively. The samples were stored in humidity chamber for 72 h at RH = 75%. In terms of mechanical properties of tensile strength, as the SPFs loading increases from 0 to 30 wt%, the tensile strength of biocomposite also increases from 2.42 to 5.31 MPa, respectively. The improvement of tensile strength was due to the effectiveness of SPF in acting as good reinforcement with SPS biopolymer. However, after the biocomposite being exposed to 75% RH for 72 h, the tensile strength was dropped drastically from 0.42 to 1.73 MPa. The same phenomena were also witnessed for the impact strength of SPF/SPS biocomposite, where the impact strength increases with the increase of fiber loading and decreases after being exposed to 75% RH for 72 h. This was due to the presence of high hydroxyl group presented in starch biopolymer, which tends to show low moisture resistance and leads to the degradation of fiber–matrix interface region. Subsequently, this may resulted to reduction of dimensional variation, poor interfacial bonding of SPS, and poor mechanical properties of SPF/SPS biocomposites.

Sahari et al. [94] carried out a study on degradation characteristics of SPF/SPS biocomposites. In their study, SPFs were incorporated into SPS with glycerol (ratio of 70:30) using hot press at a temperature of 130°C for 30 min at a load of 10 tons. Later, samples were treated in weathering chamber to determine the environmental effect on the biocomposites. The tensile test was then performed after the weathering test. The results show that the tensile strength of the SPS and SPF/SPS biocomposites after being exposed to the environmental effect was reduced by 78.09% (from 2.42 to 0.53 MPa) and 53.67% (from 5.31 to 2.46 MPa), respectively. These phenomena might be due to the degradation of both SPF and SPS as the presence of oxygen and UV causes oxidative degradation of the polymer and it can change the morphology of the polymeric material by means of chemical cross-linking or chain scission, which resulted in low mechanical properties of SPS and biocomposites [94].

14.3.2 Micro-Size Sugar Palm Fibre-Reinforced Polymer Composites

Surface treatment on natural fiber is compulsory in order to improve interfacial properties of natural fiber and polymer composite [95]. Sanyang et al. [34] conducted investigation on effect of sugar palm-derived cellulose reinforcement on the mechanical and water barrier properties of SPS biocomposite films. SPS-reinforced sugar palm cellulose composite films (SPS-C) were fabricated using different SPC contents (1–10 wt%) using a solution casting method. The mechanical properties of the composite films showed increased tensile strength and modulus, whereas the elongation at break decreased with the increase in SPC loading. Incorporation of 1–10 wt% SPC into SPS biopolymer decreased the elongation at break for the biocomposite films (from 40.99% to 32.8%) and increased the tensile modulus and tensile strength values of the biocomposite films (from 31.38 to 92.33 MPa and from 10.5 to 19.68 MPa, respectively). This might be attributed to the high compatibility between the SPS biopolymer and sugar palm cellulose [34].

14.3.3 Nanosize Sugar Palm Fibre-Reinforced Polymer Composites

Other related details on mechanical properties of nanosized SPF-reinforced SPS can be found in the study by Ilyas et al. [23]. Ilyas et al. [23] also discussed the potential of hydrolysis treatment for surface modification of SPF used to reinforce

SPS composites. The experiment was conducted to investigate the effect of different concentrations (0–1.0 wt%) of sugar palm nanocrystalline cellulose (SPNCC)-reinforced SPS on morphological, mechanical, and physical properties of the bionanocomposites film. The SPS/SPNCCs with 1.0 wt% had undergone an increment in both the tensile strength and Young's modulus when compared with the neat SPS film, from 4.80 to 11.47 MPa and 53.97 to 178.83 MPa, respectively. Meanwhile, the elongation at break (%) indicates a decrease in value (from 38.1% to 24.01%) because of the reduction in the molecular mobility of the SPS biomatrix, making film stiffer (Figure 14.2).

Ilyas et al. [24] conducted a studied on effect of sugar palm nanofibrillated cellulose concentrations on morphological, mechanical, and physical properties of biodegradable films based on agro-waste sugar palm (*Arenga pinnata (Wurmb.) Merr*) starch. Bionanocomposites were prepared by mixing SPS and sorbitol/glycerol with different sugar palm nanofibrillated cellulose loading (0–1.0 wt%) using solution casting method. Results showed that as the sugar palm nanofibrillated cellulose (SPNFC) concentration increased from 0.1 to 1.0 wt%, the tensile strength and tensile modulus of SPS/SPNFCs nanocomposite films were increased from 6.80 to 10.68 MPa and 59.07 to 121.26 MPa, respectively. These might be due to the compatible interaction between the SPNFCs and SPS polymer matrices, which facilitated adequate interfacial adhesion because of their chemical similarities [30,96]. However, the elongation at break for the nanocomposite films decreased from 38.1% to 25.38%, as the SPNFC concentration was increased from 0 to 1.0 wt% (Figure 14.3). This was due to the addition of SPNFC

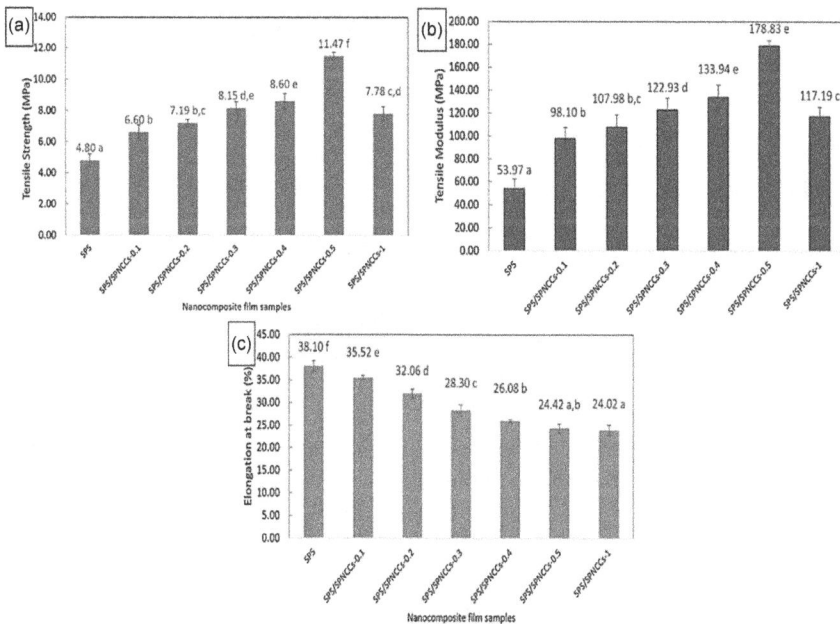

FIGURE 14.2   Effect of SPNCC loading on the (a) tensile strength, (b) tensile modulus, and (c) elongation at break (%) of SPS-SPNCC nanocomposite films [23].

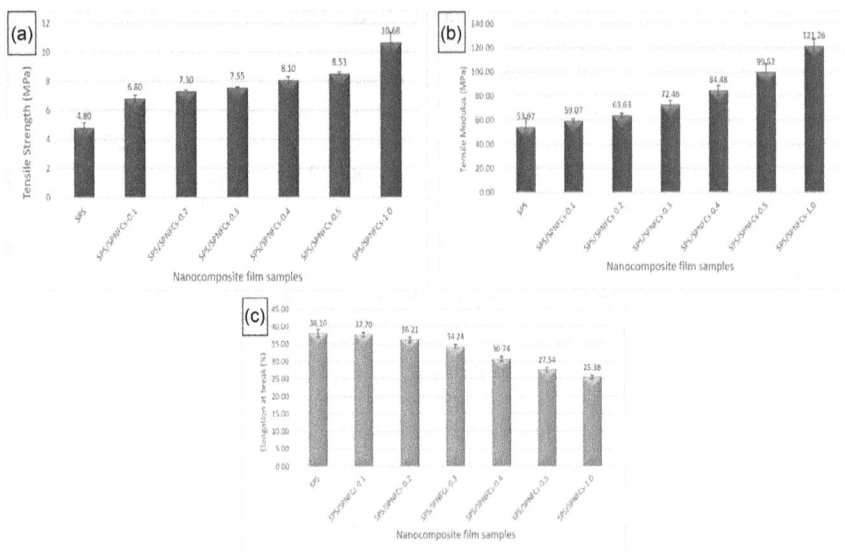

FIGURE 14.3 Effect of SPNFC concentrations on the (a) tensile strength, (b) tensile modulus, and (c) elongation at break (%) of SPS-SPNFC nanocomposite films [24].

concentration, which restricts the molecular mobility and ductility of the SPS matrix, making the composite materials stiffer. Ilyas et al. [24] concluded that there were three main factors that had potentially affected the mechanical performances of the nanocomposite material: (1) the processing method, (2) the dimension and morphology of the nanofiller, and (3) the micro/nanostructure of the matrix and matrix/filler interface.

14.4 CONCLUSIONS AND REMARKS

Tensile properties of SPF-reinforced polymer composites were reviewed. The fact that the strength of the biocomposite specimens is high as glass fibre composites has proven the potential use of SPFs as reinforcement in composites. The results confirmed that more studies are needed to determine the beneficial and cost-effective applications of the fibers in composites. This study has also confirmed that the vacuum resin impregnation using PF as polymer matrix has the highest tensile strength compared to other polymer matrix. It is expected that with proper treatments, the mechanical properties of the SPFs and the adhesion of SPF–polymer resin can be improved. Therefore, more studies on various effects of fabrications, polymer matrix used, ratio, and fibre treatment are required.

ACKNOWLEDGMENTS

The authors would like to thank Universiti Putra Malaysia and Ministry of Education, Malaysia, for the financial support through the Graduate Research Fellowship (GRF) scholarship, Universiti Putra Malaysia Grant scheme Hi-CoE (6369107), and Fundamental Research Grant Scheme FRGS/1/2017/TK05/UPM/01/1 (5540048).

REFERENCES

1. R. A. Ilyas and S. M. Sapuan, "The preparation methods and processing of natural fibre bio-polymer composites," *Current Organic Synthesis*, vol. 16, no. 8, pp. 1068–1070, 2020.
2. R. A. Ilyas and S. M. Sapuan, "Biopolymers and biocomposites: chemistry and technology," *Current Analytical Chemistry*, vol. 16, pp. 1–4, 2020.
3. P. Peças, H. Carvalho, H. Salman, and M. Leite, "Natural fibre composites and their applications: a review," *Journal of Composites Science*, vol. 2, no. 4, p. 66, 2018.
4. M. R. M. Asyraf, M. R. Ishak, S. M. Sapuan, N. Yidris, and R. A. Ilyas, "Woods and composites cantilever beam: a comprehensive review of experimental and numerical creep methodologies," *Journal of Materials Research and Technology*, vol. 9, no. 3, pp. 6759–6776, 2020.
5. A. Atiqah, M. Jawaid, S. M. Sapuan, M. R. Ishak, M. N. M. Ansari, and R. A. Ilyas, "Physical and thermal properties of treated sugar palm/glass fibre reinforced thermoplastic polyurethane hybrid composites," *Journal of Materials Research and Technology*, vol. 8, no. 5, pp. 3726–3732, 2019.
6. R. A. Ilyas, S. M. Sapuan, M. L. Sanyang, M. R. Ishak, and E. S. Zainudin, "Nanocrystalline cellulose as reinforcement for polymeric matrix nanocomposites and its potential applications: a review," *Current Analytical Chemistry*, vol. 14, no. 3, pp. 203–225, 2018.
7. E. Syafri *et al.*, "Effect of sonication time on the thermal stability, moisture absorption, and biodegradation of water hyacinth (*Eichhornia crassipes*) nanocellulose-filled bengkuang (*Pachyrhizus erosus*) starch biocomposites," *Journal of Materials Research and Technology*, vol. 8, no. 6, pp. 6223–6231, 2019.
8. R. Jumaidin, R. A. Ilyas, M. Saiful, F. Hussin, and M. T. Mastura, "Water transport and physical properties of sugarcane bagasse fibre reinforced thermoplastic potato starch biocomposite," *Journal of Advanced Research in Fluid Mechanics and Thermal Sciences*, vol. 61, no. 2, pp. 273–281, 2019.
9. R. Jumaidin *et al.*, "Characteristics of cogon grass fibre reinforced thermoplastic cassava starch biocomposite: water absorption and physical properties," *Journal of Advanced Research in Fluid Mechanics and Thermal Sciences*, vol. 62, no. 1, pp. 43–52, 2019.
10. R. Jumaidin, M. A. A. Khiruddin, Z. Asyul Sutan Saidi, M. S. Salit, and R. A. Ilyas, "Effect of cogon grass fibre on the thermal, mechanical and biodegradation properties of thermoplastic cassava starch biocomposite," *International Journal of Biological Macromolecules*, vol. 146, pp. 746–755, 2019.
11. A. M. N. Azammi *et al.*, "Characterization studies of biopolymeric matrix and cellulose fibres based composites related to functionalized fibre-matrix interface," In K. L. Goh, S. Thomas, R. T. De Silva, and Aswathi M.K., (eds.), *Interfaces in Particle and Fibre Reinforced Composites*, 1st ed. London: Elsevier, 2020, pp. 29–93.
12. H. Abral *et al.*, "Highly transparent and antimicrobial PVA based bionanocomposites reinforced by ginger nanofiber," *Polymer Testing*, vol. 81, p. 106186, 2020.
13. H. Abral *et al.*, "Transparent and antimicrobial cellulose film from ginger nanofiber," *Food Hydrocolloids*, vol. 98, p. 105266, 2020.
14. H. Abral *et al.*, "Effect of ultrasonication duration of polyvinyl alcohol (PVA) gel on characterizations of PVA film," *Journal of Materials Research and Technology*, vol. 9, no. 2, pp. 2477–2486, 2020.
15. M. R. Ishak, S. M. Sapuan, Z. Leman, M. Z. A. A. Rahman, U. M. K. K. Anwar, and J. P. Siregar, "Sugar palm (*Arenga pinnata*): its fibres, polymers and composites," *Carbohydrate Polymers*, vol. 91, no. 2, pp. 699–710, 2013.
16. S. M. Sapuan, M. R. Ishak, Z. Leman, R. A. Ilyas, and M. R. M. Huzaifah, "Development of products from sugar palm trees (*Arenga pinnata* Wurb. Merr): a community project," in *INTROPica*, Serdang, Selangor, 2017, pp. 12–13.

17. S. M. Sapuan *et al.*, "Development of sugar palm-based products: a community project," In S. M. Sapuan, J. Sahari, M.R. Ishak, and M.L. Sanyang (Eds.), *Sugar Palm Biofibers, Biopolymers, and Biocomposites*, 1st ed. Boca Raton, FL: CRC Press/ Taylor & Francis Group, 2018, pp. 245–266.
18. M. L. Sanyang, S. M. Sapuan, M. Jawaid, M. R. Ishak, and J. Sahari, "Recent developments in sugar palm (*Arenga pinnata*) based biocomposites and their potential industrial applications: a review," *Renewable and Sustainable Energy Reviews*, vol. 54, pp. 533–549, 2016.
19. R. A. Ilyas, S. M. Sapuan, and M. R. Ishak, "Isolation and characterization of nanocrystalline cellulose from sugar palm fibres (*Arenga Pinnata*)," *Carbohydrate Polymers*, vol. 181, pp. 1038–1051, 2018.
20. M. S. N. Atikah *et al.*, "Degradation and physical properties of sugar palm starch/sugar palm nanofibrillated cellulose bionanocomposite," *Polimery*, vol. 64, no. 10, pp. 27–36, 2019.
21. R. A. Ilyas *et al.*, "Sugar palm (*Arenga pinnata* (Wurmb.) Merr) cellulosic fibre hierarchy: a comprehensive approach from macro to nano scale," *Journal of Materials Research and Technology*, vol. 8, no. 3, pp. 2753–2766, 2019.
22. R. A. Ilyas, S. M. Sapuan, M. R. Ishak, and E. S. Zainudin, "Sugar palm nanofibrillated cellulose (*Arenga pinnata* (Wurmb.) Merr): effect of cycles on their yield, physicchemical, morphological and thermal behavior," *International Journal of Biological Macromolecules*, vol. 123, pp. 379–388, 2019.
23. R. A. Ilyas, S. M. Sapuan, M. R. Ishak, and E. S. Zainudin, "Development and characterization of sugar palm nanocrystalline cellulose reinforced sugar palm starch bionanocomposites," *Carbohydrate Polymers*, vol. 202, pp. 186–202, 2018.
24. R. A. Ilyas *et al.*, "Effect of sugar palm nanofibrillated cellulose concentrations on morphological, mechanical and physical properties of biodegradable films based on agro-waste sugar palm (*Arenga pinnata* (Wurmb.) Merr) starch," *Journal of Materials Research and Technology*, vol. 8, no. 5, pp. 4819–4830, 2019.
25. R. A. Ilyas, S. M. Sapuan, M. R. Ishak, and E. S. Zainudin, "Effect of delignification on the physical, thermal, chemical, and structural properties of sugar palm fibre," *BioResources*, vol. 12, no. 4, pp. 8734–8754, 2017.
26. J. Sahari, S. M. Sapuan, Z. N. Ismarrubie, and M. Z. Rahman, "Physical and chemical properties of different morphological parts of sugar palm fibres," *Fibres and Textiles in Eastern Europe*, vol. 91, no. 2, pp. 21–24, 2012.
27. H. E. Moore, "A new subfamily of palms—the caryotoideae," *Principes*, vol. 4, pp. 102–117, 1960.
28. J. Mogea, B. Seibert, and W. Smits, "Multipurpose palms: the sugar palm (*Arenga pinnata* (Wurmb) Merr.)," *Agroforestry Systems*, vol. 13, no. 2, pp. 111–129, 1991.
29. R. A. Ilyas, S. M. Sapuan, M. R. Ishak, and E. S. Zainudin, "Sugar palm nanocrystalline cellulose reinforced sugar palm starch composite: degradation and water-barrier properties," in *IOP Conference Series: Materials Science and Engineering*, 2018, vol. 368, no. 1.
30. R. A. Ilyas, S. M. Sapuan, M. R. Ishak, and E. S. Zainudin, "Water transport properties of bio-nanocomposites reinforced by sugar palm (*Arenga pinnata*) nanofibrillated cellulose," *Journal of Advanced Research in Fluid Mechanics and Thermal Sciences Journal*, vol. 51, no. 2, pp. 234–246, 2018.
31. J. Sahari, S. M. Sapuan, E. S. Zainudin, and M. A. Maleque, "Effect of water absorption on mechanical properties of sugar palm fibre reinforced sugar palm starch (spf/sps) biocomposites," *Journal of Biobased Materials and Bioenergy*, vol. 6, pp. 1–5, 2012.
32. J. Sahari, S. M. Sapuan, E. S. Zainudin, and M. A. Maleque, "Mechanical and thermal properties of environmentally friendly composites derived from sugar palm tree," *Materials and Design*, vol. 49, no. 2, pp. 285–289, 2013.

33. M. L. Sanyang, Y. Muniandy, S. M. Sapuan, and J. Sahari, "Tea tree (*Melaleuca alternifolia*) fiber as novel reinforcement material for sugar palm biopolymer based composite films," *BioResources*, vol. 12, no. 2, pp. 3751–3765, 2017.

34. M. L. Sanyang, S. M. Sapuan, M. Jawaid, M. R. Ishak, and J. Sahari, "Effect of sugar palm-derived cellulose reinforcement on the mechanical and water barrier properties of sugar palm starch biocomposite films," *BioResources*, vol. 11, no. 2, pp. 4134–4145, 2016.

35. M. L. Sanyang, S. M. Sapuan, M. Jawaid, M. R. Ishak, and J. Sahari, "Development and characterization of sugar palm starch and poly(lactic acid) bilayer films," *Carbohydrate Polymers*, vol. 146, pp. 36–45, 2016.

36. R. Jumaidin, S. M. Sapuan, M. Jawaid, M. R. Ishak, and J. Sahari, "Thermal, mechanical, and physical properties of seaweed/sugar palm fibre reinforced thermoplastic sugar palm starch/agar hybrid composites," *International Journal of Biological Macromolecules*, vol. 97, pp. 606–615, 2017.

37. R. Jumaidin, S. M. Sapuan, M. Jawaid, M. R. Ishak, and J. Sahari, "Effect of seaweed on physical properties of thermoplastic sugar palm starch/agar composites," *Journal of Mechanical Engineering and Sciences*, vol. 10, no. 3, pp. 2214–2225, 2016.

38. R. Jumaidin, S. M. Sapuan, and R. A. Ilyas, "Physio-mechanical properties of thermoplastic starch composites : a review," in *Prosiding Seminar Enau Kebangsaan 2019*, 2019, pp. 104–108.

39. D. R. Adawiyah, T. Sasaki, and K. Kohyama, "Characterization of arenga starch in comparison with sago starch," *Carbohydrate Polymers*, vol. 92, no. 2, pp. 2306–2313, 2013.

40. T. H. Jatmiko, C. D. Poeloengasih, D. J. Prasetyo, and Hernawan, "Modelling of moisture adsorption for sugar palm (*Arenga pinnata*) starch film," *AIP Conference Proceedings*, vol. 1823, p. 020007, 2017.

41. T. H. Jatmiko, C. D. Poeloengasih, D. J. Prasetyo, and V. T. Rosyida, "Effect of plasticizer on moisture sorption isotherm of sugar palm (*Arenga pinnata*) starch film," *AIP Conference Proceedings*, p. 080004, 2016.

42. C. D. Poeloengasih *et al.*, "A physicochemical study of sugar palm (*Arenga pinnata*) starch films plasticized by glycerol and sorbitol," *AIP Conference Proceedings*, p. 080003, 2016.

43. W. Apriyana, C. D. Poeloengasih, Hernawan, S. N. Hayati, and Y. Pranoto, "Mechanical and microstructural properties of sugar palm (*Arenga pinnata* Merr.) starch film: effect of aging," *AIP Conference Proceedings*, vol. 1755, 2016.

44. M. R. Mansor, S. M. Sapuan, M. A. Salim, M. Z. Akop, and M. M. Tahir, "Modeling of kenaf reinforced sugar palm starch biocomposites mechanical behaviour using Halpin-Tsai model," *Recent Advances in Environment, Ecosystems and Development*, pp. 94–99, 2015.

45. H.-T. Liao and C.-S. Wu, "New biodegradable blends prepared from polylactide, titanium tetraisopropylate, and starch," *Journal of Applied Polymer Science*, vol. 108, no. 4, pp. 2280–2289, 2008.

46. P. C. Belibi, T. J. Daou, J. M. B. Ndjaka, B. Nsom, L. Michelin, and B. Durand, "A comparative study of some properties of cassava and tree cassava starch films," *Physics Procedia*, vol. 55, pp. 220–226, 2014.

47. M. Baratter, E. F. Weschenlfelder, F. Stoffel, M. Zeni, and L. T. Piemolini-barreto, "Analysis and evaluation of cassava starch-based biodegradable trays as an alternative packaging to fresh strawberry (*Fragaria ananassa cv San Andreas*)," *American Journal of Polymer Science and Technology*, vol. 3, no. 4, pp. 76–81, 2017.

48. B. Ayana, S. Suin, and B. B. Khatua, "Highly exfoliated eco-friendly thermoplastic starch (TPS)/poly (lactic acid)(PLA)/clay nanocomposites using unmodified nanoclay," *Carbohydrate Polymers*, vol. 110, pp. 430–439, 2014.

49. B. Montero, M. Rico, S. Rodríguez-Llamazares, L. Barral, and R. Bouza, "Effect of nanocellulose as a filler on biodegradable thermoplastic starch films from tuber, cereal and legume," *Carbohydrate Polymers*, vol. 157, pp. 1094–1104, 2017.

50. A. Edhirej, S. M. Sapuan, M. Jawaid, and N. I. Zahari, "Preparation and characterization of cassava bagasse reinforced thermoplastic cassava starch," *Fibers and Polymers*, vol. 18, no. 1, pp. 162–171, 2017.

51. M. J. Halimatul, S. M. Sapuan, M. Jawaid, M. R. Ishak, and R. A. Ilyas, "Water absorption and water solubility properties of sago starch biopolymer composite films filled with sugar palm particles," *Polimery*, vol. 64, no. 9, pp. 27–35, 2019.

52. K. González, A. Retegi, A. González, A. Eceiza, and N. Gabilondo, "Starch and cellulose nanocrystals together into thermoplastic starch bionanocomposites," *Carbohydrate Polymers*, vol. 117, pp. 83–90, 2015.

53. R. Thakur *et al.*, "Amylose-lipid complex as a measure of variations in physical, mechanical and barrier attributes of rice starch- ι -carrageenan biodegradable edible film," *Food Packaging and Shelf Life*, 2017.

54. M. Bootklad and K. Kaewtatip, "Biodegradation of thermoplastic starch/eggshell powder composites," *Carbohydrate Polymers*, vol. 97, no. 2, pp. 315–320, 2013.

55. Y. Lu, L. Weng, and X. Cao, "Morphological, thermal and mechanical properties of ramie crystallites—reinforced plasticized starch biocomposites," *Carbohydrate Polymers*, vol. 63, no. 2, pp. 198–204, 2006.

56. E. Teixeira, A. A. S. Curvelo, A. C. Corrêa, J. M. Marconcini, G. M. Glenn, and L. H. C. Mattoso, "Properties of thermoplastic starch from cassava bagasse and cassava starch and their blends with poly (lactic acid)," *Industrial Crops and Products*, vol. 37, no. 1, pp. 61–68, 2012.

57. X. Tang, S. Alavi, and T. J. Herald, "Effects of plasticizers on the structure and properties of starch-clay nanocomposite films," *Carbohydrate Polymers*, vol. 74, no. 3, pp. 552–558, 2008.

58. X. Cao, Y. Chen, P. R. Chang, and M. A. Huneault, "Preparation and properties of plasticized starch/multiwalled carbon nanotubes composites," *Journal of Applied Polymer Science*, vol. 106, no. 2, pp. 1431–1437, 2007.

59. Z. Leman, S. M. Sapuan, M. Azwan, M. M. H. M. Ahmad, and M. A. Maleque, "The effect of environmental treatments on fiber surface properties and tensile strength of sugar palm fiber-reinforced epoxy composites," *Polymer - Plastics Technology and Engineering*, vol. 47, no. 6, pp. 606–612, 2008.

60. S. N. A. Safri, M. T. H. Sultan, M. Jawaid, and M. S. Abdul Majid, "Analysis of dynamic mechanical, low-velocity impact and compression after impact behaviour of benzoyl treated sugar palm/glass/epoxy composites," *Composite Structures*, vol. 226, no. July, p. 111308, 2019.

61. N. S. Z. Munawar, M. R. Ishak, R. M. Shahroze, M. Jawaid, and M. Y. M. Zuhri, "An investigation of the morphological and tensile properties of vacuum resin impregnated sugar palm fibers with various thermosetting resins," *BioResources*, vol. 14, no. 3, pp. 5212–5223, 2019.

62. M. Chandrasekar *et al.*, "Flax and sugar palm reinforced epoxy composites: effect of hybridization on physical, mechanical, morphological and dynamic mechanical properties," *Materials Research Express*, vol. 6, no. 10, p. 105331, 2019.

63. J. Sahari, S. M. Sapuan, Z. N. Ismarrubie, and M. Z. a. Rahman, "Investigation on bending strength and stiffness of sugar palm fibre from different parts reinforced unsaturated polyester composites," *Key Engineering Materials*, vol. 471–472, no. October, pp. 502–506, 2011.

64. S. Misri, Z. Leman, S. M. Sapuan, and M. R. Ishak, "Mechanical properties and fabrication of small boat using woven glass/sugar palm fibres reinforced unsaturated polyester hybrid composite," *IOP Conference Series: Materials Science and Engineering*, vol. 11, p. 012015, 2010.

65. M. N. Norizan, K. Abdan, and R. A. Ilyas, "Effect of water absorption on treated sugar palm yarn fibre/glass fibre hybrid composites," in *Prosiding Seminar Enau Kebangsaan 2019*, 2019, pp. 78–81.

66. M. R. Ishak, Z. Leman, S. M. Sapuan, M. Z. A. Rahman, and U. M. K. Anwar, "Impregnation modification of sugar palm fibres with phenol formaldehyde and unsaturated polyester," *Fibers and Polymers*, vol. 14, no. 2, pp. 250–257, 2013.

67. A. Ticoalu, T. Aravinthan, and F. Cardona, "Experimental investigation into gomuti fibres / polyester composites," In *21st Australasian Conference on the Mechanics of Structures and Materials (ACMSM 21)*. Melbourne, Australia: CRC Press/Balkema, 2010, pp. 451–456.

68. A. N. Oumer and D. Bachtiar, "Modeling and experimental validation of tensile properties of sugar palm fiber reinforced high impact polystyrene composites," *Fibers and Polymers*, vol. 15, no. 2, pp. 334–339, 2014.

69. A. E. Hadi, D. Bachtiar, J. P. Siregar, and M. R. M. Rejab, "Mechanical properties of untreated and treated sugar palm fibre reinforced polypropylene composites," in *AIP Conference Proceedings*, 2019, p. 020013.

70. D. Bachtiar, M. S. Salit, E. Zainudin, K. Abdan, and K. Z. H. M. Dahlan, "Effects of alkaline treatment and a compatibilizing agent on tensile properties of sugar palm fibrereinforced high impact polystyrene composites," *BioResources*, vol. 6, no. 4, pp. 4815–4823, 2011.

71. A. A. Mohammed, D. Bachtiar, M. R. M. Rejab, and J. P. Siregar, "Effect of microwave treatment on tensile properties of sugar palm fibre reinforced thermoplastic polyurethane composites," *Defence Technology*, vol. 14, no. 4, pp. 287–290, 2018.

72. A. M. Radzi, S. M. Sapuan, M. Jawaid, and M. R. Mansor, "Water absorption, thickness swelling and thermal properties of roselle/sugar palm fibre reinforced thermoplastic polyurethane hybrid composites," *Journal of Materials Research and Technology*, vol. 8, no, 5, pp. 3988–3994, 2019.

73. F. Agrebi, N. Ghorbel, B. Rashid, A. Kallel, and M. Jawaid, "Influence of treatments on the dielectric properties of sugar palm fiber reinforced phenolic composites," *Journal of Molecular Liquids*, vol. 263, pp. 342–348, 2018.

74. J. Sahari, S. M. Sapuan, E. S. Zainudin, and M. A. Maleque, "Thermo-mechanical behaviors of thermoplastic starch derived from sugar palm tree (*Arenga pinnata*)," *Carbohydrate Polymers*, vol. 92, no. 2, pp. 1711–1716, 2013.

75. M. H. Alaaeddin, S. M. Sapuan, M. Y. M. Zuhri, E. S. Zainudin, and F. M. AL-Oqla, "Physical and mechanical properties of polyvinylidene fluoride - short sugar palm fiber nanocomposites," *Journal of Cleaner Production*, vol. 235, pp. 473–482, 2019.

76. N. Razali, S. M. Sapuan, and N. Razali, "Mechanical properties and morphological analysis of roselle/sugar palm fiber reinforced vinyl ester hybrid composites," In S. M. Sapuan, H. Ismail, and E. S. Zainudin (Eds.), *Natural Fibre Reinforced Vinyl Ester and Vinyl Polymer Composites*. Duxford, UK: Woodhead Publishing, An Imprint of Elsevier, 2018, pp. 169–180.

77. M. R. M. Huzaifah, S. M. Sapuan, Z. Leman, M. R. Ishak, and R. A. Ilyas, "Effect of soil burial on water absorption of sugar palm fibre reinforced vinyl ester composites," in *6th Postgraduate Seminar on Natural Fiber Reinforced Polymer Composites 2018*, 2018, pp. 52–54.

78. I. M. Ammar, M. R. M. Huzaifah, S. M. Sapuan, M. R. Ishak, and Z. B. Leman, "Development of sugar palm fiber reinforced vinyl ester composites," In S. M. Sapuan, H. Ismail, and E. S. Zainudin (Eds.), *Natural Fibre Reinforced Vinyl Ester and Vinyl Polymer Composites*. Duxford, UK: Woodhead Publishing, An Imprint of Elsevier, 2018, pp. 211–224.

79. A. Atiqah, M. Jawaid, M. R. Ishak, and S. M. Sapuan, "Effect of alkali and silane treatments on mechanical and interfacial bonding strength of sugar palm fibers with thermoplastic polyurethane," *Journal of Natural Fibers*, vol. 15, no. 2, pp. 251–261, 2018.

80. A. M. Radzi, S. M. Sapuan, M. Jawaid, and M. R. Mansor, "Effect of alkaline treatment on mechanical, physical and thermal properties of roselle/sugar palm fiber reinforced thermoplastic polyurethane hybrid composites," *Fibers and Polymers*, vol. 20, no. 4, pp. 847–855, 2019.
81. A. M. Radzi, S. M. Sapuan, M. Jawaid, and M. R. Mansor, "Influence of fibre contents on mechanical and thermal properties of roselle fibre reinforced polyurethane composites," *Fibers and Polymers*, vol. 18, no. 7, pp. 1353–1358, 2017.
82. M. J. Halimatul, S. M. Sapuan, M. Jawaid, M. R. Ishak, and R. A. Ilyas, "Effect of sago starch and plasticizer content on the properties of thermoplastic films: mechanical testing and cyclic soaking-drying," *Polimery*, vol. 64, no. 6, pp. 32–41, 2019.
83. A. Edhirej, S. M. Sapuan, M. Jawaid, and N. I. Zahari, "Cassava/sugar palm fiber reinforced cassava starch hybrid composites: physical, thermal and structural properties," *International Journal of Biological Macromolecules*, vol. 101, pp. 75–83, 2017.
84. R. A. Ilyas et al., "Thermal, biodegradability and water barrier properties of bio-nanocomposites based on plasticised sugar palm starch and nanofibrillated celluloses from sugar palm fibres," *Journal of Biobased Materials and Bioenergy*, vol. 14, pp. 1–13, 2020.
85. R. A. Ilyas et al., "Sugar palm (*Arenga pinnata* [Wurmb.] Merr) starch films containing sugar palm nanofibrillated cellulose as reinforcement: water barrier properties," *Polymer Composites*, vol. 41, no. 2, pp. 459–467, 2020.
86. M. D. Hazrol, S. M. Sapuan, M. Y. M. Zuhri, and R. A. Ilyas, "Electrical properties of sugar palm nanocellulose fibre reinforced sugar palm starch biopolymer composite," in *Prosiding Seminar Enau Kebangsaan 2019*, 2019, pp. 57–62.
87. M. D. Hazrol, S. M. Sapuan, R. A. Ilyas, M. L. Othman, and S. F. K. Sherwani, "Electrical properties of sugar palm nanocrystalline cellulose, reinforced sugar palm starch nanocomposites," vol. 55, no. 5, pp. 363–370, 2020.
88. H. Y. Sastra, J. P. Siregar, S. M. Sapuan, and M. M. Hamdan, "Tensile properties of *Arenga pinnata* fiber-reinforced epoxy composites," *Polymer-Plastics Technology and Engineering*, vol. 4, no. 1, pp. 149–155, 2007.
89. M. J. Suriani, M. M. Hamdan, H. Y. Sastra, and S. M. Sapuan, "Study of interfacial adhesion of tensile specimens of *Arenga pinnata* fiber reinforced composites," *Multidiscipline Modeling in Materials and Structures*, vol. 3, no. 2, pp. 213–224, 2007.
90. A. Afzaluddin, M. Jawaid, M. S. Salit, and M. R. Ishak, "Physical and mechanical properties of sugar palm/glass fiber reinforced thermoplastic polyurethane hybrid composites," *Journal of Materials Research and Technology*, vol. 8, no. 1, pp. 950–959, 2019.
91. A. Edhirej, S. M. Sapuan, M. Jawaid, and N. I. Zahari, "Tensile, barrier, dynamic mechanical, and biodegradation properties of cassava/sugar palm fiber reinforced cassava starch hybrid composites," *BioResources*, vol. 12, no. 4, pp. 7145–7160, 2017.
92. I. M. Ammar, M. R. M Huzaifah, S. M. Sapuan, Z. Leman, and M. R. Ishak, "Mechanical properties of environment-friendly sugar palm fibre reinforced vinyl ester composites at different fibre arrangements," *EnvironmentAsia*, vol. 12, no. 1, pp. 25–35, 2019.
93. N. M. Nurazzi, A. Khalina, S. M. Sapuan, and R. A. Ilyas, "Mechanical properties of sugar palm yarn/woven glass fiber reinforced unsaturated polyester composites: effect of fiber loadings and alkaline treatment," *Polimery*, vol. 64, no. 10, pp. 12–22, 2019.
94. J. Sahari, M. S. Salit, E. S. Zainudin, and M. A. Maleque, "Degradation characteristics of SPF/SPS biocomposites fabrication of SPF/SPS biocomposites," *Fibres and Textiles in Eastern Europe*, vol. 22, no. 5107, pp. 96–98, 2014.
95. A. Blaga, "GRP composite materials in construction: properties, applications and durability," *Industrialization Forum*, vol. 9, no. 1, pp. 27–32, 1978.
96. R. A. Ilyas et al., "Production, processes and modification of nanocrystalline cellulose from agro-waste: a review," In B. Movahedi (Ed.), *Nanocrystalline Materials*. London: IntechOpen Limited, 2019, pp. 3–32.

15 Extraction and Characterization of Malaysian Cassava Starch, Peel, and Bagasse, and Selected Properties of the Composites

M.I.J. Ibrahim
University of Sabha, Libya
Universiti Putra Malaysia

A. Edhirej
University of Sabha, Libya

S.M. Sapuan, M. Jawaid,
N.Z. Ismarrubie, and R.A. Ilyas
Universiti Putra Malaysia

CONTENTS

15.1 INTRODUCTION

Cassava is ranked fourth among principal crops grown in the world. It is majorly produced in African, Asian, and Latin American countries. It is a root crop that is enriched with starch, serving as a major source of calories in the tropical regions. In addition, cassava has its utility as an industrial raw material for the manufacture of various products. The starch production characteristic of cassava has the following advantages in comparison with other grain crops: high purity level, condensing properties, flavorless taste, excellent consistency, cost-effectiveness, and high concentration of starch [1,2]. In the 1970s, Malaysia was one of the largest starch exporters having 20,913 ha of starch producing crops. The major importers were the United Kingdom, Japan, Canada, and the USA. In 2010, Malaysia had 2,769 ha of cassava plantation, which produced approximately 37,187 tons of cassava crop in one year [2].

However, this massive production of cassava crops produced large amounts of cassava waste, which supposedly had detrimental effects on the environment [3]. Cassava serves to be among the major food crop in tropical and subtropical countries. Approximately 200 million people can acquire 500 cals. per day from cassava. It is able to produce significant quantities of the crop even in minimalist climatic and soil conditions having remarkable tolerance to drought and infertile soils. Moreover, the plant is immune to diseases and pest attacks. Owing to these characteristics, cassava is among the most popular crops for farmers [4]. The roots are convenient to plant and have economic production costs. Hence, they are quite suitable to be used as either food or industrial raw material. However, farmers prefer to produce edible varieties of the crop more than starch varieties, as the former provides double profits [5]. In the same way, owing to its accessible regeneration capacity in comparison with other agricultural residues, cassava bagasse serves as an abundant solar energy reservoir. For example, in comparison with sugar cane bagasse, cassava bagasse is more advantageous, as it requires no pre-treatment and can respond well to the action of micro-organisms [6]. Owing to its ability to convert its starch to bio-ethanol and bio-degradable plastics, cassava is gaining extensive popularity. It is also being considered to use it as animal feeds. Large-scale industries are setting up to plant cassava crops considering its advantageous utilities [5]. Cassava has numerous advantages such as renewable source of energy, ecofriendly, economical, manageable weight, high specific mechanical performance, convenient recyclability and ease of separation, and biodegradability [7,8]. Owing to these characteristics, the crop attracts interest in various research ventures. Recently, the use of natural fibers as reinforcing materials in polymers and composites has been a significant area of interest. Lignocellulosic are preferred more than inorganic fillers owing to their renewable nature, biodegradability, lesser energy consumption, convenient availability around the globe, economical and density, and high specific strength and modulus [9–11]. This study considers the physical, thermal, and morphological properties as well as the chemical composition of cassava starch, peel, and bagasse obtained from starch extraction of cassava root planted in Malaysia and to explore their potential to be used in the development of biocomposites. From the literature review and to the best of our knowledge, there is no study reported on physical, thermal, and morphological properties as well as the chemical composition of cassava starch, peel, and bagasse from Malaysian cassava by any researchers until now.

15.2 MATERIALS AND METHODS

15.2.1 MATERIALS

Cassava starch was extracted from fresh cassava root roots following the procedure described in [12], which were collected from a local market. Cassava peel and bagasse were obtained from the same extraction process and were dried, ground, sieved, and characterized.

15.2.2 EXTRACTION PROCESS

Cassava starch was obtained from cassava tubers as per the following method described in [12]. The tubers, as shown in Figure 15.1a, were washed, peeled, and shredded (Figure 15.1b). The product was mixed with distilled water and filtered through a clean cloth (Figure 15.1c). The white precipitate (starch) was separated, sun-dried, and then analyzed. Figure 15.1d shows the obtained cassava starch, whereas Figure 15.1e and f shows cassava peel and bagasse, which were obtained from the same process and dried at 50° to constant weight, shredded, sifted through a 300 μm mesh sieve, and characterized to be used as a filler of thermoplastic cassava starch matrices. A significant component of cassava root is starch, which has found to be around 68.62% of the total dry, and cassava tubers yield mainly two by-products: peel and bagasse. Peel represented an 11.6% dry base, which is within the range 5%–15% reported in [13]. Cassava bagasse is a solid fibrous residue (around 17% of the tuber) that remains after the flour or starch has been extracted [13] in this study, which was found to be 19.78%. Industrial production of cassava starch involves the elimination of soluble sugars and the separation of fibers resulting in a purified starch and a solid residue called cassava bagasse. The cassava bagasse is mainly composed of water (70–80 wt%), residual starch, and cellulose fibers. The cellulose fiber content ranges between 15 and 50 wt% of the total solid residue (dry weight basis), and the remainder is residual starch [14].

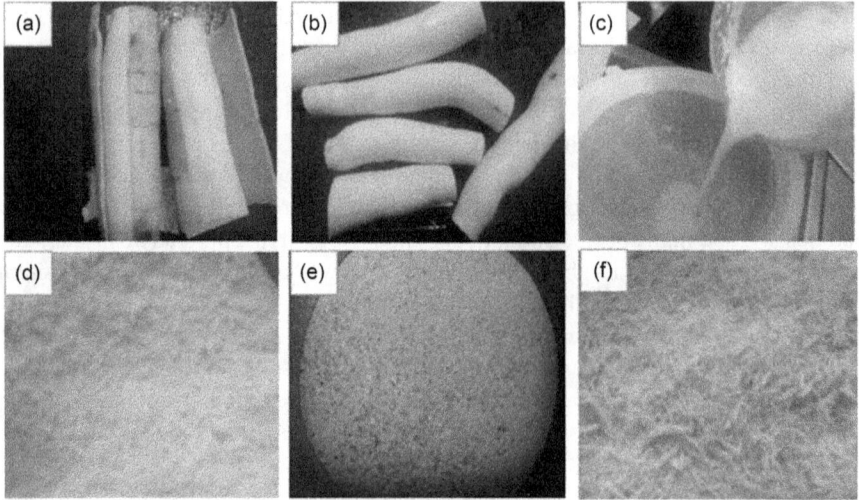

FIGURE 15.1 Photographs of cassava root and its product. (a) Cassava root, (b) peeled root, (c) grated root with water, (d) cassava starch, (e) cassava peel, and (f) cassava bagasse.

15.2.3 Characterization of Cassava Starch, Peel, and Bagasse

15.2.3.1 Chemical Composition

Versino and García [15] described the methods used to study the dry matter, ash, moisture, crude protein, starch, carbohydrate, and energy content of cassava starch. Moreover, the methods used to investigate acid detergent fiber (ADF), neutral detergent fiber (NDF), lignin (LIG), ash, cellulose, and hemicellulose of cassava peel and bagasse were described in previous work [15]. NDF and ADF were used to determine the chemical composition of cassava fiber. This is the most practiced method for calculating the major fiber constituents, i.e., cellulose, hemicelluloses, and lignin. The proportions of cellulose and hemicelluloses were calculated using the following equations, respectively:

$$\text{Cellulose} = \text{ADF} - \text{lignin} \qquad (15.1)$$

$$\text{Hemicelluloses} = \text{NDF} - \text{ADF} \qquad (15.2)$$

15.2.3.2 Physical Properties

15.2.3.2.1 Particle Size Distribution

Mastersizer 2000 E (Malvern Instruments) having a built-in Q spec Dry powder Feeder was used to observe particle size distribution cassava starch, peel, and bagasse.

15.2.3.2.2 Density (ρ)

The density of cassava bioproducts was calculated by using helium gas through a gas pycnometer. The density ρ is a basic physical property of matter, which is defined as the ratio of its mass m to its volume V as shown the following in equation:

$$\rho = \frac{m}{v} = \text{g/cm}^3 \qquad (15.3)$$

The psychometric density of solids is calculated to analyze the volume of a known mass of powder by computing the volume of gas evacuated under given conditions. The density of cassava bioproducts was determined using a gas pycnometer. Helium gas is used as the displacing fluid since it penetrates the most delicate pores assuring maximum accuracy. For this reason, helium is recommended since its small atomic dimension enables entry into crevices and pores.

15.2.3.2.3   Water Content (WC)

The weight loss was determined to calculate the moisture content of cassava starch, peel, and bagasse. Powder samples were weighed (w_1), dried at 105°C for 24 h, and weighed again (w_2). Water content was calculated as the percentage of initial powder weight lost during drying and that on a wet basis, as shown in the following equation:

$$\text{WC}(\%) = \frac{w_1 - w_2}{w_1} \times 100 \qquad (15.4)$$

15.2.3.2.4   Water Holding Capacity (WHC)

The water absorption of cassava starch, peel, and bagasse samples were determined as per the method explained in [16]. The samples (3.0 g) were dissolved in 25 mL of distilled water and placed in pre-weighed centrifuge tubes. After being kept in the centrifuge tubes for 25 min at 3,000 rpm, the dispersions were mixed and were left at room temperature for 1 h. The supernatants were removed, and the residue was dehydrated in an oven for 25 min at 50°C, to determine the moisture content of the samples. The water absorption capacity was denoted as grams of water-bound per gram of the sample on a dry basis [16]:

$$\text{WHC}(\%) = \left[\left(M_{\text{final}} - M_{\text{initial}} \right) / M_{\text{initial}} \right] \times 100 \qquad (15.5)$$

15.2.3.3   Thermal Properties

15.2.3.3.1   Differential Scanning Calorimetry (DSC)

Differential scanning calorimetry (DSC) tests were conducted in a differential scanning calorimeter after heating the sample from room temperature to 280°C at 10 °C/min. The following factors were obtained through these thermo-grams: onset (T_o), conclusion (T_c), melting temperatures (T_m), and enthalpy rate to this thermal transition (ΔH_{gel}).

15.2.3.3.2   Thermal Gravimetric Analyzer (TGA)

Thermal gravimetric analysis was conducted by using 10 mg of starch and filler. The following conditions were provided for the analysis: a heating rate of 10 °C/min, and temperature ranging from room temperature to 500°C.

15.2.3.4 Scanning Electron Microscopy (SEM)

The surface morphology of the samples was studied by using scanning electron microscopy (SEM) (Hitachi S-3400N) functioning at an acceleration voltage of 20 kV. The apparatus was framed on bronze stumps through a double-sided tape and coated with a gold layer (40–50 μm), under high vacuum mode.

15.2.3.5 Structural Properties

15.2.3.5.1 Fourier Transform Infrared Spectroscopy (FTIR)

An IR spectrometer (Bruker Vector 22) was used to determine the FTIR spectra of the natural starch and fillers by using potassium bromide (KBr) discs made from powdered samples.

15.2.3.5.2 X-Ray Diffraction

Rigaku D/max 2500 X-ray powder diffractometer (Rigaku, Tokyo, Japan) was used to study X-ray diffraction patterns of the cassava starch, peel, and bagasse. The relative crystallinity index (R_c) was calculated according to Equation (15.6) proposed in previous work [17], and based on the calculus of amorphous area (A_a) and crystalline areas (A_c), it is calculated with the aid of the ORIGIN 8.0 program:

$$R_c = A_c / \left(A_c + A_a \right) \tag{15.6}$$

15.3 RESULT AND DISCUSSION

15.3.1 Chemical Composition of Cassava Starch, Peel, and Bagasse

Table 15.1 provides the chemical composition of cassava starch. It shows that the cassava has high concentrations of starch and carbohydrates, whereas the concentrations of ash and protein are deficient. In this study, the carbohydrate content in starch is found to be 86.47 g/100 g. This amount was similar to the value provided in a previous work [18]. The physical, mechanical, and thermal properties of natural fiber are greatly affected by its chemical composition. The natural fiber majorly constitutes of cellulose, hemicelluloses, lignin, and ash. Cellulose serves as the main structural component providing strength to the walls of the stem of the plant as well as to the fiber [19]. Table 15.2 shows the chemical composition of cassava fiber (peel and bagasse). It shows that the cassava peel and bagasse contain low concentrations of the cellulose and high concentrations of hemicellulose content. Nonetheless, the cellulose and hemicellulose content of cassava bagasse is higher than that of cassava peel. On the other hand, both cassava bagasse and peel show a similar concentration of lignin and the ash. The lignin content was calculated to determine the proportion of the resistant components of the fibrous residue, which plays a significant role in providing strength to the fiber walls [20].

Cassava fiber has more advantages than residues of other crops owing to a lower concentration of ash, such as rice straw and wheat straw, which have 17.5% and 11.0% ash contents, respectively.

TABLE 15.1
Chemical Composition of Cassava Starch

No	Chemical Composition of Cassava Starch	
1	Moisture g/100 g	12.66
2	Crude protein g/100 g	0.555
3	Ash g/100 g	0.31
4	Starch g/100 g	58.815
5	Carbohydrate g/100 g	86.475
6	Energy Kcal/100 g	348

TABLE 15.2
Chemical Composition of Cassava Peel and Bagasse

No	Content	Peel (%)	Bagasse (%)
1	ADF	10.37	13.16
2	NDF	33.75	42.42
3	LIG	2.87	3.12
4	Ash	3.38	3.36
5	Cellulose	7.5	10.04
6	Hemicelluloses	23.38	29.26

15.3.2 PHYSICAL PROPERTIES

15.3.2.1 Particle Size Distribution

Figure 15.2a presents the particle size distribution of cassava starch. From the figure, it is clear that the highest percentage (42%) is for the starch size 10–20 μm, followed by 25% for the size 10 μm. The majority of cassava starch particle size is less than 30 μm with 93%; this result agrees with the findings of [21]. Versino, López, and García [22] reported that the size of the Argentina cassava starch is smaller than 53 μm. Cassava peel and bagasse have similar particle size distribution. Figure 15.2b and c shows that cassava peel and bagasse exhibit large particle size (250–300 μm) with 33% and 39%, respectively. However, bagasse has a higher amount of the smallest size (0–50 μm) because it contains less amount of non-extracted starch.

15.3.2.2 Density

The densities of cassava starch, peel, and bagasse were calculated from the average of five replicates, as shown in Tables 15.1 and 15.2. The densities of cassava starch, peel, and bagasse are quite similar, i.e., 1.48, 1.48, and 1.45 g/cm³, respectively.

15.3.2.3 Moisture Content (MC)

The moisture content of natural fiber is an important criterion that needs to be considered in choosing natural fiber as a reinforcement material. This is because moisture

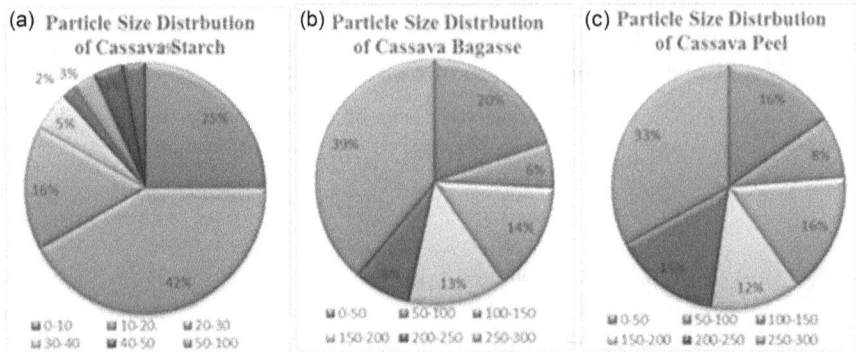

FIGURE 15.2   Particle size distribution (PSD) of (a) cassava starch, (b) bagasse, and (c) peel.

content affects dimensional stability, electrical resistivity, tensile strength, porosity, and swelling behavior of natural fiber in a composite material [1]. The result of the moisture content of cassava fiber indicated that peel fiber was the lowest with 12.23%, as shown in Tables 15.1 and 15.2. Therefore, bagasse fiber shows a similar content of starch.

15.3.2.4   Water Holding Capacity (WHC)

The water absorption is a critical criterion for several applications of TPS products. The starches absorb water due to their hygroscopic nature. The results of the water absorption test are presented in Tables 15.1 and 15.2 for cassava starch, peel, and bagasse. From the result, cassava starch absorbed a lower amount of water (52.97%) as compared to the fiber, which indicates that fibers are highly hydrophilic [23]. Cassava peel exhibited higher retention capacity of water than bagasse due to the reduction of cellulose content in the peel, and this result is in agreement with the work reported in [24]; these results are in agreement with their chemical composition shown in Tables 15.1 and 15.2.

15.3.3   Thermal Properties

15.3.3.1   Thermal Gravimetric Analyzer (TGA)

To study the variations in the thermal stability and to evaluate the weight loss, the raw material during heating, TGA curves for cassava starch, peel, and bagasse were used. It was shown that weight loss of the natural starch and fibers was made in two stages once reached the temperature below 500°C; in the first stage, water loss occurs at the temperature below 100°C, and in the second stage, the material starts decaying. The thermo-gravimetric indicates that the maximum decaying of material occurred at the temperature range of 277°C–353°C. At 500°C, 87.42% of the native starch was decayed. This result was in agreement with [25], which reported that at 500°C, 85.4% of the yellow ginger starch was decomposed. Figure 15.3a shows that the highest rates of thermal decomposition occurred at the temperature of 255.13°C for peel, 290.85°C for bagasse, and 293.25°C for starch. Similar results were reported in previous work [26] for Colombian cassava starch and [22] for Argentinian cassava starch. Taken as

(a)

(b)

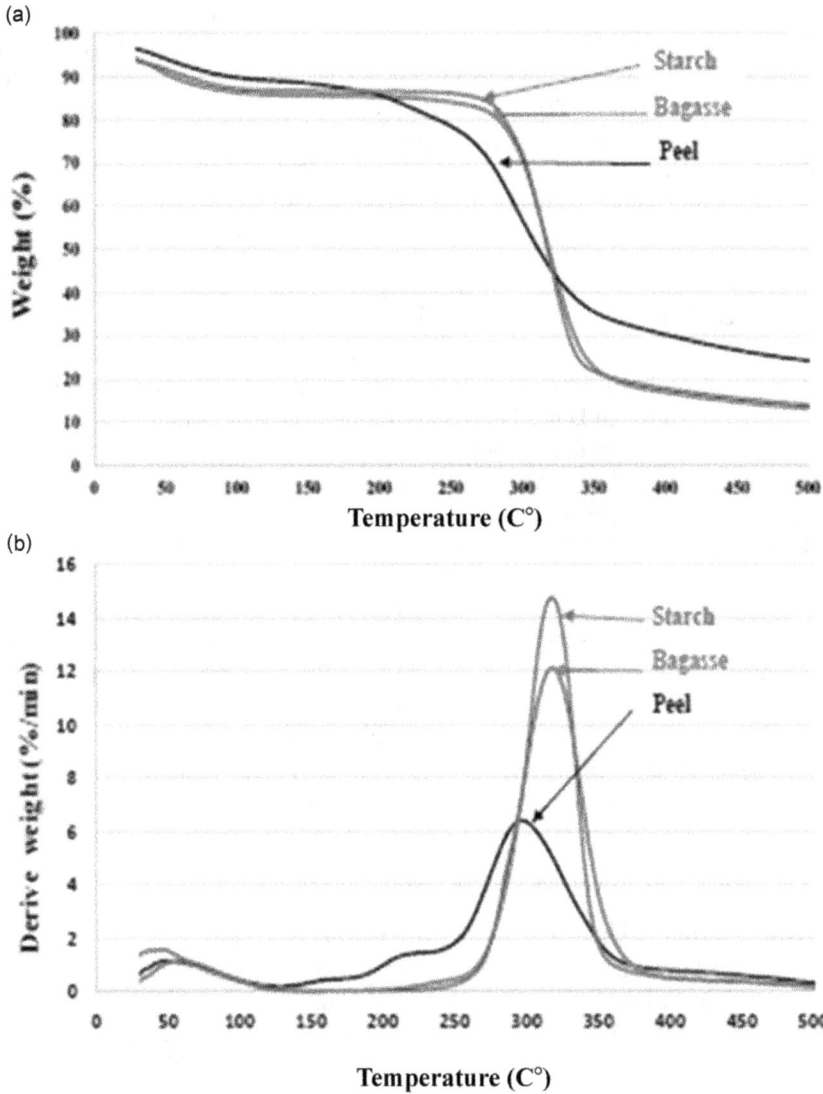

FIGURE 15.3 (a) TGA and (b) DTG curves of cassava starch, peel, and bagasse.

a whole, cellulose initiates to decompose at a temperature of 315°C and totally splits into its components at 400°C for all different fibers. From the available literature, cellulose will begin to destroy at a high temperature of 315°C [27]. When the needed temperature is attained, the process of decaying begins, and the mass loss rate is quick. The one hundredth of weight loss for peel and bagasse fiber is 78.06% and 88.56% correspondingly at 500°C. Figure 15.3b shows the DTG curves of cassava starch, peel, and bagasse. It can be seen that the DTG curves of cassava starch and bagasse are similar. The dehydration of all materials, in the range of 50°C–200°C, was small, where no peak is seen in this temperature range. This result is in agreement with the observation of [28]. Table 15.3 shows the degradation temperature of cassava starch and fiber, as well as DTG peaks. The DTG peak of cassava starch and bagasse is around 323°C and 322°C, respectively, and this result is slightly higher than value (316°C) reported in [28] for Brazilian cassava starch. The peel shows a peak of around 298°C. The thermal degradation of the fiber begins with the degradation of hemicelluloses (200°C–260°C), cellulose (240°C–350°C), and lignin (280°C–500°C) [28].

15.3.3.2 Differential Scanning Calorimetry (DSC)

The DSC was used to calculate various features of transitions such as "onset" temperature (T_o), "conclusion" temperature (T_c), peak temperature (T_p), and gelatinization enthalpy (ΔH_{gel}) through DSC. The gelatinization peak temperature (T_p) of DSC of cassava starches was obtained from Figure 15.4. The first gelatinization peak (T_o) and second gelatinization peak (T_c) and ascribed to starch were described to peel and bagasse. As displayed in Figure 15.4, the T_p of cassava peel was discovered to be of a higher degree as compared to that of bagasse. Thermal characteristic studies using thermo-gravimetric analysis and DSC showed that cassava starch and fiber were thermally stable. From Figure 15.4, T_o and T_p of cassava starch was around 143.02°C and 172.87°C, respectively. These values were slightly higher than the ones documented in [29], which reported that the T_p of Brazilian cassava starch occurred at 169.2°C. The cassava peel shows higher thermal stability behavior with 146.49°C and 177.76°C for T_o and T_p, respectively, as compared to the bagasse, which presents 130.81°C and 169.23°C for T_o and T_p. This result is in agreement with the report of [22], which discussed the thermal properties of Argentinian cassava peel and bagasse.

15.3.4 MORPHOLOGICAL AND STRUCTURAL PROPERTIES

15.3.4.1 Scanning Electron Microscopy (SEM)

Figure 15.5b which is a magnified micrograph of Figure 15.5a illustrated that the cassava starch granules were rounded and oval-shaped, and the majority were between 5 and 30 μm which is in agreement with the study on the particle size distribution and the work reported in [29] for Brazilian cassava starch, where the starch granules surrounded by the remaining cell walls of the parenchyma tissue can be observed.

TABLE 15.3

Thermal Degradation of Cassava Starch, Peel, and Bagasse

Material	Degradation Temperature (°C)			Mass Residue (%) at 500°C	DTG Peak Temperature (°C)
	T_{on}	$T_{25\%}$	$T_{50\%}$		
Starch	293.26	296.5	317.5	12.58	323.4
Peel	255.13	265.6	311.1	21.94	298.8
Bagasse	290.85	294.8	318.2	11.44	322.6

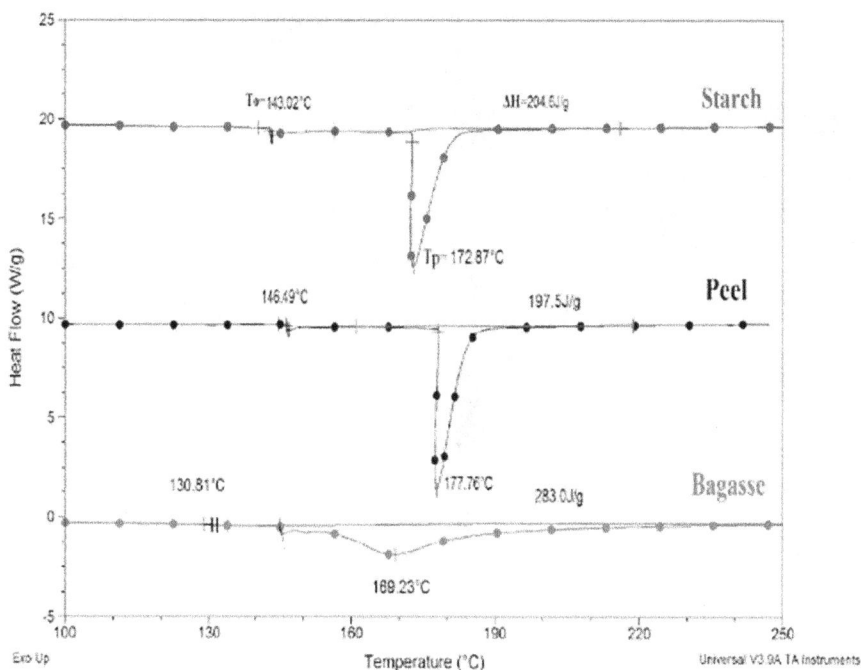

FIGURE 15.4 DSC traces showing the peak temperature (T_p), onset temperature (T_o), and gelatinization enthalpy (ΔH) of cassava starch, peel, and bagasse.

Figure 15.5d and f, which is a magnified one of Figure 15.5c and e, shows clearly the shape and size of bagasse and peel particles. The micrographs illustrated a typical view of the cassava bagasse, showing a fiber with starch granules interspersed on the surface. Bagasse granules exhibited observable polygonal shape differences that are attributed to the bimodal granule size distribution of cassava starch. The cassava peel granules exhibited both round and polygonal shapes as shown in Figure 15.5f, and this was similar to the result reported in [20]. In this, it was possible to study the starch granules encompassed by the residual parenchyma tissue cell walls (Figure 15.5c and d). Describe the SEM morphological characterization of the fibrous residue. The view obtained from the micrographs displayed a fiber having starch granules covering the surface of the cassava bagasse (Figure 15.5d). The observed shape of the bagasse granules was found to be polygonal. The bimodal granule size dispersant of the cassava starch was the cause of the variations present in the bagasse granules. Cassava peel granules were found to exhibit round and polygonal shapes, as illustrated in Figure 15.5f. The results were found to concord with the prior work given in [15,20,22].

15.3.4.2 Fourier Transform Infrared Spectroscopy (FTIR)

The main differences between the spectra of both by-products (peel and bagasse) are the width and position of the band around 1,642, and 3,400 cm^{-1} for bagasse have shifted to 1,633 and 3,420 cm^{-1} for the peel (Figure 15.6). These results can be related to differences in chemical composition, as has been previously described, mainly to

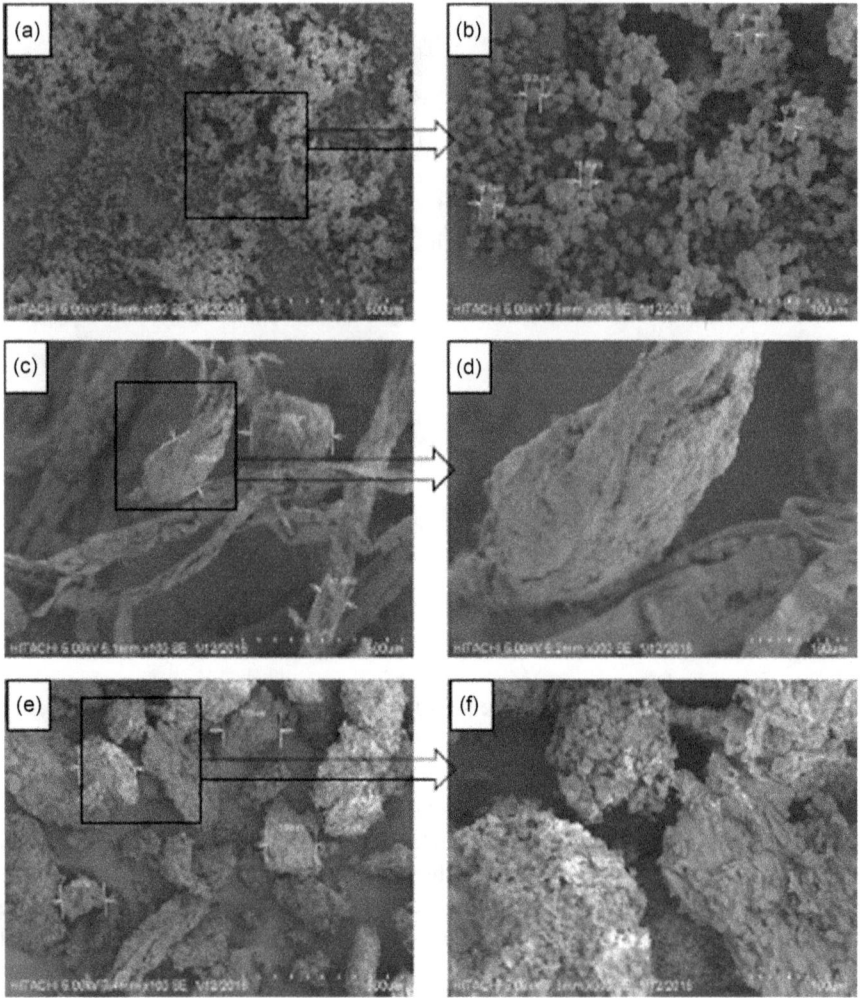

FIGURE 15.5 SEM of cassava starch (a, b), bagasse (c, d), and peel (e, f).

protein and lignin contents. Cassava TPS composites presented bands corresponding to the distinctive functions of starch. The broad peak observed around 3,400 cm^{-1} is associated with the stretching of the O–H groups. Bands identified at 2,929, and 2,886 cm^{-1} are attributed to C–H stretching. The peak at 1,642 cm^{-1} was assigned to the bending mode of the absorbed water. The peaks around 1,420 cm^{-1} are assigned to O–H bonding. Peaks at 1,153 and 1,041 cm^{-1} were attributed to the C–O bond stretching of the C–O–H group in cassava starch and the C–O bond stretching of the C–O–C group in the anhydroglucose ring, respectively [30,31]. Similar peaks were observed in [28], which recorded peaks at 3381, 2930, and 1649 cm^{-1} for Brazilian cassava starch and [22] for Argentinian cassava starch.

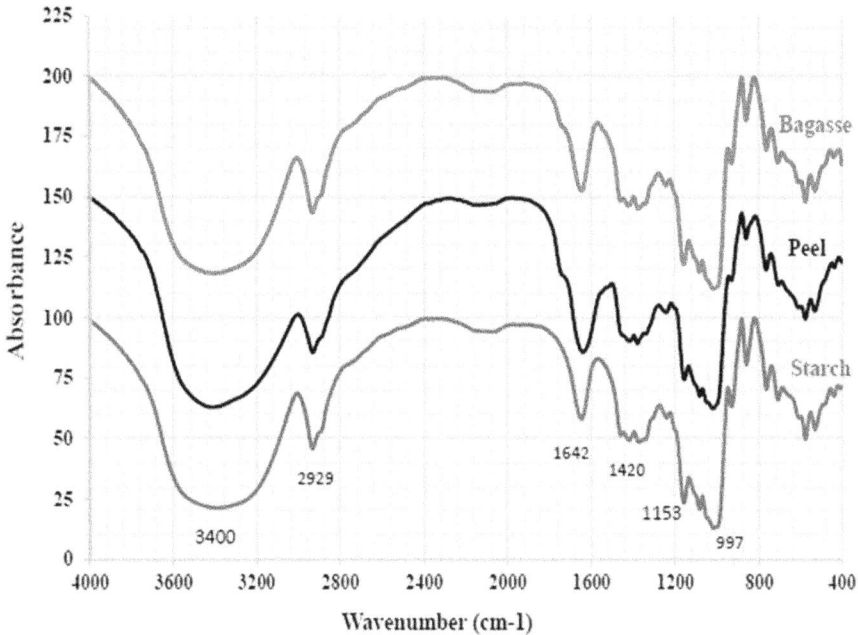

FIGURE 15.6 FTIR spectra of cassava starch, peel, and bagasse.

15.3.4.3 X-Ray Diffraction (XRD)

The framework of the native cassava starch was observed to be semi-crystalline with A-type (18.3° and 23.1°) and B-type at 15.1°, 17.4°, and 26.1° crystallinity peaks. The C-type crystalline structure was found to contain a combination of A- and B-type structures, as illustrated in Figure 15.7, and these results are in agreement with [28,32]. During the process of extraction, the structure of the starch granule may have been subjected to partial damage as a result of the use of the shear utilized for the mechanical abrading of the cassava bagasse. The retrieved profile of the X-ray diffractogram, in which two shoulders of large size were found in the range of 15°–18° (A-type crystallinity) and 19°–22° (B-type crystallinity), was found to be in accordance with this principle. As in prior reports, the cellulose type I crystal was also found to be a component of the diffractogram pattern [33]. An X-ray diffraction curve of the cassava starch is shown in Figure 15.7. It is observed that there were major diffraction peaks located at various positions (angles) of 2 (15.1°, 17.2°, 18.3°, and 23.1°) in the diffractogram of the matrix. The values observed here are similar to those of the cassava starching [34], i.e., 15°, 17°, 18°, 22°–23°, and 26.9° and 15.2° (strong), 17.4° (strong), 20° (weak), 23.3° (strong) and 26.2° (weak) [35]. This pattern is following a crystallinity pattern of the type A-B known as Ca (C in the proximity of A) [36]. This pattern displays almost 90% of type A crystallinity and 10% of type-B crystallinity. Similar results have reported in [28,37]. The cassava starch's relative crystallinity, as recorded in this study, was found to be 28.3%. This value was similar to the results of [38] (27.8%), which was C-type crystallinity.

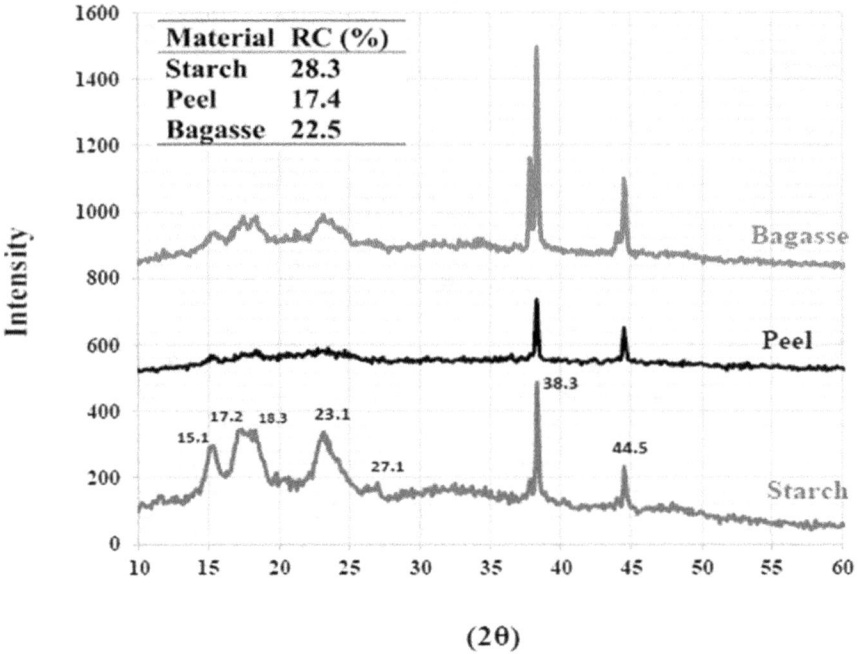

FIGURE 15.7   X-ray diffractogram of cassava starch, peel, bagasse, and its relative crystal-linity (RC).

A broad value range (28.8%–37.4%) was observed for the relative crystallinity in various types of cassava starch in [39]. As observed in the X-ray diffraction patterns (Figure 15.7), the bagasse showed a semi-crystalline structure. Bagasse presented peaks at $2\theta = 15.4°$, $17.5°$, $18.4°$, and $23.2°$, and according to [39], these peaks are related to cassava starch present in the bagasse. From Figure 15.7, it can be observed that cassava bagasse had higher relative crystallinity as compared to cassava peel. Furthermore, the crystallinity indexes of bagasse and peel were found to be 22.5% and 17.4%, respectively. This value was slightly higher than the that given in [40], in which the relative crystallinity of Brazilian cassava bagasse is 21.08%.

15.5   CONCLUSION

Thermoplastic starch and fibers derived from cassava root were successfully extracted. From the chemical composition, a higher cellulose and hemicellulose concentration was found in Bagasse as compared to peel, and the main component of cassava root was starch with 68.62%. The water absorption of peel (389.23%) is higher than bagasse (258.05%); this is due to the reduction of cellulose content. The thermal result showed that the decomposition temperature of cassava starch was 293.25°C. The morphological study indicated that cassava starch was rounded and oval-shaped, bagasse exhibited polygonal shape and peel exhibited round and polygonal shape. The gelatinization temperatures were 177.76°C, 172.87°C, and

169.23°C for the peel, starch, and bagasse, respectively. However, the endothermic gelatinization enthalpy was 283, 204.6, and 197.5 J/g for the bagasse, starch, and peel, respectively. The relative crystallinity of starch 28.3%, bagasse, shows higher crystallinity than that of peel. From this study, we concluded that starch and fiber derived from cassava root have the potential to be used as sustainable biocomposite for various applications.

ACKNOWLEDGMENT

The authors thank Universiti Putra Malaysia for financial support through the Universiti Putra Malaysia Grant scheme vote number (960700).

REFERENCES

1. M. Jawaid and H. A. Khalil, "Cellulosic/synthetic fiber reinforced polymer hybrid composites: a review," *Carbohydrate Polymers,* vol. 86, pp. 1–18, 2011.
2. A. M. Aripin, A. S. M. Kassim, Z. Daud, and M. Z. M. Hatta, "Cassava peels for alternative fiber in pulp and paper industry: chemical properties and morphology characterization," *International Journal of Integrated Engineering,* vol. 5, pp. 30–33, 2013.
3. N. Reddy and Y. Yang, "Biofibers from agricultural byproducts for industrial applications," *Trends in Biotechnology,* vol. 23, pp. 22–27, 2005.
4. D. Dai, Z. Hu, G. Pu, H. Li, and C. Wang, "Energy efficiency and potentials of cassava fuel ethanol in Guangxi region of China," *Energy Conversion and Management,* vol. 47, pp. 1686–1699, 2006.
5. E. NurulNahar and S. Tan, "Cassava mini-cuttings as a source of planting material," *Journal of Tropical Agriculture and Food Science,* vol. 40, pp. 145–151, 2012.
6. A. Pandey, C. R. Soccol, P. Nigam, V. T. Soccol, L. P. Vandenberghe, and R. Mohan, "Biotechnological potential of agro-industrial residues. II: cassava bagasse," *Bioresource Technology,* vol. 74, pp. 81–87, 2000.
7. W. N. Gilfillan, D. M. Nguyen, P. A. Sopade, and W. O. Doherty, "Preparation and characterisation of composites from starch and sugar cane fibre," *Industrial Crops and Products,* vol. 40, pp. 45–54, 2012.
8. A. Edhirej, S. M. Sapuan, M. Jawaid, and N. I. Zahari, "Cassava: its polymer, fiber, composite, and application," *Polymer Composites,* vol. 38, no. 3, pp. 555–570, 2015.
9. S. Hashmi, U. Dwivedi, and N. Chand, "Graphite modified cotton fiber reinforced polyester composites under sliding wear conditions," *Wear,* vol. 262, pp. 1426–1432, 2007.
10. L. Castillo, O. López, C. López, N. Zaritzky, M. A. García, S. Barbosa, *et al.,* "Thermoplastic starch films reinforced with talc nanoparticles," *Carbohydrate Polymers,* vol. 95, pp. 664–674, 2013.
11. R. Bodirlau, C.-A. Teaca, and I. Spiridon, "Influence of natural fillers on the properties of starch-based biocomposite films," *Composites Part B: Engineering,* vol. 44, pp. 575–583, 2013.
12. J. G. Akpa and K. K. Dagde, "Modification of cassava starch for industrial uses," *International Journal of Engineering and Technology,* vol. 2, pp. 913–919, 2012.
13. S. Aro, V. Aletor, O. Tewe, and J. Agbede, "Nutritional potentials of cassava tuber wastes: a case study of a cassava starch processing factory in south-western Nigeria," *Livestock Research for Rural Development,* vol. 22, p. Article# 213, 2010.
14. E. d. M. Teixeira, D. Pasquini, A. A. Curvelo, E. Corradini, M. N. Belgacem, and A. Dufresne, "Cassava bagasse cellulose nanofibrils reinforced thermoplastic cassava starch," *Carbohydrate Polymers,* vol. 78, pp. 422–431, 2009.

15. F. Versino and M. A. García, "Cassava (*Manihot esculenta*) starch films reinforced with natural fibrous filler," *Industrial Crops and Products,* vol. 58, pp. 305–314, 2014.

16. H. Yaich, H. Garna, S. Besbes, M. Paquot, C. Blecker, and H. Attia, "Chemical composition and functional properties of *Ulva lactuca* seaweed collected in Tunisia," *Food Chemistry,* vol. 128, pp. 895–901, 2011.

17. K. Frost, D. Kaminski, G. Kirwan, E. Lascaris, and R. Shanks, "Crystallinity and structure of starch using wide angle X-ray scattering," *Carbohydrate polymers,* vol. 78, pp. 543–548, 2009.

18. F. M. Pelissari, M. M. Andrade-Mahecha, P. J. do Amaral Sobral, and F. C. Menegalli, "Comparative study on the properties of flour and starch films of plantain bananas (*Musa paradisiaca*)," *Food Hydrocolloids,* vol. 30, pp. 681–690, 2013.

19. N. Reddy and Y. Yang, "Properties and potential applications of natural cellulose fibers from cornhusks," *Green Chemistry,* vol. 7, pp. 190–195, 2005.

20. M. C. Doporto, C. Dini, A. Mugridge, S. Z. Viña, and M. A. García, "Physicochemical, thermal and sorption properties of nutritionally differentiated flours and starches," *Journal of Food Engineering,* vol. 113, pp. 569–576, 2012.

21. O. Moreno, C. Pastor, J. Muller, L. Atarés, C. González, and A. Chiralt, "Physical and bioactive properties of corn starch–Buttermilk edible films," *Journal of Food Engineering,* vol. 141, pp. 27–36, 2014.

22. F. Versino, O. V. López, and M. A. García, "Sustainable use of cassava (*Manihot esculenta*) roots as raw material for biocomposites development," *Industrial Crops and Products,* vol. 65, pp. 79–89, 2015.

23. D. I. Munthoub and W. Rahman, "Tensile and water absorption properties of biodegradable composites derived from cassava skin/polyvinyl alcohol with glycerol as plasticizer," *Sains Malaysiana,* vol. 40, pp. 713–718, 2011.

24. N. Razali, M. S. Salit, M. Jawaid, M. R. Ishak, and Y. Lazim, "A study on chemical composition, physical, tensile, morphological, and thermal properties of roselle fiber: effect of fiber maturity," *BioResources,* vol. 10, pp. 1803–1824, 2015.

25. L. Zhang, W. Xie, X. Zhao, Y. Liu, and W. Gao, "Study on the morphology, crystalline structure and thermal properties of yellow ginger starch acetates with different degrees of substitution," *Thermochimica Acta,* vol. 495, pp. 57–62, 2009.

26. D. P. Navia and H. S. Villada, "Thermoplastic cassava flour," *Thermoplastic Elastomers,* pp. 23–38, 2012.

27. H. Yang, R. Yan, H. Chen, D. H. Lee, and C. Zheng, "Characteristics of hemicellulose, cellulose and lignin pyrolysis," *Fuel,* vol. 86, pp. 1781–1788, 2007.

28. M. G. Lomelí-Ramírez, S. G. Kestur, R. Manríquez-González, S. Iwakiri, G. B. de Muniz, and T. S. Flores-Sahagun, "Bio-composites of cassava starch-green coconut fiber: part II—structure and properties," *Carbohydrate Polymers,* vol. 102, pp. 576–583, 2014.

29. E. d. M. Teixeira, A. A. Curvelo, A. C. Corrêa, J. M. Marconcini, G. M. Glenn, and L. H. Mattoso, "Properties of thermoplastic starch from cassava bagasse and cassava starch and their blends with poly (lactic acid)," *Industrial Crops and Products,* vol. 37, pp. 61–68, 2012.

30. K. Kaewtatip and J. Thongmee, "Effect of kraft lignin and esterified lignin on the properties of thermoplastic starch," *Materials and Design,* vol. 49, pp. 701–704, 2013.

31. J. Prachayawarakorn, S. Chaiwatyothin, S. Mueangta, and A. Hanchana, "Effect of jute and kapok fibers on properties of thermoplastic cassava starch composites," *Materials and Design,* vol. 47, pp. 309–315, 2013.

32. J. J. van Soest, S. Hulleman, D. De Wit, and J. Vliegenthart, "Crystallinity in starch bioplastics," *Industrial Crops and Products,* vol. 5, pp. 11–22, 1996.

33. P. Kampeerapappun, D. Aht-ong, D. Pentrakoon, and K. Srikulkit, "Preparation of cassava starch/montmorillonite composite film," *Carbohydrate Polymers,* vol. 67, pp. 155–163, 2007.

34. E. d. M. Teixeira, D. Róz, A. Luzia, A. J. F. de Carvalho, and A. A. da Silva Curvelo, "Preparation and characterisation of thermoplastic starches from cassava starch, cassava root, and cassava bagasse," in *Macromolecular Symposia*, 2005, pp. 266–275.

35. Y. Yuan, L. Zhang, Y. Dai, and J. Yu, "Physicochemical properties of starch obtained from Dioscorea nipponica Makino comparison with other tuber starches," *Journal of Food Engineering*, vol. 82, pp. 436–442, 2007.

36. S. N. Moorthy, "Physicochemical and functional properties of tropical tuber starches: a review," *Starch-Stärke*, vol. 54, pp. 559–592, 2002.

37. H. Zobel, "Starch crystal transformations and their industrial importance," *Starch-Stärke*, vol. 40, pp. 1–7, 1988.

38. S. Srichuwong, T. C. Sunarti, T. Mishima, N. Isono, and M. Hisamatsu, "Starches from different botanical sources I: contribution of amylopectin fine structure to thermal properties and enzyme digestibility," *Carbohydrate Polymers*, vol. 60, pp. 529–538, 2005.

39. E. Nuwamanya, Y. Baguma, N. Emmambux, and P. Rubaihayo, "Crystalline and pasting properties of cassava starch are influenced by its molecular properties," *African Journal of Food Science*, vol. 4, pp. 008–015, 2010.

40. F. Debiagi, B. M. Marim, and S. Mali, "Properties of cassava bagasse and polyvinyl alcohol biodegradable foams," *Journal of Polymers and the Environment*, vol. 23, pp. 269–276, 2015.

16 Characterization of Corn Fiber-Filled Cornstarch Biopolymer Composites

M.I.J. Ibrahim
University of Sabha, Libya
Universiti Putra Malaysia

S.M. Sapuan, E.S. Zainudin, M.Y.M. Zuhri,
A. Edhirej, and R.A. Ilyas
Universiti Putra Malaysia

CONTENTS

16.1 INTRODUCTION

The growing environmental devastation ascribed to the disposal of plastic packaging waste has led to an urgent need to develop environmentally friendly packaging materials to save our ecosystem [1]. In order to resolve the current environmental predicament initiated by non-biodegradable plastics, organic biopolymers have been studied to be a potential alternative to petroleum-based plastics. Thermoplastic starch (TPS) is one of the most commonly presented biopolymers for replacing conventional plastic applications. Besides its wide availability, it is renewable, secure handling, biodegradable, and affordable [1]. Therefore, starch recently has attracted researchers and materials engineers' attention as a promising eco-friendly material as well as a potential substitute to non-biodegradable petroleum plastic materials [2]. Despite its multi-attractive properties, TPS-based films for packaging applications have been reported to have insufficient tensile strength and weak water barrier characteristics [3–5]. Such disadvantages sharply reduce their wide-ranging applications, particularly for food packaging functions. Several studies have been conducted by material researchers to upgrade the mechanical performance and improve the water propensity of starch-based products without changing their biodegradability character [6–8]. The incorporation of natural cellulose fibers throughout the preparation of starch composite films is an efficient approach for developing the operational properties of packaging films, as documented by various researchers [3,5,9]. Based on previous studies, the corn plant is deemed to be a terrific resource of natural cellulosic fibers for the reason that it has a high level of cellulose content. Corn is a tropical plant owning multipurpose for human life, such as food agent and biofuel source. It also serves as a potential supplier of starch and cellulosic fiber for producing environmental composite materials.

The residues of agricultural harvests, such as pineapple leaf, sugar palm corn stover (husks, stalk, and leaves), oil palm, cassava peel, and bagasse, are annually produced in large amounts, and they could be found at low cost and large quantity, and can be used as a recyclable source of biomasses. Among these vast capacities of plant leftover, only a minor quantity is utilized as household fuel or animal nourishment, whereas the majority of the remains are usually incinerated; consequently, there is a destructive impact on the environment due to atmospheric contamination. As a result, the utilization of these residues as reinforcing fillers for biocomposite materials is a valuable solution [10]. The stover of corn plant tree usually contains 15% husk, 35% leaves and cobs, and 50% stalk (stem). Most of the stover is discarded as waste, although it has excellent potential to be discovered as natural fiber [11]. Cornstalk is indicated to the vertical stem that grows from the root of corn plant; its lignocellulosic fiber usually includes a high-level concentration of cellulose and a minimal amount of lignin and ash [12]. The distinctive characteristics of cornstalk fiber, such as durability, moderate tensile flexibility, strength, low-density, and extendibility, will provide distinctive properties for corn fiber products [13].

The principal aim of this chapter was to prepare and characterize eco-compatible plasticized films based on cornstarch and cornstalk as reinforcing filler with the proposition of an integrated methodology to the use of whole cornstalk wastes. This research chapter was also designed to evaluate the utilization of corn plants, which is an inexpensive source of cornstarch and fibers, as a feedstock for making fiber-reinforced TPS. The use of cornstalk as reinforcement agents for thermoplastic cornstarch adds value to the wasted by-products and increases the suitability of cornstarch composite films as environmentally friendly food packaging material. It is worth to mention that the cornstalk particles used as supporting filler in the current research were not chemically handled or adapted, which will lead to the development of more environmentally friendly material and cheaper production processes.

16.2 MATERIALS AND METHODS

16.2.1 MATERIALS

The isolation of cornstarch was conducted following the procedures introduced by [14]. Corn grains (1 kg) were steeped in distilled water (4 L) and maintained for 12 h at 4°C. Afterward, the water was evacuated, and the moist grains were ground in a lab electric mixer (wet milling) until the minimum likely fraction was achieved. The crushed specimens were sifted through 75 μm mesh sieve and then were left to sediment for 8 h. The supernatant solution was removed, and the sedimented particles were suspended in distilled water in order to be centrifuged for 20 min at 3,000 rpm. The obtained cornstarch was dehydrated at 50°C for 12 h in an oven; the dried cornstarch was blended and sieved for achieving uniform particle size distribution. The extraction of cornstalk fiber was performed consistent with the method introduced by [15]. Raw cornstalks were cleaned thoroughly and chopped into a specific size; then, they were placed in an oven at 60°C for 12 h to be completely dehydrated. The outer skins of the dried corn stems were separated manually. Lastly, cornstalk fiber appears in light white. Table 16.1 shows the composition and physical properties of cornstarch and cornstalk. The fructose plasticizers were provided by Evergreen Engineering and Resources SDN-BHD, Malaysia.

TABLE 16.1
The Chemical Composition of Corn Starch and Cornstalk

	Content	Amount	Units
	1. Amylose	24.64	(g/100 g)
	2. Amylopectin	75.36	(g/100 g)
Corn Starch	3. Crude fats	7.13	(g/100 g)
	4. Crude protein	7.70	(g/100 g)
	5. Ash	0.62	(g/100 g)
	6. Moisture content	10.45	(g/100 g)
	7. Density	1.4029	(g/cm³)
	Content	**Amount**	**Units**
	1. Cellulose	10.80	(g/100 g)
	2. Hemicellulose	60.30	(g/100 g)
Cornstalk Fiber	3. Lignin	1.97	(g/100 g)
	4. Ash	10.70	(g/100 g)
	5. Moisture content	11.10	(g/100 g)
	6. Density	1.4164	(g/cm³)

16.2.2 Preparation of Composite Films

The preparation of starch-based films reinforced by stalk fiber was performed using the solution casting method. A film-forming solution includes 100 mL distilled water and 5 g of cornstarch (CS). Fructose with 0.25 g/g dry starch concentration was used as a plasticizer; this percentage was selected based on our previous study about plasticizers selection [16]. Cornstalk as reinforcing filler was used at concentrations of 2%, 4%, 6%, and 8% w/w dry starch, with a particle size of less than 300 μm. The mixture was heated to 85°C in a thermal bath for 20 min with constant stirring. After that, the film-forming solution was kept in a desiccator in a vacuum to eliminate the formation of air bubbles during heating. Following that, the solution was discharged regularly in circular plates with a 140 mm diameter. The plates were then dried up in an air-flow oven at 45°C. The dry films were removed from the plates and stored at the ambient condition within plastic bags for a week prior to characterization. The highest percentage of cornstalk that used in the fabrication of starch films is 8% (w/w dry basis); beyond this limit, the formed films seemed to be weak, cracked, incoherent, and difficult to take off from the casting plates. The films produced were labeled according to the concentration of cornstalk fiber as St 2%, St 4%, St 6%, and St 8%, whereas the film without stalk fiber labeled as control.

16.2.3 Film Thickness and Density

The thicknesses of the films were achieved using a digital vernier caliper (Mitutoyo Co., Japan). Five random measurements were recorded for each film, and the final thickness was selected from the average value, whereas the density of the film was determined directly from the film volume and weight. The considered values were the averages of five measurements. The following equation was used to calculate film density:

$$\rho = m/v = g/m^3 \tag{16.1}$$

16.2.4 Water Content (WC)

Water (moisture) content of films was concluded using the weight loss technique. Initial weight (w_1) of the film sample was measured using a digital balance. The sample was then dried on a lab oven at 105°C for 12 h, and directly the final weight (w_2) was measured via the same digital balance. Finally, WC of films was calculated as the proportion of initial weight to dry weight according to the equation:

$$WC(\%) = \left(\frac{w_1 - w_2}{w_1} \right) \times 100 \tag{16.2}$$

16.2.5 Film Water Solubility (WS)

WS of films was measured in triplicate. At first, film samples were dried at 85°C for 12 h in an air-circulation laboratory oven, and then, the initial weight was calculated. The drying temperature was selected to prevent the volatilization of the plasticizer [17]. The film samples were then immersed in 50 mL of distilled water in a lab beaker under continuous stirring for 6 h at room temperature. After that period, the remaining fragments of samples were collected and dried at 85°C until a constant weight was gained. The water solubility (WS) can be determined using the following equation:

$$WS(\%) = \left[(W_{initial} - W_{final})/W_{initial} \right] \times 100 \qquad (16.3)$$

16.2.6 Water Absorption (WA)

The water absorption assay was carried out in accordance with ASTM D 570-98. The film samples were first dehydrated for 12 h at 60°C and then cooled in a desiccator, and directly weighed ($M_{initial}$). The samples then submerged in filtered water at ambient conditions. The samples were removed regularly from the water after a certain immersion period, wiped with a soft cloth, and weighed again (M_{final}). The variations between the initial and final weights were used to calculate the WA using the following equation:

$$WA(\%) = \left[(M_{final} - M_{initial})/M_{initial} \right] \times 100 \qquad (16.4)$$

16.2.7 Scanning Electron Microscope (SEM)

The surface morphological properties of the samples were inspected by the SEM instrument type (Hitachi S-3400N, Nara, Japan). Prior to the scanning, the sample was painted with a thin gold coating to conduct electricity. A 20-kV voltage was then applied in a high-level vacuum to produce a bundle of electrons. The generated electrons deliver signals to visualize the surface structure of the sample and create high-resolution images.

16.2.8 Thermal Properties

The thermostability properties of the specimens were achieved by a thermogravimeter analyzer (Q500 V20.13 Build 39, Bellingham, USA). The film sample was located in a platinum vessel under nitrogen gas vacuum and exposed to a temperature ranging from room temperature to 450°C at a constant rate of 10 °C/min. TGA measures the percentage of weight loss as affected by the rate of temperature increase.

16.2.9 Fourier Transform Infrared Spectroscopy (FTIR)

An infrared spectrometer device type (Bruker Vector 22, Lancashire, UK) was used to track the FT-IR spectrum of samples and determine the presence of functional groups. The experiment was carried out using 16 scans per sample, over a frequency range of 4,000 to 400 cm^{-1}, with a spectral resolution of 4 cm^{-1}.

16.2.10 X-Ray Diffraction (XRD)

The XRD diffraction testing of the films was performed using a 2,500 X-ray diffractometer (Rigaku, Tokyo, Japan). The angle of scattering (2θ) was ranged from 5° to 60°, with a speed rate of 0.02° per angle. The operational current and voltage during the test were set at 35 mA and 40 kV, respectively. The relative crystallinity (R_c) of samples was calculated based on the calculus of the amorphous area (A_a) and the crystalline area (A_c) by the equation:

$$R_c = \left(A_c / \left(A_c + A_a \right) \right) \times 100 \qquad (16.5)$$

16.2.11 TENSILE PROPERTIES

A universal testing machine type, Instron 3365, with a loading cell of 5 kg was used to measure the tensile characteristics of film samples. The film samples were tested, and the tensile stress (TS) and Young's modulus along with the elongation at break were determined according to the standard ASTM D882. The film specimens were slashed into strips with sizes of 10 mm × 70 mm. The samples were fixed between two tensile clutches, and the effective length was set to be 30 mm. The films were dragged with a crosshead speed of 2 mm/min. Throughout the stretching process, the values of force (N) and deformation (mm) were documented. Calculations were taken from five different samples. The tensile properties were computed as the mean value of the achieved results.

16.3 RESULT AND DISCUSSION

16.3.1 THICKNESS AND DENSITY OF FILMS

An insignificant difference was observed in the thicknesses and densities among different films; the thickness was expanded, whereas the density was reduced upon fiber addition with minimal variations, as shown in Table 16.2. The higher the stalk concentrations in the corn film, the thicker the film. The incorporation of a higher content of filler resulted in thicker and rougher films. According to [18], this could be ascribed to the existence of larger sizes of stalk fiber particles within the starch matrix, which resulted in a less homogenous structure and could promote structural disintegration. Films with a higher concentration of fiber showed lower density values. A similar result was reported by [19] for cornhusk composites.

TABLE 16.2
The Physical Properties of Cornstarch-Based Composites

Film	Thickness (μm)	Density (g/cm³)	Moisture Content (%)	Water Solubility (%)
Control	195 ± 25.4	1.559 ± 0.08	11.13 ± 0.11	21.38 ± 0.43
St 2%	205 ± 25.4	1.550 ± 0.07	11.63 ± 0.12	21.38 ± 0.52
St 4%	350 ± 25.4	1.465 ± 0.05	11.19 ± 0.10	19.32 ± 0.41
St 6%	430 ± 25.4	1.495 ± 0.06	11.13 ± 0.19	19.61 ± 0.24
St 8%	505 ± 25.4	1.475 ± 0.11	10.66 ± 0.13	19.91 ± 0.38

16.3.2 Moisture Content and Water Solubility

The moisture content of cornstarch-based film mixed with various loadings of the stalk is displayed in Table 16.2. As compared to control film, the addition of stalk fiber concentration from 2% to 8% (w/w) observed an insignificant change in the moisture content of films; this could be attributed to the similar densities of matrix and filler as shown in Table 16.1. Furthermore, it is well known that dry cornstalk immersed in water is acted like a sponge, which means that it tends to soak up a high quantity of water in a short period. Therefore, the water solubility of films was also improved. Moreover, it affects and modifies the contact between the starch biopolymer chains, which may develop or decrease the suspension of the starch [20]. Solubility in water is an important character for the TPS-based films, especially for applications that may require water insolubility to enhance water resistance and develop product integrity [21]. It was noticed that the samples did not remain intact once submerged in water. The behavior above influenced by the stalk fiber concentration. The solubility declined with the addition of reinforcing filler concentration, although these films exhibited the highest equilibrium moisture content.

16.3.3 Water Absorption (WA)

Water absorption is a substantial parameter for TPS-based films because water acts as a plasticizer [22]. Due to the presence of hydroxyl groups in the structure of the film (as shown in FTIR), the films revealed a high propensity for water uptake. Figure 16.1 demonstrates the measurements of water retention values as a function of immersion time for cornstarch/stalk composite films. It can be observed that after 30 min of water immersion, all films experienced a high level of water retention that was due to the hydrophilic character of cornstarch and stalk fiber. From Figure 16.1, it is clear that the higher concentrations of stalk fiber tend to retain a lower amount of water uptake, which is attributed in general to the characteristics of fibers that are being hygroscopic naturally. Films containing 8% stalk appeared a lower affinity to

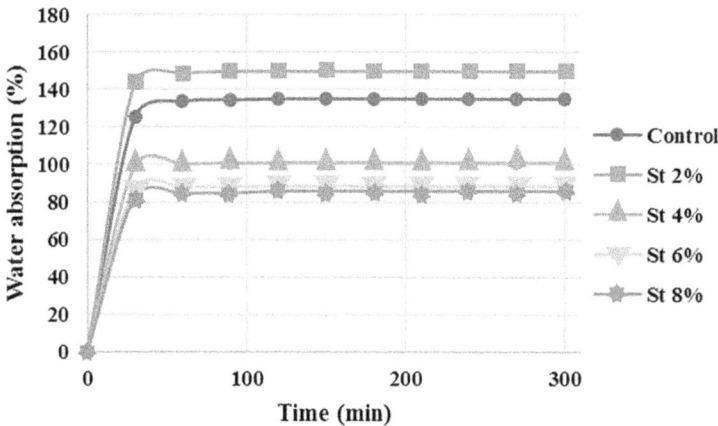

FIGURE 16.1   Water absorption of cornstarch/cornstalk fiber composite films.

absorb water compared to other films. The highest water absorption rate was 150% observed for the film containing 2% stalk, whereas the lower value was 85% for the film containing 8% stalk. Water absorption leans to decrease with an increasing amount of stalk fiber.

16.3.4 MORPHOLOGICAL PROPERTIES

The surface fracture images of the cornstarch/stalk fiber-reinforced composites are displayed in Figure 16.2. The typical appearance showed consistent and homogeneous structures and evidently covered by the polymer matrix; this was mainly due to the function of fiber in forming solid interaction links with the TPS polymer

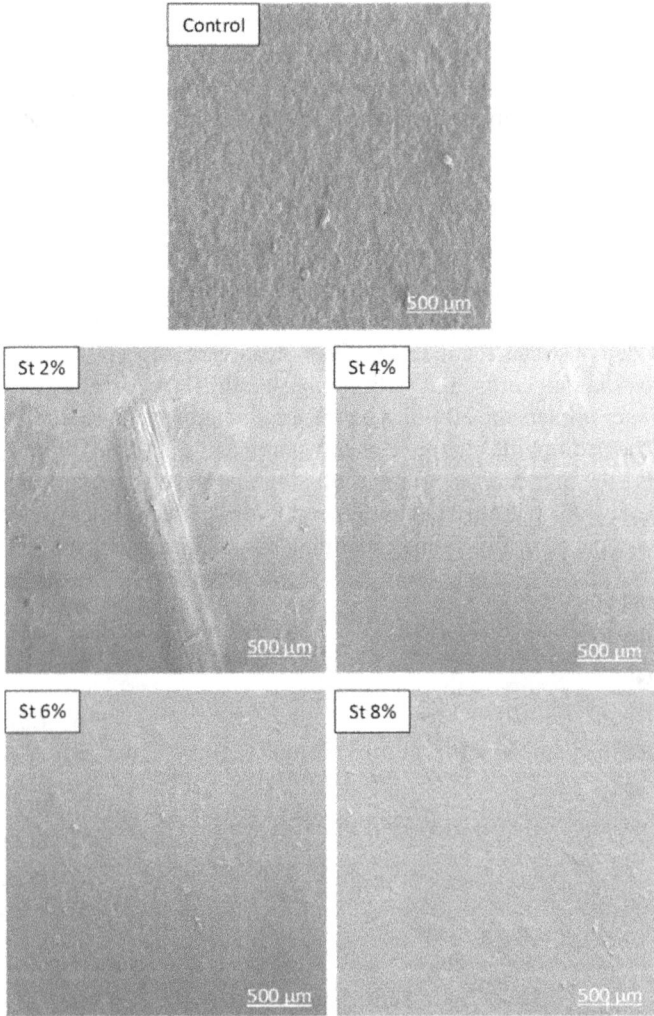

FIGURE 16.2   SEM images of cornstarch/cornstalk fiber composite films.

matrix because both matrix and reinforcement are biological compounds with similar polarity [23]. It should be observed that the film with St 2% demonstrated a less coherent surface, and the starch matrix did not clearly cover the stalk fiber; also the presence of pores was detected in the structure. However, further increasing the concentration of stalk fiber from % to 6% and then to 8%, the voids disappeared and the fracture surfaces became coarser and more rigid. Moreover, there was evidence of excellent adhesion between stalk fiber and starch matrix, since no microcracks or longitudinal gaps in the structure were noticed. In film formation, matrix homogeneity is a positive indicator of its structural homogeneity and consistency [24, 25]. In this regard, it can be concluded that films reinforced with higher stalk fiber will show an efficient stress transfer between filler and the polymer matrix and, therefore, presented better tensile characteristics. In general, consistent distribution and intelligible fiber to polymer interactions were noticed that might be assigned to the hydrophilic similarity of both matrix and reinforcement. A similar surface fracture was detected when cassava bagasse was reinforced thermoplastic cassava starch biocomposites [26].

16.3.5 THERMAL PROPERTIES

The thermostability of starch-based composites reinforced with cornstalk fiber at a multiloading is demonstrated in Figure 16.3. TGA and DTG curves were utilized to verify the thermal properties of samples in terms of the decomposition temperatures and residues of material after the highest rate of degradation. Based on the data obtained, the thermal degradation and weight loss of cornstarch/stalk fiber composites occurred in three heating stages, as indicated in Table 16.3. Each stage is linked with a leading peak in the DTG graph and corresponds to a particular mass loss in the TGA diagram. The initial mass loss appeared at less than 100°C and continued to about 200°C that was mostly due to the removal of water particles by evaporation and the volatilization of fructose particles [27–29]. Samples with higher moisture content experienced higher heating rates and thus more weight loss, and this is consistent with moisture content results (Table 16.2). Further heating

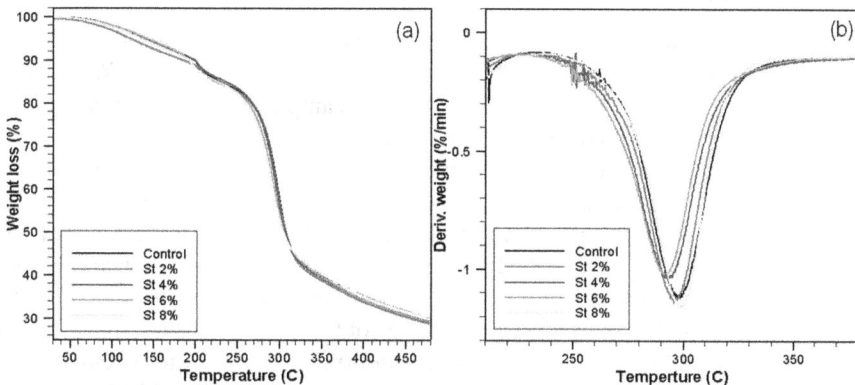

FIGURE 16.3 (a) TGA and (b) DTG curves of cornstarch/cornstalk fiber composites.

TABLE 16.3

Degradation Temperatures Cornstarch/Cornstalk Fiber Composites

Film Sample	Degradation Temperature (°C)			Mass Residue (%)	Weight Loss (%)
	Phase 1 (plasticizer + water)	Phase 2 (starch)	Phase 3 (fiber)		
Control	184	277	–	29.63	50.95
CS-St2%	206.65	279.43	299.89	28.73	55.44
CS-St4%	200.07	277.58	297.47	29.24	56.18
CS-St6%	184.97	273.92	293.04	29.14	55.96
CS-St8%	192.72	271.98	295.42	30.07	55.03

caused the second loss in weight started with the degradation of the water-soluble amylopectin in starch structure [30]. However, the highest rate of degradation was attributed to the breakdown of the major constituents of fiber, namely hemicellulose, cellulose, and lignin. Lomelí-Ramírez et al. [31] stated that in the lignocellulosic fibers, the thermal decomposition starts with the disintegrating of hemicellulose at 200°C–260°C, cellulose at 240°C–350°C, and finally lignin at 280°C–500°C that depends on the plant type and the percentages of fabric components. Lastly, after the lignin is entirely decomposed, the residual component is inorganic such as silica (silicon dioxide, SiO_2), which can be considered as a char (mass residues).

In general, the composite films showed close onset degradation temperatures ranging from 293°C to 299°C, higher than the decomposition temperature of the control film, which started its maximum decaying at 277°C. Furthermore, the mass residues following the last breakdown of the composite showed small variations compared to the neat control film; such observations indicate that the incorporation of stalk fiber as a reinforcing agent does not enhance the thermal stability of cornstarch-based composites. A similar finding was discovered by Edhirej et al. [32] who reinforced cassava starch hybrid composites with cassava bagasse and SPF.

16.3.6 FTIR Analysis

The FTIR spectrum curves of the control film and cornstarch/cornstalk fiber composite films with various loading of stalk fiber are displayed in Figure 16.4. It was clear that the FTIR spectra curves of all film samples exhibited characteristic absorption peaks associated with fiber and polymer components. Since the materials used were extracted from the same biological sources, their chemical composition contains certain constituents, which are lignin, cellulose, and hemicellulose for the fibers and amylose and amylopectin for the starch. The transmittance bands at the peaks 930–1,030 cm^{-1} are ascribed to O–C stretching within the anhydroglucose ring [33]. The bending of water fragments in cornstarch generated the double bands at around 1,500–1,600 cm^{-1} [34]. The stretching of C–H groups showed up the sharp bands at 2,850–3,000 cm^{-1} [35]. The broad absorption bands with extreme intensity at 3,200–3,500 cm^{-1} are assigned to the vibrational stretching of O–H hydroxyl groups within the fiber and starch structure [36,37]. However, there is no evidence of chemical reactions, as no new peaks appeared. Thus, the interactions between the reinforcing fibers and the starch matrix molecules were determined by detecting the shift of the band position, following the fiber loading. For example, it can be seen from Figure 16.4 that the O–H stretching peaks at approximately 3,200–3,500 cm^{-1} for CS film were shifted to lower intensity bands after the reinforcement. The shifting of band position is an indicator of increasing intermolecular hydrogen bonding in the hybrid structure, which provided stronger interaction and more compatibility [36]. A similar interpretation was reported by Jumaidin et al. [38] when they reinforced sugar palm starch-based composites with seaweed.

16.3.7 DIFFRACTION ANALYSIS

The X-ray diffraction curve of the non-reinforced film, as well as cornstarch-based reinforced films at different stalk fiber loadings, is introduced in Figure 16.5. It was clear that all film samples offered similar pattern behavior to the control film; the only difference was in the increased intensity of the main peaks following fiber loading. Regarding the control film without lignocellulosic fiber, the crystalline structure was constructed from gelatinization and retrogradation of starch molecules as effected by heating, resulted in sharp 2θ peaks diffracted at angles 17.57°, 20.15°, and 22.48°, which represents the typical A-type pattern of the natural plant starches [39,40]. In general, the incorporation of fibers, as expected, increased the intensity of the main peaks, which in turn enhanced the crystallinity index of the hybrid composites. According to Bledzki and Gassan [41], the observed correlation between fiber loading and increased crystallinity might be explained through the function of cellulose in improving crystallinity by creating a cross-linked network

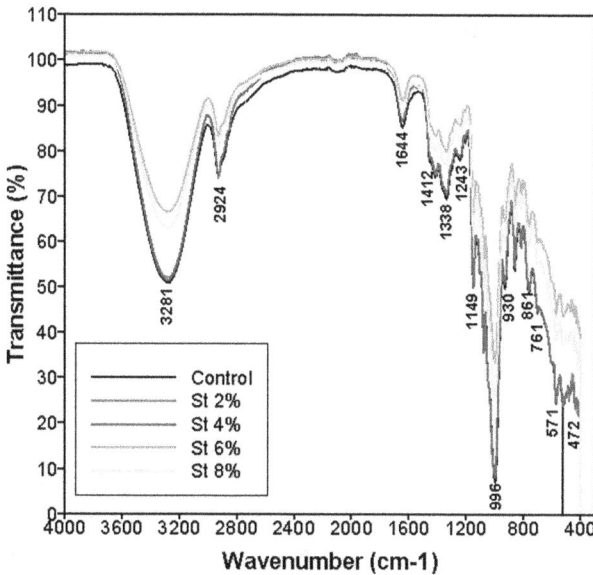

FIGURE 16.4 FTIR spectrum of cornstarch/cornstalk fiber composite films.

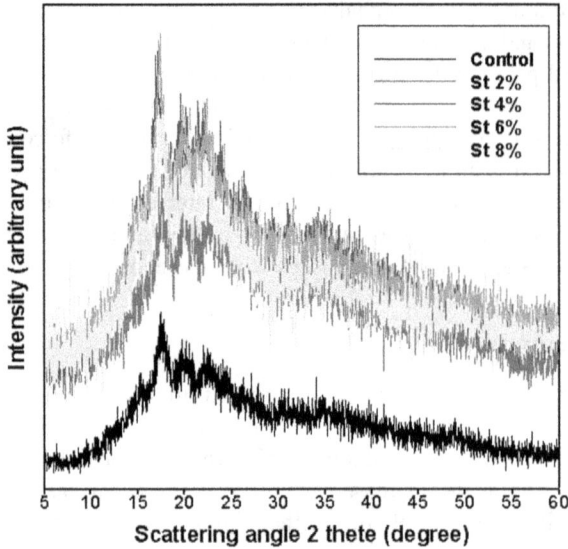

FIGURE 16.5   XRD curve of cornstarch/cornstalk fiber composite films.

and hindering disintegration of the composite by forming covalent bonding between fiber and matrix. The crystallinity values of cornstarch films as affected by different fiber loadings are displayed in Table 16.4. Natural fibers are oriented materials; therefore, improvement in the crystallinity of starch-based composites is expected with the increase of fiber content [42].

16.3.8   TENSILE PROPERTIES

The effect of cornstalk fiber loading on the tensile strength, tensile modulus, and elongation at break of cornstarch-based composite films was concluded, and the results are introduced in Figure 16.6. As anticipated, it is noticeable that the tensile strength and tensile modulus of composite films increased as the stalk fiber concentration increased from 2 to 8 wt%. The tensile strength and Young's modulus of the pure cornstarch film (control) were 6.8 and 61.35 MPa, respectively. The addition of cornstalk as

TABLE 16.4
Crystallinity Index of Cornstarch/Cornstalk Fiber Composite Films

Film Sample	Crystallinity Index (%)
Control	15
St 2%	18
St 4%	18.1
St 6%	16.4
St 8%	28.7

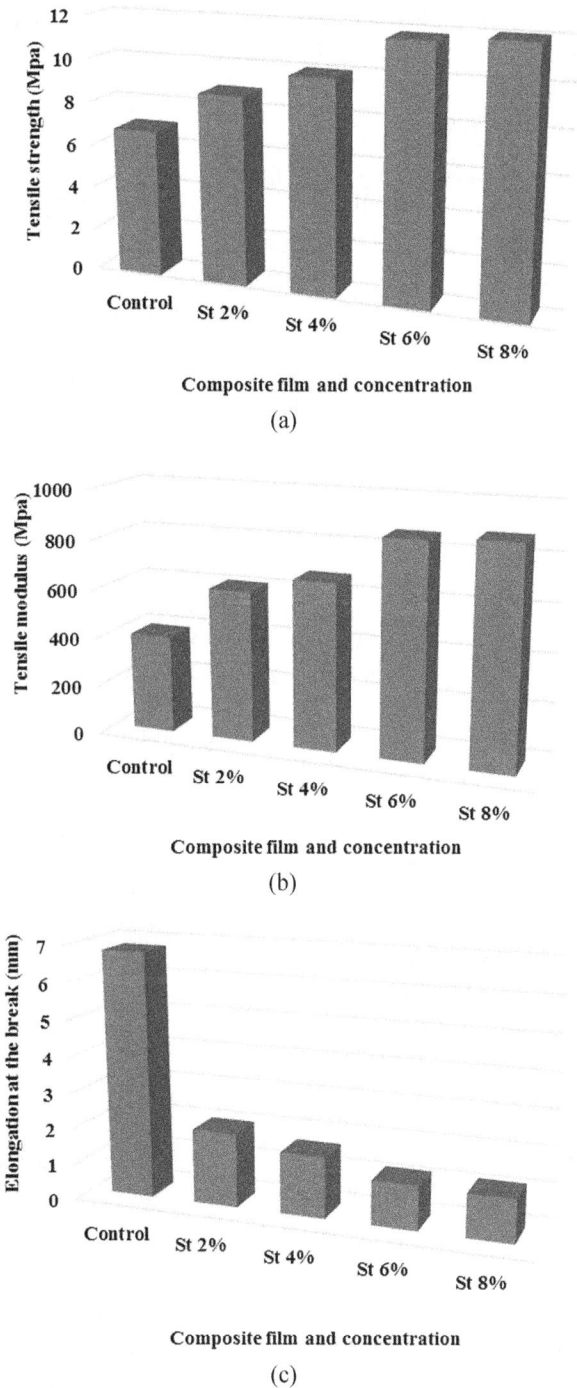

FIGURE 16.6 Tensile properties of the composite films. (a) Tensile strength, (b) tensile modulus, and (c) elongation of the break.

reinforcement from 2 to 8 wt% significantly increased the tensile strength and tensile modulus values of the composite films. Thus, at the highest stalk fiber loading (8 wt%), the tensile strength of starch-based composite films improved by 75%, whereas the tensile modulus was 90% higher than that of the control film. The observed tensile behavior can be attributed to the beneficial interaction between the starch and stalk matrices, which facilitated sufficient interfacial adhesion because of their chemical similarities. Moreover, the improvement of tensile properties was also linked with the increase of the crystallinity index, as shown in Table 16.4. Similar results were conveyed by other researchers [7,43]. On the other hand, the elongation at break for the composite films decreased as the stalk concentration increased from 2 to 8 wt% showed inverse behavior to the increase in tensile strength and tensile modulus, which demonstrated a reduction in elongation at break of approximately 85%. The introduction and increase of stalk decreased the molecular mobility of the cornstarch matrix, making the composite materials stiffer and less flexible [44]. Therefore, cornstalk composite films became more resistant to break, stiffer, and less stretchable than the pure cornstarch films.

16.4 CONCLUSION

Biodegradable composite films were formed from cornstarch/cornstalk fiber through solution casting and dehydration method using 25% w/w fructose as a plasticizer. The effect of cornstalk loading on the tensile, physical, morphological, thermal, and structural properties of cornstarch-based films was successfully evaluated. The concentration of cornstalk fiber affected physical properties. For instance, higher content of stalk increased film thickness and decreased film density. Furthermore, the higher concentration of stalk slightly decreased moisture content and water solubility of the films. The addition of stalk fiber considerably improved the overall mechanical performance of the films as well as thermal stability. The addition of natural fibers (cornstalk) as a reinforcing agent is an exciting option to address the properties of the resulting composite film, which allows for a more extensive application range of environmentally friendly starch-based materials. Furthermore, it should be noted that cornstalk particles used as fibers in this study were not chemically handled or modified, which would lead to the development of more environmentally friendly and cheaper production processes and materials.

ACKNOWLEDGMENTS

The authors would like to thank Universiti Putra Malaysia and the Ministry of Education, Malaysia, for the financial support through the Universiti Putra Malaysia Grant (9600700).

REFERENCES

1. M. Ibrahim, S. Sapuan, E. Zainudin, and M. Zuhri, "Potential of using multiscale corn husk fiber as reinforcing filler in cornstarch-based biocomposites," *International Journal of biological macromolecules,* vol. 139, pp. 596–604, 2019.
2. N. Savadekar and S. Mhaske, "Synthesis of nano cellulose fibers and effect on thermoplastics starch based films," *Carbohydrate Polymers,* vol. 89, pp. 146–151, 2012.

3. E. d. M. Teixeira, D. Pasquini, A. A. Curvelo, E. Corradini, M. N. Belgacem, and A. Dufresne, "Cassava bagasse cellulose nanofibrils reinforced thermoplastic cassava starch," *Carbohydrate Polymers,* vol. 78, pp. 422–431, 2009.

4. C. Bilbao-Sainz, J. Bras, T. Williams, T. Sénechal, and W. Orts, "HPMC reinforced with different cellulose nano-particles," *Carbohydrate Polymers,* vol. 86, pp. 1549–1557, 2011.

5. C.-A. Teacă, R. Bodîrlău, and I. Spiridon, "Effect of cellulose reinforcement on the properties of organic acid-modified starch microparticles/plasticized starch bio-composite films," *Carbohydrate Polymers,* vol. 93, pp. 307–315, 2013.

6. M. Sanchez-Garcia, E. Gimenez, and J. M. Lagarón, "Morphology and barrier properties of solvent cast composites of thermoplastic biopolymers and purified cellulose fibers," *Carbohydrate Polymers,* vol. 71, pp. 235–244, 2008.

7. A. B. Dias, C. M. Müller, F. D. Larotonda, and J. B. Laurindo, "Mechanical and barrier properties of composite films based on rice flour and cellulose fibers," *LWT-Food Science and Technology,* vol. 44, pp. 535–542, 2011.

8. C. M. Müller, J. B. Laurindo, and F. Yamashita, "Effect of cellulose fibers addition on the mechanical properties and water vapor barrier of starch-based films," *Food Hydrocolloids,* vol. 23, pp. 1328–1333, 2009.

9. A. M. Slavutsky and M. A. Bertuzzi, "Water barrier properties of starch films reinforced with cellulose nanocrystals obtained from sugarcane bagasse," *Carbohydrate polymers,* vol. 110, pp. 53–61, 2014.

10. J. Sahari, S. Sapuan, E. Zainudin, and M. Maleque, "Mechanical and thermal properties of environmentally friendly composites derived from sugar palm tree," *Materials and Design,* vol. 49, pp. 285–289, 2013.

11. S. Sokhansanj, A. Turhollow, J. Cushman, and J. Cundiff, "Engineering aspects of collecting corn stover for bioenergy," *Biomass and Bioenergy,* vol. 23, pp. 347–355, 2002.

12. C. Mendes, F. Adnet, M. Leite, C. G. Furtado, and A. Sousa, "Chemical, physical, mechanical, thermal and morphological characterization of corn husk residue," *Cellulose Chemistry and Technology,* vol. 49, pp. 727–735, 2015.

13. N. Reddy and Y. Yang, "Properties and potential applications of natural cellulose fibers from cornhusks," *Green Chemistry,* vol. 7, pp. 190–195, 2005.

14. A. Ali, T. A. Wani, I. A. Wani, and F. A. Masoodi, "Comparative study of the physico-chemical properties of rice and corn starches grown in Indian temperate climate," *Journal of the Saudi Society of Agricultural Sciences,* vol. 15, pp. 75–82, 2016.

15. P. Baranitharan and G. Mahesh, "Alkali treated maize fibers reinforced with epoxy poly matrix composites," *Magnesium,* vol. 15, p. 150, 2014.

16. M. Ibrahim, S. Sapuan, E. Zainudin, and M. Zuhri, "Physical, thermal, morphological, and tensile properties of cornstarch-based films as affected by different plasticizers," *International Journal of Food Properties,* vol. 22, pp. 925–941, 2019.

17. M. Jouki, N. Khazaei, M. Ghasemlou, and M. HadiNezhad, "Effect of glycerol concentration on edible film production from cress seed carbohydrate gum," *Carbohydrate Polymers,* vol. 96, pp. 39–46, 2013.

18. F. Versino and M. A. García, "Cassava (*Manihot esculenta*) starch films reinforced with natural fibrous filler," *Industrial Crops and Products,* vol. 58, pp. 305–314, 2014.

19. T. Wan, R. Huang, Q. Zhao, L. Xiong, L. Luo, X. Tan, *et al.,* "Synthesis and swelling properties of corn stalk-composite superabsorbent," *Journal of Applied Polymer Science,* vol. 130, pp. 698–703, 2013.

20. A. Edhirej, S. Sapuan, M. Jawaid, and N. I. Zahari, "Preparation and characterization of cassava starch/peel composite film," *Polymer Composites,* vol. 39, pp. 1704–1715, 2018.

21. M. Bertuzzi, M. Armada, and J. Gottifredi, "Physicochemical characterization of starch based films," *Journal of Food Engineering,* vol. 82, pp. 17–25, 2007.

22. A. Edhirej, S. M. Sapuan, M. Jawaid, and N. I. Zahari, "Effect of various plasticizers and concentration on the physical, thermal, mechanical, and structural properties of cassava-starch-based films," *Starch-Stärke,* vol. 69, p. 1500366, 2017.

23. Z. Ahmad, N. H. A. Razak, N. S. M. Roslan, and N. Mosman, "Evaluation of kenaf fibers reinforced starch-based biocomposite film through water absorption and biodegradation properties," *Journal of Engineering Science,* vol. 10, p. 31, 2014.

24. M. C. Galdeano, S. Mali, M. V. E. Grossmann, F. Yamashita, and M. A. García, "Effects of plasticizers on the properties of oat starch films," *Materials Science and Engineering: C,* vol. 29, pp. 532–538, 2009.

25. S. Mali, M. V. E. Grossmann, M. A. Garcia, M. N. Martino, and N. E. Zaritzky, "Microstructural characterization of yam starch films," *Carbohydrate Polymers,* vol. 50, pp. 379–386, 2002.

26. A. Edhirej, S. Sapuan, M. Jawaid, and N. I. Zahari, "Preparation and characterization of cassava bagasse reinforced thermoplastic cassava starch," *Fibers and Polymers,* vol. 18, pp. 162–171, 2017.

27. P. Suppakul, B. Chalernsook, B. Ratisuthawat, S. Prapasitthi, and N. Munchukangwan, "Empirical modeling of moisture sorption characteristics and mechanical and barrier properties of cassava flour film and their relation to plasticizing–antiplasticizing effects," *LWT-Food Science and Technology,* vol. 50, pp. 290–297, 2013.

28. K. M. Dang and R. Yoksan, "Development of thermoplastic starch blown film by incorporating plasticized chitosan," *Carbohydrate Polymers,* vol. 115, pp. 575–581, 2015.

29. Q. Huang, J. Zhao, M. Liu, J. Chen, X. Zhu, T. Wu, *et al.*, "Preparation of polyethylene polyamine@ tannic acid encapsulated MgAl-layered double hydroxide for the efficient removal of copper (II) ions from aqueous solution," *Journal of the Taiwan Institute of Chemical Engineers,* vol. 82, pp. 92–101, 2018.

30. M. I. Ibrahim, S. M. Sapuan, E. S. Zainudin, and M. Y. M. Zuhri, "Extraction, chemical composition, and characterization of potential lignocellulosic biomasses and polymers from corn plant parts," *BioResources,* vol. 14, pp. 6485–6500, 2019.

31. M. G. Lomelí-Ramírez, S. G. Kestur, R. Manríquez-González, S. Iwakiri, G. B. de Muniz, and T. S. Flores-Sahagun, "Bio-composites of cassava starch-green coconut fiber: part II—structure and properties," *Carbohydrate Polymers,* vol. 102, pp. 576–583, 2014.

32. A. Edhirej, S. Sapuan, M. Jawaid, and N. I. Zahari, "Cassava/sugar palm fiber reinforced cassava starch hybrid composites: physical, thermal and structural properties," *International Journal of Biological Macromolecules,* vol. 101, pp. 75–83, 2017.

33. J. Fang, P. Fowler, J. Tomkinson, and C. Hill, "The preparation and characterization of a series of chemically modified potato starches," *Carbohydrate Polymers,* vol. 47, pp. 245–252, 2002.

34. J. Sahari, S. Sapuan, E. Zainudin, and M. Maleque, "Thermo-mechanical behaviors of thermoplastic starch derived from sugar palm tree (*Arenga pinnata*)," *Carbohydrate Polymers,* vol. 92, pp. 1711–1716, 2013.

35. J. Prachayawarakorn, N. Limsiriwong, R. Kongjindamunee, and S. Surakit, "Effect of agar and cotton fiber on properties of thermoplastic waxy rice starch composites," *Journal of Polymers and the Environment,* vol. 20, pp. 88–95, 2012.

36. Y. Wu, F. Geng, P. R. Chang, J. Yu, and X. Ma, "Effect of agar on the microstructure and performance of potato starch film," *Carbohydrate Polymers,* vol. 76, pp. 299–304, 2009.

37. Q. Huang, J. Zhao, M. Liu, Y. Li, J. Ruan, Q. Li, *et al.*, "Synthesis of polyacrylamide immobilized molybdenum disulfide (MoS2@ PDA@ PAM) composites via mussel-inspired chemistry and surface-initiated atom transfer radical polymerization for removal of copper (II) ions," *Journal of the Taiwan Institute of Chemical Engineers,* vol. 86, pp. 174–184, 2018.

38. R. Jumaidin, S. M. Sapuan, M. Jawaid, M. R. Ishak, and J. Sahari, "Thermal, mechanical, and physical properties of seaweed/sugar palm fiber reinforced thermoplastic sugar palm starch/agar hybrid composites," *International Journal of Biological Macromolecules,* vol. 97, pp. 606–615, 2017.

39. Q. Zhou, M. W. Rutland, T. T. Teeri, and H. Brumer, "Xyloglucan in cellulose modification," *Cellulose,* vol. 14, pp. 625–641, 2007.

40. L. Zhang, W. Xie, X. Zhao, Y. Liu, and W. Gao, "Study on the morphology, crystalline structure and thermal properties of yellow ginger starch acetates with different degrees of substitution," *Thermochimica Acta,* vol. 495, pp. 57–62, 2009.

41. A. Bledzki and J. Gassan, "Composites reinforced with cellulose based fibers," *Progress in Polymer Science,* vol. 24, pp. 221–274, 1999.

42. X. Ma, J. Yu, and J. F. Kennedy, "Studies on the properties of natural fibers-reinforced thermoplastic starch composites," *Carbohydrate Polymers,* vol. 62, pp. 19–24, 2005.

43. M. Pereda, G. Amica, I. Rácz, and N. E. Marcovich, "Preparation and characterization of sodium caseinate films reinforced with cellulose derivatives," *Carbohydrate Polymers,* vol. 86, pp. 1014–1021, 2011.

44. M. L. Sanyang, S. Sapuan, M. Jawaid, M. R. Ishak, and J. Sahari, "Effect of sugar palm-derived cellulose reinforcement on the mechanical and water barrier properties of sugar palm starch biocomposite films," *BioResources,* vol. 11, pp. 4134–4145, 2016.

Index

For Product Safety Concerns and Information please contact our EU
representative GPSR@taylorandfrancis.com
Taylor & Francis Verlag GmbH, Kaufingerstraße 24, 80331 München, Germany